Wireless Communication with Artificial Intelligence

This reference text discusses advances in wireless communication, design challenges, and future research directions to design reliable wireless communication.

The text discusses emerging technologies, including wireless sensor networks, Internet of Things (IoT), cloud computing, mm-Wave, Massive MIMO, cognitive radios (CR), visible light communication (VLC), wireless optical communication, signal processing, and channel modeling. The text covers artificial intelligence-based applications in wireless communication, machine-learning techniques and challenges in wireless sensor networks, and deep learning for channel and bandwidth estimation during optical wireless communication.

This book will be useful for senior undergraduate, graduate students, and professionals in the fields of electrical engineering, and electronics and communication engineering.

Wireless Communications and Networking Technologies: Classifications, Advancement and Applications

Series Editor: D.K. Lobiyal, R.S. Rao and Vishal Jain

The series addresses different algorithms, architecture, standards and protocols, tools, and methodologies that could be beneficial in implementing next-generation mobile networks for communication. Aimed at senior undergraduate students, graduate students, academic researchers, and professionals, the proposed series will focus on the fundamentals and advances of wireless communication and networking, as well as mobile ad-hoc network (MANET), wireless sensor network (WSN), wireless mess network (WMN), vehicular ad-hoc networks (VANET), vehicular cloud network (VCN), vehicular sensor network (VSN) reliable cooperative network (RCN), mobile opportunistic network (MON), delay tolerant networks (DTN), flying ad-hoc network (FANET) and wireless body sensor network (WBSN).

Cloud Computing Enabled Big-Data Analytics in Wireless Ad-hoc Networks
Sanjoy Das, Ram Shringar Rao, Indrani Das, Vishal Jain and Nanhay Singh

Smart Cities
Concepts, Practices, and Applications
Krishna Kumar, Gaurav Saini, Duc Manh Nguyen, Narendra Kumar and Rachna Shah

Wireless Communication
Advancements and Challenges
Prashant Ranjan, Ram Shringar Rao, Krishna Kumar and Pankaj Sharma

Wireless Communication with Artificial Intelligence
Emerging Trends and Applications
Anuj Singal, Sandeep Kumar, Sajjan Singh and Ashish Kr. Luhach

Computational Intelligent Security in Wireless Communications
Suhel Ahmad Khan, Rajeev Kumar, Omprakash Kaiwartya, Raees Ahmad Khan and Mohammad Faisal

For more information about this series, please visit: https://www.routledge.com/Wireless%20Communications%20and%20Networking%20Technologies/book-series/WCANT

Wireless Communication with Artificial Intelligence

Emerging Trends and Applications

Edited by
Anuj Singal, Sandeep Kumar, Sajjan Singh,
and Ashish Kr. Luhach

CRC Press
Taylor & Francis Group
Boca Raton London New York

CRC Press is an imprint of the
Taylor & Francis Group, an **informa** business

First edition published 2023

by CRC Press
6000 Broken Sound Parkway NW, Suite 300, Boca Raton, FL 33487-2742

and by CRC Press
4 Park Square, Milton Park, Abingdon, Oxon, OX14 4RN

CRC Press is an imprint of Taylor & Francis Group, LLC

Library of Congress Cataloguing-in-Publication Data
A catalog record for this title has been requested

ISBN: 978-1-032-13712-4 (hbk)
ISBN: 978-1-032-13713-1 (pbk)
ISBN: 978-1-003-23052-6 (ebk)

DOI: 10.1201/9781003230526

Typeset in Times
by MPS Limited, Dehradun

Contents

Preface

This book aims to provide more relevant information regarding Emerging Trends in Wireless Communication using Artificial Intelligence. This information is beneficial for students, academics, and researchers from industry who wish to know more about the recent trends of wireless communication. In the past few years, wireless communication technologies have made significant advancements. Also, the demands of high-speed wireless communication, high data rate, high efficiency, and more reliable networking have been increasing steadily. The massive growth of the Internet and the advances in wired and wireless network technologies fulfill this demand. However, this increased demand raises many critical security challenges that must be addressed to allow for the broad deployment of new technologies. So, the book provides a wide range of emerging wireless communication technologies and facilitates researchers, engineers, and students to share the latest research results and the advanced research methods of the field. This book describes the need, design challenges, advancements, and future research direction to design reliable wireless communication.

Further, it covers many emerging wireless communication technologies like Wireless Sensor Network, Internet of Things (IoT), Cloud Computing, mm-Wave, Massive MIMO, Cognitive Radios (CR), Visible Light Communication (VLC), Optical wireless communication, signal processing, and channel modeling, etc. This will give direction for future research to scientists and researchers working in emerging wireless communication technologies. The book highlights the importance of developing these technologies actively. The future of wireless networks seems to be faster, wiser, and more efficient. Implementation strategies and future research directions meet the ongoing design and application requirements of several modern and real-time applications. The book would be a better choice than most available books, which were published long ago and hence don't address the current needs of advancement in wireless communication with Artificial Intelligence, topics seldom elaborated on in old books. Features of this book are that it

- Reviews the challenges, advancements and future research direction to design reliable wireless communication.
- Highlights new techniques and applications in wireless communication with practical implementation.
- Facilitates future research development and collaborations.
- Promotes mutual understanding of researchers in different disciplines of wireless communication.
- Reviews learning methods in emerging trends in wireless communication with artificial intelligence.
- Covers real-time applications.

We express our appreciation to all of the contributing authors who helped us tremendously with their contributions, time, critical thoughts, and suggestions to put together this peer-reviewed edited volume. The editors are also thankful to Taylor and Francis Publisher and their team members for the opportunity to publish this volume. Lastly, we thank our family members for their love, support, encouragement, and patience during the entire period of this work.

Dr. Anuj Singal
Dr. Sandeep Kumar
Dr. Sajjan Singh
Dr. Ashish Kr. Luhach

Editors

Dr. Anuj Singal earned a B.Tech degree in Electronics and Communication from the University of GGSIPU (Delhi), India, in 2008 and the M.E degree in Electronics and Communication from Thappar University (Patiala), India. He completed a Ph.D. in Electronics and Communication Engineering from Department of ECE, GJU S&T, Hisar. His current research interests include wireless communication, optical communication, antenna, machine learning, among others. Presently, he is an Assistant Professor in the Electronics and Communication Engineering Department of GJU S&T, Hisar. He has published 36 research papers in various International/National Journals (including IEEE, Elsevier, and Springer) and proceedings of the reputed International/National Conferences (including Springer and IEEE). He has reviewed many SCI and Scopus index international journals. He has supervised five M. Tech Scholars.

Dr. Sandeep Kumar is presently a Professor in the Department of Computer Science & Engineering, Koneru Lakshmaiah Educational Foundation, Vaddeswaram, Andhra Pradesh, India. He has good academics and research experience in various areas of Electronics and Communication. His areas of research include Embedded systems, Image processing, Biometrics and Machine Learning. He has successfully filed seventeen National and seven International patents. He has received three invitations to be a guest in Scopus Indexed IEEE/Springer Conferences. He has been invited ten times to be an expert in various colleges/universities in India. He has published 115 research papers in various International/National Journals (including IEEE and Springer) and proceedings of the reputed International/National Conferences (including Springer and IEEE). He has been awarded "Best Paper Presentation" in Nepal and India in 2017 & 2018 respectively. He has been awarded "Best Performer Award" in Hyderabad, India, in 2018. He has been awarded also "Young Researcher Award" in Thailand, 2018. He has also been awarded "Best Excellence Award" in Delhi, 2019. He is an active member of 17 professional international societies. He has been nominated to the board of editors/reviewers of 30 peer-reviewed and refereed journals (including IEEE, Springer). He has conducted three International conferences and six workshops. He has also attended 24 seminars, workshops, and short-term courses in IITs. He has supervised 23 M.Tech scholars and has been supervising four more Ph.D. scholar and three M.Tech scholars. His four books have been published internationally - two by Taylor & Francis, USA and two by SP & Wiley in the United States.

Dr. Sajjan Singh earned his B.E degree in Electronics & Communication Engineering from M.D.U. Rohtak, India in 2004, M.Tech degree in Microelectronics & VLSI Design from University Institute of Engineering & Technology, Kurukshetra University, Kurukshetra, India in 2007 and Ph.D degree in the area of wireless communication at the Lingaya's University, Faridabad, India in 2015. At present, he is Professor and Head of the Electronics & Communication Engineering Department at CGC Technical Campus, Mohali, Punjab. He has published 47 research papers in various reputed research international and national journals. He has guided more than 20 M.Tech thesis and is guiding one Ph.D student. He has been reviewer of many SCI and Scopus index international journals such as the *International Journal of Electronics*. He has organized many national level conferences and workshops sponsored through AICTE/ISTE and PTU. His research interests include almost all aspects of wireless communication with special emphasis on OFDM, MIMO-OFDM, SC-FDMA, Channel Modeling and IOT.

Dr. Ashish Kr. Luhach is an Associate Professor in the Department of Electrical & Communication Engineering, The PNG University of Technology, Papua New Guinea. He has academic and research experience in various areas of Wireless Adhoc network and Artificial Network. His areas of research include image processing, cloud computing, and machine Learning. He has been invited many times as an expert in various colleges/universities in India. He has published 60 research papers in various International/National Journals (including IEEE and Springer) and proceedings of the reputed International/National Conferences (including Springer and IEEE). He has been nominated to the board of editors/reviewers of peer-reviewed and refereed journals (including IEEE, Springer). He has conducted many international conferences and workshops.

Contributors

Shilpy Agrawal
Department of Electronics and Communication Engineering
G. D. Goenka University Gurugram
Haryana, India

Rohit Anand
Department of ECE
G. B. Pant Engineering College
New Delhi, India

T. Anil kumar
Department of ECE
Lendi Institute of Engineering and Technology
Jonnada, Vizianagaram District
Andhra Pradesh, India

Anuarg
Department of Mechanical & Automation Engineering
G. B. Pant Engineering College
New Delhi, India

Renuka Arora
Assistant Professor
Ganga Institute of Technology and Management
Kablana, Jhajjar (Haryana), India

M. Rajan Babu
Department of ECE
Lendi Institute of Engineering and Technology
Andhra Pradesh, India

Gaurav Bathla
Department of CSE
Chandigarh University
Punjab, India

Rishu Bhatia
Associate Professor
Ganga Institute of Technology and Management
Kablana, Jhajjar (Haryana), India

Aayush Chibber
Deparment of ECE
G. B. Pant Engineering College
New Delhi, India

Khvati Chopra
Department of Computer Science of Engineering
Jamia Hamdard, New Delhi, India

Vipul Dixit
Department of Electronics and Communication Engineering
PDPM Indian Institute of Information Technology Design and Manufacturing
Jabalpur, India

Amit Kumar Garg
Professor
Deenbandhu Chhotu Ram University of Science and Technology
Sonipat, India

K. Gurucharan
ECE Department
Lendi Institute of Engineering and Technology
Andhra Pradesh, India

Rakesh Joon
Associate Professor
Ganga Institute of Technology and Management
Kablana, Jhajjar (Haryana), India

Harpreet Kaur
Department of ECE
GNI Technical Campus
Hyderabad, India

Manpreet Kaur
Department of CSE
GNI Technical Campus
Hyderabad, India

Sarabpreet Kaur
Department of CSE
CEC Jhanjeri
Mohali, India

Deepak Kedia
Department of Electronics and Communication Engineering
Guru Jambheshwar University of Science & Technology
Hisar, Haryana, India

M. Khari
Department of CSE
NSUT East Campus
Delhi, India

S. S. Kiran
ECE Department
Lendi Institute of Engineering and Technology
Andhra Pradesh, India

B. Kiranmai
ECE Department
Lendi Institute of Engineering and Technology
Andhra Pradesh, India

A. Kumar
Department of Electrical
NSUT East Campus
Delhi, India

A. Naresh Kumar
Department of Electrical and Electronics Engineering
Institute of Aeronautical Engineering
Hyderabad, India

Atul Kumar
Department of Electronics and Communication Engineering
PDPM Indian Institute of Information Technology Design and Manufacturing
Jabalpur, India

K. Kumar
Department of ECE
NSUT East Campus
Delhi, India

Ramnish Kumar
E.C.E. Department
GJU S&T
Hisar, Haryana, India

Rohit Kumar
Department of CSE
Chandigarh University
Punjab, India

E. Kusuma Kumari
ECE Department
Srivasavi Engineering College
Tadepalligudem, Andhra Pradesh, India

Harsha Vardan Maddiboyina
R & D Department
Planet Sigma Embedded Systems Pvt. Ltd.
Hyderabad, India

B. T. P. Madhav
Lendi Institute of Engineering and Technology
India

Khushboo Pachori
Department of ECE
Oriental Institute of Science and Technology
Bhopal, India

Rutvik Patel
Department of CSE
Institute of Computer Science and Technology (Ganpat University)
Ahmedabad, Gujarat, India

V. A. Sankar Ponnapalli
Department of Electronics and Communication Engineering
Sreyas Institute of Engineering and Technology
Hyderabad, India

M. Venkateswara Rao
Lendi Institute of Engineering and Technology
India

T. J. V. Subrahmanyeswara Rao
ECE Department
Sasi Institute of Engineering & Technology
Tadepalligudem, Andhra Pradesh, India

Rasveen
Department of Electronics and Communication Engineering
G. D. Goenka University Gurugram
Haryana, India

Arvind Rehalia
Department of Instrumentation and Control
Bharati Vidyapeeth College of Engineering
New Delhi, India

Sanjima
Ganga Institute of Technology & Management
India

Aaryan Sharma
Department of Instrumentation and Control
Bharati Vidyapeeth College of Engineering
New Delhi, India

Dinesh Sharma
CE Department
Chandigarh College of Engineering & Technology
Chandigarh, India

Purnima K. Sharma
ECE Department, Srivasavi Engineering College
Tadepalligudem, Andhra Pradesh, India

Anuj Singal
Department of Electronics and Communication Engineering
Guru Jambheshwar University of Science & Technology
Hisar, Haryana, India

Jagtar Singh
Kurukshetra University
N.C. College of Engineering
Israna, Paniapt, Haryana, India

Dinesh Kumar Verma
PhD Scholar
Deenbandhu Chhotu Ram University of Science and Technology
Sonipet, India

1 Advanced Wireless Communications for Future Technologies-6G and beyond 6G

S. S. Kiran, M. Rajan Babu, B. Kiranmai, and K. Gurucharan
Department of ECE, Lendi Institute of Engineering & Technology, Andhra Pradesh, India

CONTENTS

DOI: 10.1201/9781003230526-1

1.1 INTRODUCTION

In the past few years, wireless communication technologies have made significant advancements. The current 5G innovations that contain standard methods for interpreting like system softwarization and optimization, mm-wave, and ultra-densification through the implementation of novel cellular networks have benefitted product technology solutions, intellectuals, academic researchers, standards bodies, and end-users significantly. VAR, e-commerce, mobile payments, processor interactions, and superior wireless broadband, among other emerging technologies and verticals, have shown 5G's tremendous opportunities, which continue to adapt and evolve to a broad range of new use cases. However, as societal requirements change, the number of new use cases that 5G cannot adequately service has significantly increased. For instance, volumetric teleportation, the next era of VAR, demands a Tbps-level transmission rate and a fraction of a second delay, most of which are difficult to attain even with 5G's millimeter wave (mmWave) frequency bands. Accelerated manufacturing industry and the transition from the Fourth Industrial Revolution to the eventual Industry-X.0 standard will drive 5G network density even ahead of the 106 km^2 metric, requiring a reassessment of existing network management techniques.

Furthermore, as connection density increases, so will expectations for greater energy efficiency, which 5G is not intended to meet. As a result, the research community has shifted its focus on the aforementioned significant challenges. For instance, we think the existing innovations in terahertz band transmissions, smart surfaces and surroundings, and network automation may promise advanced wireless communications for future technologies.

To accomplish this objective, a combination of social demands and emerging technologies that assist in meeting those criteria are the most significant factors for a millennial to jump beyond conventional mobile networks. Taken mutually, essential factors provide a strong argument for a vigorous debate on the future era of advanced wireless technologies, often referred to as 6G networks. We anticipate that 6G will facilitate a profoundly smart, reliable, adaptable, and safe domestic cellular network and the unification of space communications to achieve total wireless accessibility. This paper outlines our long-term vision of wireless communications, spotlighting new use cases and defining the crucial technological solutions to make 6G a reality. We will begin by explicitly describing the performance indicators that will help improve the quality of 6G technologies. While the "International Telecommunication Union's Telecommunication" research many formal measuring standards, early data has recently emerged in the public domain. The following will describe the performance indicators.

The Capacity of System: This KPI category focuses on metrics that relate to system throughput. These include things like max peak spectral efficiency, observed data rate, observed spectral efficiency, area traffic throughput, optimum channel bandwidth, and power density. In this context, the perceived data rate and spectral efficiency measurements related to 95% of all user locations should have the same values [1].

Latency of System: The end-to-end latency measures and delay jitter are included in this category of KPIs. It is worth noting that jitter is a unique key performance indicator for 6G that keeps track of the system's latency fluctuations and is not present in 5G.

System Administration: This KPI category focuses on measures relating to network management and optimization, energy efficiency, dependability, and mobility, to name a few. 5G, on the other hand, does not include an objective key performance indicator for energy efficiency; 6G does include a 1 Tb/J energy efficiency target.

Radio Design Paradigms and Novel Spectrum Use: While 5G guaranteed widespread acceptance of millimeter wave spectrum, the need for higher-quality data rates, as a result, wider channel bandwidths in 6G would demand the use of terahertz (THz) and sub-terahertz range. Simultaneously, the inclusion of new spectrum bands will need creative, innovative radio design capable of sensing and communicating over the whole electromagnetic range.

Architectures of Novel Networks: Traditional mobile-based wireless communication architectures are still unable to expand to fulfill the demands of 6G in terms of area traffic capacity and connection density. Instead, 6G and Beyond 6G will need to incorporate advanced communications systems into the core framework of the ecosystem.

Advanced Automation and Deep Intelligence: Manual network configuration will be unfeasible because of the rigorous spectrum reliability, efficiency, and latency requirements of 6G and Beyond 6G. Instead, wireless sensor networks and automation will take the spotlight, helping develop a more self-sufficient network.

Extending Network Coverage beyond Terrestrial Networks: 6G and Beyond 6G will need near-Earth and deep-space connections to attain total wireless ubiquity. Many explanations have been developed and still need research to attain this ambitious goal. All advanced technologies include the following particulars.

i. A network with abundant spectrum resources operating in the THz band
ii. Intelligent Communication Environments – ICE
iii. Artificial-Intelligence-System (AIS)
iv. "Very Large Scale Network" Automation
v. Energy-Saving Ambient Backscatter Communications,
vi. The "Internet of Space Things" – (IoT) is enabled by "CubeSats and UAVs,"
vii. Multi-User MIMO Communication Networks of Cell-Free – MuMCNF

We also pay attention to three key aspects. Although new technologies are anticipated to influence the future of advanced communications, they are not yet mature adequate for 6G and beyond 6G. The "Internet of Nano-Things," "Internet of Bio Nano-Things," and "Quantum Communications" are among them. Furthermore, robust security solutions, besides these key technologies, will be critical for the success of 6G.

Furthermore, while prior art on 6G and Beyond 6G wire-free systems has been more publicly available in recent years, we think this is the case. The bulk of prior publications in this field has concentrated on a few issues, perhaps impeding a complete analysis of important factors that will define wireless communications' future. Furthermore, discussions of systems other than 6G, including the "Internet of Nano-Things," "Internet of Bio-Nano-Things," and "Quantum Communications" – QC, are essential but notably missing from current literature.

This chapter aims to give readers a detailed look into the future generations of wireless communications. The research on 6G wireless networks and research beyond 6G is continuously evolving and breaking new ground. This publication aims to give a complete assessment of use cases and essential supporting technologies and a comprehensive roadmap for 6G and beyond wireless systems. The overall aim is to urge the scientific community to collaborate to address the significant research problems connected with implementing 6G networks (Figure 1.1).

The remainder of this chapter is divided into the sections below. We show a wide range of use cases enabled by 6G in the second section. In addition, the third and fourth sections include information on the most critical technologies for the development of 6G and Beyond 6G, as well as a consideration of the main difficulties that each faces. Later, in the 5th section, Pervasive Artificial Intelligence Systems are addressed; these systems have better applicability in a broad range of areas in both academia and industry. The commoditization of crucial network components was then highlighted in the 6th section as one of the significant contributions of network softwarization. The 7th section then focuses on 6G radio reconfigurable trans receiver front ends, which link many devices wirelessly – subsequently discussed Ambient Backscatter Communications in the 8th section, which relates IoT and sensors. The Internet of Space Things, Multi-User MIMO Communication Systems, and potential enablers for beyond 6G systems are discussed in the 9th, 10th, and 11th sections, respectively, followed by a 6G and Beyond timetable in the 12th section. Finally, we end this chapter in the last section.

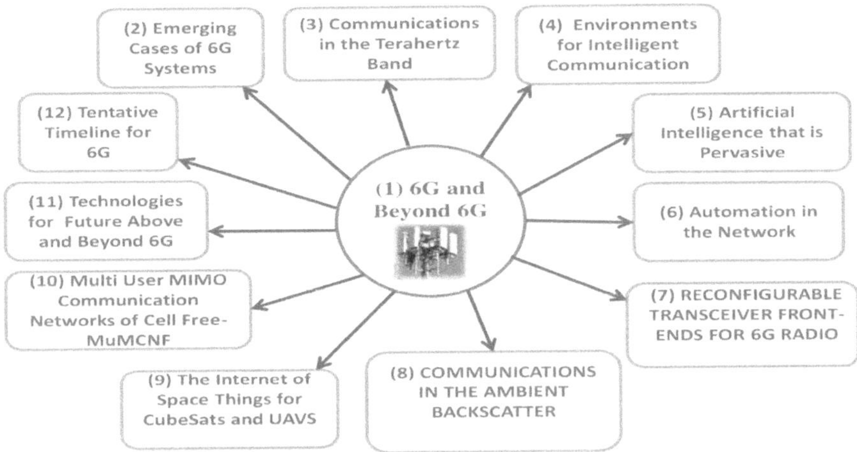

FIGURE 1.1 Main enabling technologies for advanced wireless communications systems of the 6G and beyond 6G.

1.2 EMERGING CASES OF 6G SYSTEMS:

As a result of the 5G experience, the 6G use cases will be based on knowledge. Before widespread commercial availability, the most useful new use cases enabled by 5G will include "Enhanced Mobile Broadband" – eMBB, "Ultra-Reliable Low-Latency Communication" – URLLC, and "Machine Type Communications" MTC, which are designed to support various applications. Alternatively, as stated in Section I, the "5G KPIs" are inadequate for different purposes. We can better anticipate which applications will benefit the most from 6G as we better know the performance trade-offs associated with 5G systems in contexts of Energy Efficiency, Low Latency, Reliable Coverage, Throughput, and Dependability. As illustrated in Figure 1.2, we demonstrate a range of essential use cases enabled by 6G.

Remote Healthcare in Real-Time: The quality and availability of connection are critical factors in the success of remote healthcare systems [2]. Regarding the latter, we emphasize that, with the rollout of 6G, we will see the most significant possible wireless communications quality. It will focus on throughput and latency using leading enabling technologies, including TeraHz band communications and "Network Automation Solutions." Furthermore, As stated in Section 11, developments in 6G and Beyond are anticipated to play a significant function in illness detection and conduct as a connection solution in the healthcare sector.

Cyber-Physical Systems with Autonomous Behavior: Two of today's most promising "cyber-physical systems" are "Autonomous Aerial Vehicles" (AAVs) and autonomous cars [3,4]. Connecting a large quantity of data to high-passing terrain planning, optimization of routes, traffic, and security data are transmitted between the component nodes, i.e., both automobiles and AAVs. While enormous quantities of data must be transmitted on time and without errors, it is also important to remember that these nodes usually travel at speeds of over 100 km/h. Consequently, in addition to millisecond latency and exceptional dependability, the

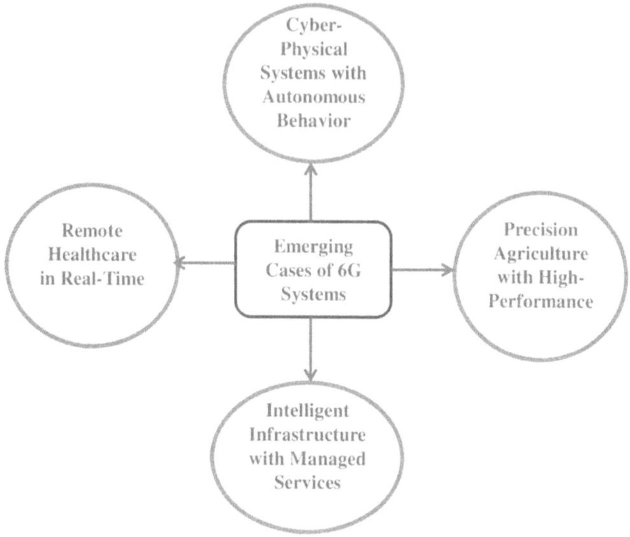

FIGURE 1.2 6G Systems are best suited for certain use cases.

connection solution that allows autonomous "Cyber-Physical Systems" to operate at very high speeds must also provide stable operation significantly faster than others, which is currently available not feasible through 5G networks [5].

Precision Agriculture with High-Performance: For decades, soil moisture monitoring has been a staple in irrigation choices in the larger area of precision agriculture. On the other hand, the absence of a reliable Wi-Fi connection continues to stymie real-time measurements and irrigation automation systems. Aside from standard automatic watering systems, high-performance precision agriculture focuses on providing data-driven insights to meet customers' unique requirements, farms, crops, and soil. On the other hand, rural connectivity limitations make accessible and timely access to such information a significant issue. Consequently, we expect 6G, which emphasizes ubiquitous wireless connection, will significantly affect agricultural technological development.

Intelligent Infrastructure with Managed Services: The use cases mentioned thus far mainly include using third-party systems to use sophisticated telecommunications infrastructure. Altogether, the development of such infrastructure is a valuable use case in and of itself. In addition to network optimization techniques, there is a requirement to exert control over wireless signal propagation. To accomplish this objective, we need to highlight that current wireless networks have always been essential, a supporting role in 5G and its predecessor systems. 6G and Beyond 6G will need precise controls over how electromagnetic waves interact with the interior and outside environments., as demonstrated by the applications described below, with the ever-increasing need for data. In this regard, we believe that the environments for intelligent communication outlined in the 4th section will show a vital task in the future generation of wireless systems' ubiquity and pervasiveness.

1.3 COMMUNICATIONS IN THE TERAHERTZ BAND

Several new wireless communications technologies have resulted in a significant increase in wireless data traffic in recent years. The need for more excellent data rates and improved coverage has followed this exponential development [6]. It has been predicted that 0.1–10 terahertz (THz) communications will play a dynamic role in developing future wireless communications technologies and development throughout the next decade.

To further scientific knowledge, the "Federal-Communications-Commission" (FCC) has opened the frequencies between 95 GHz and 160 GHz for research [7]. To get a maximum data speed of 100 Gbps, some cellular operators have chosen to use millimeter waves, a lower band of frequencies, for their 5G services. Others have chosen high millimeter-wave frequencies in the hopes of achieving 100 Gbps; others have chosen high millimeter-wave frequencies. Test results so far have been disappointing, with a peak data rate of approximately 1 Gbps1. Variables such as the staggering complexity of practical communication lines and system malfunction and interaction with other systems are the primary reasons for the significant discrepancy between the planned and actual data rates. THz bands, which lie between the mm-Wave and IR light spectrums, as illustrated in Figure 1.3, were traditionally regarded as a "no-land" man because of their abundance of ranges resources. On the other hand, THz connections have become a viable alternative for establishing indoor communications networks due to significant advances in the transceiver and antenna design. Recently, the realization of a "Wireless Network on Chip" (WNoC) employing Tera-Hz bands has made significant progress [8].

1.3.1 APPLY THZ BAND COMMUNICATIONS CASES

Unlike lower-frequency wireless networks, THz-band wireless communications offer various unusual application situations due to the unique EM Wave and photonic properties connected to this very high-frequency range. "THz-band Spectrum" may be used for the following conditions in adding to the anticipated Tbps-level connections for cellular systems:

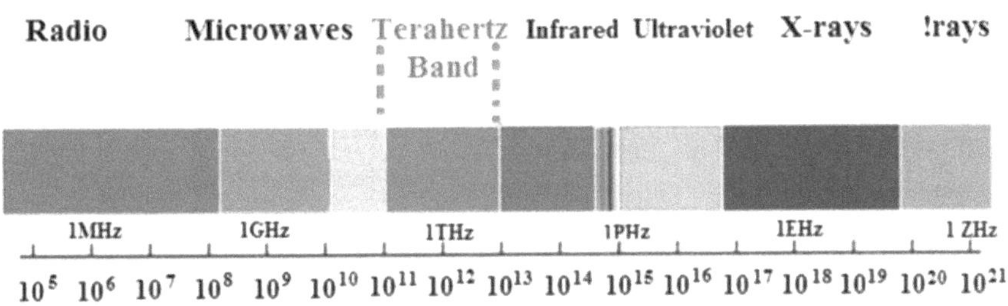

FIGURE 1.3 For "Wireless Communication" connections over short, moderate, lengthy, inside, and in the proximity, the Tera Hz band provides hundreds of GHz of useable spectrum resources.

i. Local Area Networks
ii. Data Center Networks
iii. On-Chip Wireless Network
iv. Personal Area Networks
v. Nano-Networks
vi. Satellite-to-Satellite Communications

1.3.2 The THz Band's Equipment

In TeraHz band transceivers, the need for more output power, reduced noise levels, and enhanced reception sensitivity led to breakthrough device design. THz band signal production is divided into three primary directions: electronic-based, photonics-based, and developing material-based.

GaAs and InP are attractive for photonics-based technology because they have higher electron mobilities than other III-V semiconductors. This capability, coupled with their relatively high frequencies (more than 100 GHz), makes them good candidates for applications in this area. 50 Gbps data connections may be achieved in an indoor setting using the "Uni Traveling-Carrier Photodiode" (UTCP) technique, with a data rate of 300 GHz and a fiber spacing of 300 microns [9–12].

1.3.3 The THz Band's Physical Layer Modeling

To achieve high-frequency wireless communications, precise channel models are needed to include channel characteristics, such as significant air attenuation and molecule absorption, as well as radiation impacts, for example, reflection, dispersion, and diffraction. Several attempts to offer basic knowledge of such channels have been described in recent studies. For example, an early paper in [13,14] shows the THz band channel's incredible capacity. The 300 GHz THz band, which goes from 200 to 340 GHz, has been published [15]. The single-bounce reflected routes, particularly the direct route between transmitter and reception, dominate the received power. Alternatively, energy diffraction and scattering are considerably less pronounced compared to other channel effects, such as diffraction and scattering.

1.3.4 THz Band Communications – Medium Access Control

Some features of "Medium Access Control" systems in Tera-Hz band communications include properties that address issues such as hearing impairment and blockage of the line of sight, among others [16,17]. For high-frequency MAC protocols, MAC solutions in RF systems are not utilized. Instead, transceivers using high-powered laser beams implement very specific handshakes. These razor-sharp beams may offer more significant power-radiation gain and longer transmission distances, but they can also cause deafness if misaligned. As a result, in MAC scheme design, the deafness-avoidance method is needed. A beam-training phase is used in IEEE 802.15.3c [18], as well as other solutions to estimate and guide beams toward intended devices. An *a priori* aided channel-tracking method [19] was recently suggested using angular division multiplexing [20]. The findings of such suggested methods indicate that effective beam alignment techniques may substantially increase channel throughput.

1.3.5 Open THz Band Communications Problems

THz band antenna arrays are still a significant problem for fabrication and testing. Like photolithography and electron beam lithography, several printing techniques may result in a product with several hundred plasmonic antenna components on the front. Large antenna arrays may increase signal coverage by creating multi-array radiation patterns with high-directivity main lobes, concentrating energy in desired directions. However, such angular scope is limited by such highly directed beams, resulting in low energy efficiency for each client due to poor energy efficiency in the transmitter. The "THzPrism" method has been suggested to generate numerous beams with minor frequency changes in various directions while retaining excellent distance coverage [21].

1.4 ENVIRONMENTS FOR INTELLIGENT COMMUNICATION

As a result of the proliferation of new wireless devices, services, and applications, with a growing need for quicker wireless communications, despite the problems mentioned above, in mmWave, and THz-band frequencies, the most significant barrier is the short communication distance. Although present solutions emphasize developing wireless transceiver H/W and S/W and network-optimization techniques, future solutions will focus on creating more creative and efficient technologies. Wireless communication settings may be actively used to control signal propagation in indoor and outdoor situations. Staying abreast of changes in the behavior of electromagnetic waves when interacting with interior furnishings and exterior buildings and other infrastructure is the key to controlling signal propagation in settings. Reflective, absorptive, collimated waves, as illustrated in Figure 1.4, are standard electromagnetic wave properties. The "Intelligent Communication Settings" concept is based on control algorithms that use "Deep Learning and Reinforcement learning" to customize environments dynamically. The following subsections go through these controlled wave characteristics, current research initiatives, and outstanding problems.

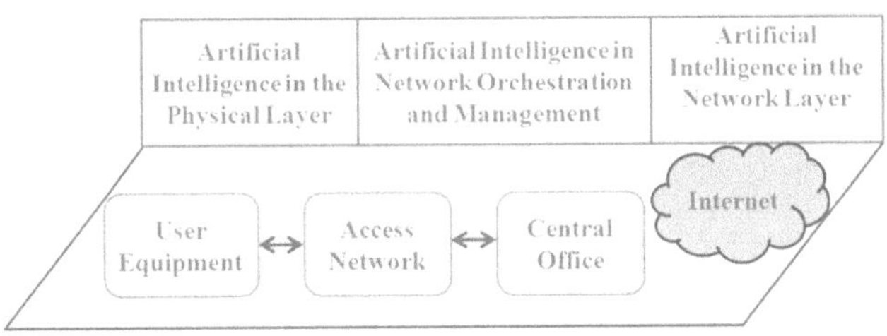

FIGURE 1.4 Artificial intelligence applications at various levels of wireless systems.

1.4.1 Intelligent Communication Environments Include the Following

A 3D structure with many layers, each unique set of capabilities, is a good metaphor for intelligent ecosystems. The EM behavior layer [22] comprises metasurfaces, two-dimensional representations of meta-materials with an adjustable impedance that controls the EM waves' reflection orientations [23–25].

Intelligent communication environments provide the following benefits over-current relays with numerous antennas, which are extensively used in wireless systems:

i. Increased spatial variety owing to the intelligent surface's broad coverage and adjustable antenna arrays,
ii. Because the computer and communication layers lie immediately under the surface layer, the processing time is decreased.
iii. The flexibility of the network is increased when waves from various directions converge, and intelligent surfaces can help redirect the waves in the correct direction.

1.4.2 Intelligent Communication Environments Have a Layered Structure

Here, we describe the tiered structure and each layer's functioning concerning the previously discussed Intelligent Communication Environments.

i. Plane of Meta-material
ii. Plane of Sensing and Actuation
iii. The Computer Plane
iv. Plane of Communication

1.4.3 Apply Intelligent Environments Case Studies

The wireless system may significantly enhance communication effectiveness by using well-coordinated tiles in Intelligent Environments.

i. Enhancements to Signal Propagation
ii. Security at the Physical Layer

1.4.4 Intelligent Environments – Open Problems

For Intelligent Environments to become a market-ready solution, several outstanding issues must be addressed.

Dimensional and Energy Consumption Trade-offs:

The Intelligent Environments are expected to be applied to interior walls, ceilings, and building facades concerning real-world applications. They will need dimensions that suit particular installation locations while still meeting connection requirements. Meanwhile, additional reflect-array components and RF chains are added to the system because of the sophisticated signal-processing circuitry. Balancing energy and dimension usage while serving customers with regular performance is difficult.

Existing Solutions Compatibility: This protocol stack is used for the different channels and user connections with the current Wi-Fi access points. Smart environments must conform with the standard IEEE 802.11 series to help increase the indoor coverage of signals.

Standardization: There was no consensus on standardizing device design, maximum emission power, and communication protocols, given that several possible approaches, including reflectarrays, metasurfaces, and selective surfaces, are being investigated. As additional ideas emerge, a work group's standardization effort is required to create a solitary framework.

1.5 ARTIFICIAL INTELLIGENCE PERVASIVE

Artificial intelligence (AI) has seen tremendous growth in recent years, resulting in its purpose in an extensive range of fields in both academia and industry. Under the AI umbrella, we concentrate on machine-learning algorithms and their applications in advanced wireless networks in this section. As a result, in this section, these two terms are used interchangeably. "Artificial intelligence" may be used at every level of a wireless network, as shown in Figure 1.4. To adjust network resources to different situations, machine-learning (ML) techniques may be used at the network layer for traffic clustering.

1.5.1 Artificial Intelligence in "Physical Layer"

Physical Layer models have traditionally been model-oriented, which involves proposing and optimizing mathematical models that follow a specific framework to meet a set of predetermined performance requirements. A channel model, as well as other parametric parameters, are expected for channel estimation. These model-based solutions typically work well if the derivation of mathematical models is reasonably straightforward or if a closed-form solution exists. Field measurements or numerical simulations can then be used to validate the models. However, model-based solutions in real-world situations are restricted in complex contexts due to nonlinearity within systems and uncontrolled interference. Another approach builds the model by learning from the data and is based on statistics or data sets. This technique is proper when the theoretic examination is willful or finding a closed-form explanation is challenging. Diffusion-based channels, prevalent in molecular communication, are incredibly reliant on the location, making them challenging to describe mathematically [26–30].

1.5.2 Artificial Intelligence in Network Orchestration and Management

• Supervised Learning: This method is commonly used to solve problems involving traffic prediction [31,32], classification [33], and slice resource prediction [34]. Reinforcement Learning is commonly used in resource management problems

[35,36]. The famous virtual network-embedding issue, where the network orchestrator optimizes virtual network functions on the underlying physical substrate, is an excellent candidate for reinforcement learning [37]. Other uses include network infrastructure elastic scaling [38], failure prevention, and configuration rollback [39]. (QoE) [40] Moreover, network security [41], where branded data for exercise is unavailable, and (ii) the application's real-time nature makes waiting for feedback impractical. Unsupervised learning can be a valuable tool in these situations [42].

1.5.3 Learning from a "Quantum Machine"

Furthermore, we point out that machine-learning techniques may be coupled with quantum computing to successfully manage vast quantities of data in various applications [43]. Because of the significant speedup, the efficiency of a "Quantum Machine Learning Algorithm" may be evaluated using complexity-theory measures. In [43], it was shown that, presently, existing quantum machine-learning methods permitted by small-scale quantum computers have shown significant promise in addressing arrangement and recession issues compared to conventional machine-learning algorithms.

1.5.4 Open Difficulties for Pervasive Artificial Intelligence

Researchers have highlighted the possibility of AI attaining great time and spectrum efficiency once integrated into 5G during the early years of standardization. Specifically, AI algorithms can aid in identifying network abnormalities, the allocation of network resources, and network administration, among other activities that would be inefficient with traditional methods [44]. These solutions, however, have yet to be formally accepted in 5G standards across the globe. The ITU launched an AI/ML in 5G competition in June 2020 to encourage academics to discover and solve real-world issues using AI/ML solutions in 5G directions that are important to them. As a result, similar initiatives are expected to be incorporated into subsequent 5G development but will become more tangible and ubiquitous in 6G. While universal artificial intelligence in wireless communication networks will lead to a paradigm change toward data-driven methods, there are still unsolved issues.

1.6 AUTOMATION IN THE NETWORK

"SDN and NFV" have usually been considered to be the primary paradigm shifts that came about as a result of 5G's [45] debut. While wired and wireless networks both saw advantages from the introduction of SDN and NFV, these technologies were responsible for complex network management, cost-effective service deployments, "Enhanced Traffic Engineering," and fine-grained network slice methods. The commoditization of essential network components like switches [46] and base stations [47–49], which enable their deployment on "Commercial off-the-Shelf" (COTS) hardware, has also aided the network's softwarization. Furthermore, the massive "Open-Source Community" that supports these projects has helped bring together a broader range of stakeholders than was previously feasible.

1.6.1 Programmable Data Planes Defined by Software

Open Flow [50,51], the most common Southbound API, is associated with SDN and has been extensively used in 5th Generation networks. However, since OpenFlow's stateless match-action abstraction depends heavily on static header field matching, it prevents real data plane programmability. We also point out that many applications rely on the system's real-time state. Consequently, a significant body of research aims to develop stateful data planes, in which particular stateful packet handling and switch tasks are delegated to data plane switches [52–54]. A device with a stateful data plane, for example, may retain some packet information and use it to method fresh packets in the equal stream. The controller still sets the basic packet forwarding rules. However, including state information at the switch level provides context for rule selection.

"Programmable Stateful Data Planes" poses several interconnected research problems:

A general broad-based description of the state is required, as well as abstractions that reveal this information. Since distributed switching devices will handle packet-level state maintenance, a state consistency method is needed. To avoid competing forwarding operations, a system like this might be implemented via the controller. Security concerns provide a significant obstacle; if the data plane switches are intended to work with packet data, a hostile actor may use faulty packets to cause state changes. Packet integrity verification will be required in this scenario, and ultra-lightweight methods will be required.

1.6.2 Decomposition and Orchestration of Services on an Automatic Basis

Several services may be provided over the same physical infrastructure [21], and the cell phone industry has made it a significant research emphasis [55,56]. The slice creation and deployment procedure, on the other hand, is mainly template-driven and needs a human setup. The current 3GPP network-slicing standard [21], for example, is mainly based on the idea of "Network Slice Templates" (NSTs). The "virtual Network Functions" (VNFs) and related service function chains that make up a network service are expressly defined by an NST. As a result, such network slicing primitives can only deploy a restricted number of network services, namely those for which a template has previously been established. Trying to tackle new kinds of network services this way is not scalable because it lacks a method for (i) accommodating new types of network services, and (ii) gaining more sophisticated services on the network to increase operational burden as a result of the creation and maintenance of templates.

For network slice automation, we offer the concept of automated service deconstruction and orchestration, which goes beyond the conventional template-driven approach, as shown in Figure 1.5. To that aim, we identify three main stakeholders, as defined by the 3GPP: communication service consumers (CSCs), "Communication Service Providers" (CSPs), and "Virtual Infrastructure Service Providers" (VISPs). CSCs ask CSPs for communications services, while CSPs create system parts and arrange them across VISP arrangements to provide the required services. The CSCs offer high-level needs like latency, throughput, and reliability as part of the slice automation process, similar to the developing Intent-Based Networking Paradigm [57,58].

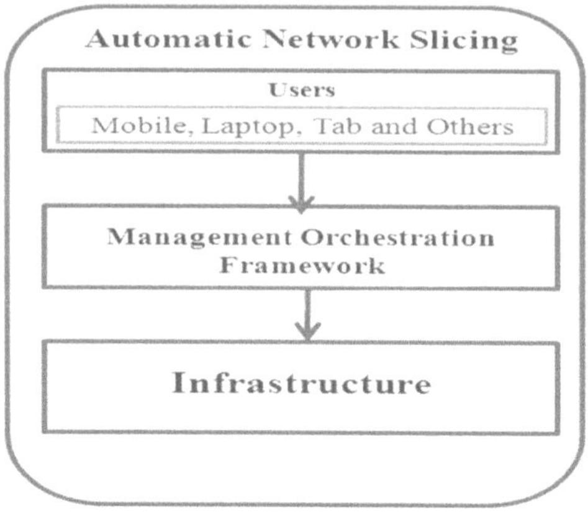

FIGURE 1.5 Framework for automated network slicing.

The request is subsequently decomposed into its component VNF-forwarding graphs by the CSC (VNF-FG). Instead of using a template to map services to VNF-FGs, CNN models abstract facility needs and generate the appropriate VNF-FG. The resource needs for the component VNFs are also included in the service-specific VNF-FG, enabling smooth deployment onto the underlying infrastructure.

1.6.3 Networks of Self-Driving Cars

The network operator has been at the heart of network operations for decades. On the other side, human operators are finding it more challenging to manage real-time networks due to the rising complexity of communications networks and the ever-increasing number of linked gadgets that create a continuous state of flux as a consequence [58]. A "Self-Driving Network" should start with a "High Level" objective or purpose as its input. Increasing on the idea of intentions, we may divide them into two categories: imperative and declarative. While the former specifies how a particular process should be completed in great detail, the latter states the end objective, lacking defining how the stated objective should be accomplished. However, the implementation of self-driving networks has several research difficulties, which will be discussed more below.

- **Accurate Intent Definitions:** As earlier stated, functional determined meaning necessity balance imperative and declarative elements. Automation is rendered useless if the goal is mainly urgent. The automation process, on the other hand, becomes excessively complicated if the goal is simply declarative. Furthermore, network architecture must enhance the quality of data supplied to control algorithms [59]. "Quality of Data" (QoD) is generally acknowledged as a precondition for "Quality of Service" (QoS) in the area of self-driving networks [59].
- **In-Band Telemetry (INT):** Researchers are looking at in-band telemetry because they need a sound system observing data without the extra expense. To include extra information within data packets, the INT approach

uses programmable data planes [60]. Change processing times, buffering occupancy rates, and even particular policy rules are examples of this kind of data. As packets travel across the network, they accumulate additional information that may be retrieved as needed, resulting in highly comprehensive and accurate network data sets. To that aim, INT's effect on network performance must be quantified [61–63].

1.7 RECONFIGURABLE TRANSCEIVER FRONT-ENDS FOR 6G RADIO

The ever-increasing need for more incredible wireless data speeds, joined with the enormous rise in the number of wirelessly linked devices, has resulted in an overcrowded electromagnetic (EM) spectrum. The higher-frequency ranges must be addressed to solve the spectrum shortage issue and increase the wireless network's capacity. From RF to TeraHz, solutions including frequency-agile, ultra-broadband reconfigurable systems must be designed, implemented, and optimized. This device, which can detect and transmit simultaneously over the whole electromagnetic spectrum (1 GHz to 10 THz), is essential to the "Next Generation Wireless Communications" NGWC infrastructure. Through this area, we aim to inspire progressive ideas to achieve all-spectrum communications. Figure 1.6 depicts All Spectrum Detecting and Communications, including Advanced CMOS with antenna, Optoelectronics and Plasmonics for 6G and Beyond 6G.

1.7.1 All-Spectrum Sensing and Access in a Dynamic Model

In recent years, a concerted research effort at the physical and connection levels has resulted in promising development for individual cognitive radios at RF frequencies (CRs). For example, the UK's Research and Innovation Program recently funded the "6G Mitola Radio" research project, which seeks to create self-regulating

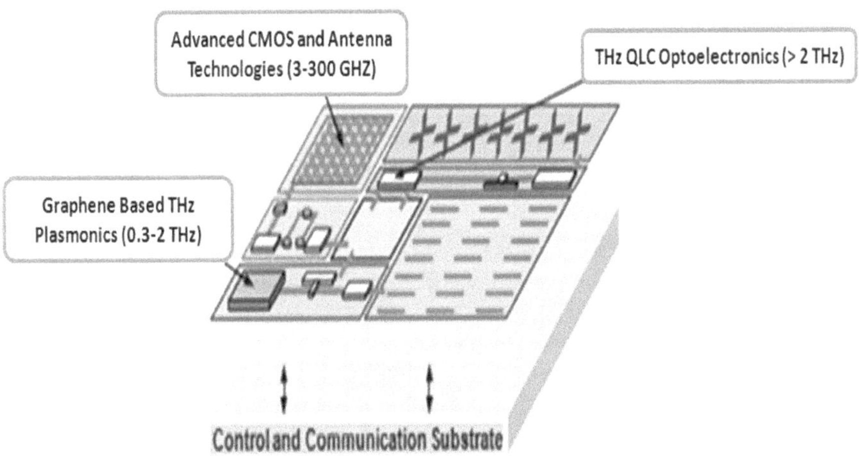

FIGURE 1.6 6G hybrid front-end for dynamic all spectrum detecting and communication.

communities for wireless communications that are fair and efficient 4. This study will enable smooth convergence across diverse wireless networks by allowing radios to make intelligent choices to improve end-user experience.

1.7.2 Design of a Multi-Band Transceiver

Any multi-band transceiver design must make the best possible choice of materials and technologies to allow all-spectrum communications. For multi-band operations, most existing solutions depend on CMOS. However, this technique is only effective in narrow bands [64–68].

1.8 COMMUNICATIONS IN THE "AMBIENT BACKSCATTER"

Sensors are required to work in various conditions and have long battery life in the IoT world. Backscatter technology is used in Radio Frequency Identification (RFID) solutions to moderate and reflect Radio Frequency signals rather than generate them, saving much energy. Due to signal attenuation over long distances, current modulated backscatter systems impose strict restrictions on the closeness of the backscatter transmitter to the RF source. Furthermore, backscatter transmitters have modulated backscatter receivers. Thus, they are passive and can transmit data only when backscatter receivers [69] make requests. As a result, new solutions in the 6G IoT network are needed to improve energy efficiency while increasing flexibility and scalability.

1.8.1 Backscatter Communications Operation Principles

"Backscatter communications," as the name advises, reflect signals to the source. The indications are dispersed across a specific angular choice of the surroundings due to the imperfect reflection. The signals may therefore be picked up by a "Backscatter Communication Receiver" within range. Monostatic, bistatic, and "Ambient Backscatter Communications" (ABC) are three architectural variations of backscatter communications [70,71]. We will go over the first two types briefly before turning our attention to the third.

1.8.2 Ambient Backscatter Communications Problems with Open Problems

Energy Efficiency and Spectral Range: Ambient backscatter communications are still in their initial research and development steps. As previously stated, good performance requires careful planning of backscatter devices. The current solutions are behind owing to the lack of predictable IoT device deployment. While maintaining transmission distance, long-range IoT devices should use ambient backscatter connections. Individual backscatter communication devices may be energy efficient, but IoT networks with hundreds or thousands of them may need system-level energy optimization.

1.9 THE INTERNET OF SPACE THINGS FOR CUBESATS AND UAVS

IoST refers to a combination of the Internet of Things and Space Things; it describes how things on Earth may be connected and communicates with one another. This expansion is required for future communication networks for the following reasons: i) The IoT is heavily reliant on the existing arrangements and thus lacks flexibility and adaptability; (ii) traditional IoT solutions cannot provide global coverage, particularly in rural regions such as the South Pole and the North Pole, due to a mismatch between construction costs and income; and (iii) traditional IoT solutions cannot provide global coverage due to a mismatch between construction costs and service revenue; and (iv) traditional.

To meet this objective, the "Internet of Things" (IoT) is a pervasive "Cyber-Physical System" with applications in watching and surveillance, backhauling in space, as well as ground-to-space data integration, air, and space. As illustrated in Figure 10, the Internet of Things comprises ground stations, client locations, and on-Earth detecting devices. The space sector includes CubeSats, UAVs, and near-Earth monitoring equipment. "Ground-to-Satellite Connections" (GSLs) enable "IoST Hubs & Cube-Sats" to conversation needs and information. "Inter Satellite links" (ISLs) transmit data to CubeSats in a similar orbit in addition to orbits neighboring to it. The UAVs also create a localized data aggregation layer by establishing connections with sensors and CubeSats. To that aim, we should mention that UAVs are anticipated to play a significant role in the forthcoming 3-GPP announcements 16 and 17 [72]. In addition, since the space segment is such an essential part of the system, research into creating tiny satellites, or CubeSats [73], for use in IoST is necessary [74].

1.9.1 Subsystem for Multi-Band Communications

Existing Cube-Sats have limited communications capability, relying on bandwidth ranging from 1–2 GHz to 26.5–40 GHz. A "Multi-Band Communications Subsystem" is necessary for this reason. There are two main disadvantages to this strategy. The conventional frequency bands, for starters, are getting more congested [75]. With current frequency bands, the Tbps-level throughput needed by IoST cannot be accomplished. IoST [76] suggested using various frequency bands ranging from RF to THz spectrum to address spectrum shortages and capacity constraints in existing satellite networks. Advances in high-frequency device improvement have enabled such frequencies [77–79]. We have created multi-frequency transceivers and antenna systems as part of the multi-band communications sub-system, as detailed below [80–82].

1.9.2 Design of the System Constellation

To provide precise worldwide coverage and adequate connection performance, an optimal constellation design is necessary for the context of IoST. Contrary to conventional LEO constellations, which may only include a few dozen satellites, more advanced LEO constellations may include hundreds or even thousands of spacecraft; for example, Astrocast, a CubeSat-based IoT system, can hold up to 64 satellites [83]. At first glance, such systems' coverage and connection leave a lot to be desired. Mega-constellations of several hundred satellites have acquired considerable momentum in the last year [84], driven by the demand for better coverage, dependable connection, and higher redundancy. "Mega-Constellations" deliver some benefits over conventional constellations, together with greater coverage density, better relationship, and increased redundancy, to name a few. IoST makes substantial use of "SDN and NFV" to increase network resource efficiency meaningfully, shorten network administration, and decrease operating costs [74], going beyond the typical bent pipe structure of "Satellite Communications Systems." IoST aims to provide CubeSats-as-a-service in a way close to the "infrastructure-as-a-service" (IaaS) model, with promising outcomes as demonstrated in [74].

1.10 MULTI-USER MIMO COMMUNICATION NETWORKS OF CELL-FREE MUMCNF

MU-MIMO Communications has been tested and used in 5G wireless networks [80] at base stations with over 100 antenna components to enhance array gain and utilize diversity. Network MIMO is a similar idea: rather than cramming more than one hundred antenna components into only Base Station, it creates a synchronized structure made up of many Base Stations with many antennas. This coordination system achieves spatial divergence by enabling a single user to be serviced by several BSs simultaneously when just one BS is connected to the user, mitigating the drawback of lousy channel conditions and eliminating inter-cell interference [82]. On the other hand, according to thorough comparison research, MU-MIMO communication methods exceed network MIMO in end-user received signal potency and total costs in setup [85]. In comparison to traditional MIMO with cell-free MU-MIMO, as shown in Figure 1.7.

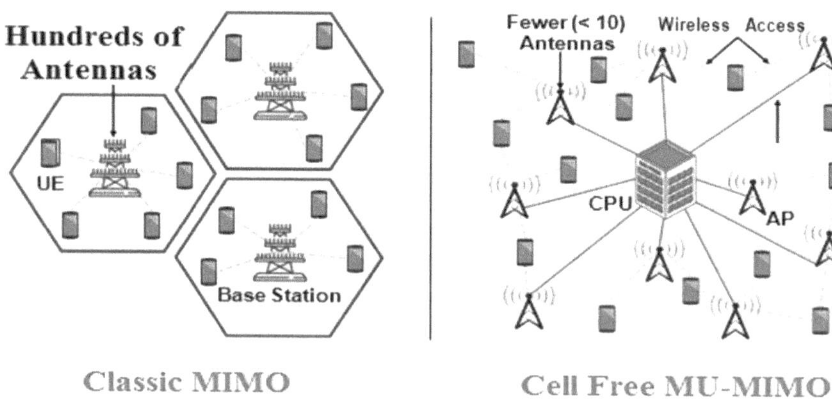

FIGURE 1.7 Comparison to traditional MIMO with cell-free MU-MIMO.

1.10.1 Cell-Free MU-MIMO Communication Systems Channel Characteristics

Inter-cell interference, small-scale fading, and other adversarial channel effects are theoretically demonstrated to disappear when the number of antenna components approaches infinity [84]. Such effects will not influence propagation channels in cell-free MIMO communications. Media under this kind of coordinated system will, in particular, fulfill the criteria of sound propagation [86]. To optimize the sum rate, favorable propagation conditions require that the channel vectors between the BSs and UEs be orthogonal. This feature is particularly noticeable in traditional extensive MIMO communications [87]. It has been shown that in cell-free MU-MIMO systems with a high number of APs (about 1000/km^2), optimal propagation conditions may be obtained.

1.10.2 Open Problems in MU-MIMO Communications without a Cell

Because the field of "Cell-Free MIMO Communications" is still developing, several unsolved issues need to be investigated further. We believe that coordination and optimization issues will significantly impact the system's performance as a whole and future deployments.

- **User Scheduling:** Channel Characterization and capacity study have been the subject of much research. On the other hand, the prior art does not address situations connecting systems with a high number of operators to service. There may be a limit on the number of APs that can serve a user while still maintaining an acceptable average throughput in such cases. According to current research, all operators will be assisted concurrently under the same frequency source block. When the amount of operators exceeds the capacity of the system to provide them all at the same time, a fair scheduling system should be considered.

1.11 TECHNOLOGIES FOR A FUTURE ABOVE AND BEYOND 6G

For the time being, we have gone through the essential wireless network drivers that will shape the future generation of "Wireless Communications." Many interesting early phase technologies have emerged that are set to alter the way we think about data transmission shortly. To that aim, we will look at three potential concepts in this section: the "Internet of Nano Things," the Internet of "Bio-Nano-Things," and "Quantum Communications."

1.11.1 Nano Things Internet

Additional spectrum resources will be required to support a multitude of wireless devices and services. Several new wireless communications models are also expected to be realized shortly, alongside the requirement for more spectrum resources to meet the diverse demands of wireless devices and services.

With the rise of wireless ubiquity, we have seen certain circumstances where electromagnetic waves do not function well. Few of them are not reachable owing to hardware constraints, for example, in hyper-saline water or via the vasculature, where transmission ranges may be shallow as the frequencies of operation rise in the application mentioned above scenarios for THz band communications. Sounds range from the nanometer (one billionth of a meter) to the meter (one ten-millionth of a meter) (i.e., 109 to 107 m in size), prompting research on nano-network communications [88,89]. Unlike traditional wire-free communication systems, the transceivers and devices used on the "Internet of Nano Things" (IoNT) operate on the nano-scale scale and are programmed differently.

1.11.2 Bio-Nano-Things Internet for Health Applications

The idea of the Internet of Bio-NanoThings is fundamental to IoNT because of its unique features and uses (IoBNT). The IoBNT, which was first presented in 2015, has gained much momentum for its attempts to integrate telecoms and healthcare solutions [90,91] synergistically. The IoBNT presents several unique difficulties and possibilities in human healthcare applications:

Interdisciplinary study in both communications and data analytics may significantly aid in modeling biological processes, such as "Cancer Cell Formations" – CCF, "Alzheimer's Disease," and the development of efficient disease management strategies. The IoBNT's holistic network design will combine components at diverse levels, such as inside cells and across tissues, organs, and systems.

1.11.3 Quantum Communications

More spectrum, transceiver front-ends, complicated communications, and higher reliability requirements are anticipated as networks go beyond 6G. Quantum computing has long been regarded as a crucial empowering technology for creating complicated systems [92]. "Quantum Systems" are instrumental in the solution of complex optimization problems. Traditional methods, such as the geographic routing algorithm, demonstrate considerable complexity to produce optimum solutions in the optimal routing issue with multiple goals, while less complicated procedures often compromise optimality. Quantum computing has been demonstrated to decrease complexity while attaining optimality in such situations.

1.12 TENTATIVE TIMELINE FOR 6G

To this point, we have considered how societal needs drive the shift from 5th Generation to 6th Generation, as well as a slew of novel and forthcoming applications that may greatly benefit from 6G. The growing technical maturity and global deployments of 5G systems are illustrated in Figure 1.8.

Simultaneously, we have spoken about how general requirements will drive the shift from 5th Generation to 6th Generation, as well as a host of fresh and developing use cases that will gain the most from 6G, culminating in critical H/W and S/W co-design demonstrations by 2025, in 2026 and beyond, full-scale 6G test-beds will be available. These test-beds will, we believe, provide the ideal setting for displaying 6G's promise and proving its applicability for applications such as multisensory

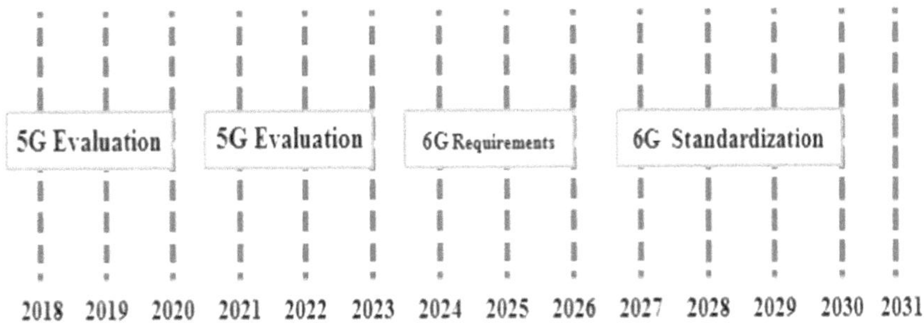

FIGURE 1.8 Timetable for 6G and beyond systems projection.

holographic teleportation, "Real-Time Remote Healthcare," manufacturing automation, and intelligent arrangement and surroundings, to mention a few.

1.13 CONCLUSION

The primary supporting methods for the next era of Advanced Wireless Communication are examined in this chapter, together with their crucial usage cases and a viewpoint on present and future research and improvement. We think that in the future, 6G and Beyond-6G will be the standard.

We believe that the emphasis on wireless ubiquitousness, or the unrestricted accessibility of high-quality wireless connectivity, will be central to 6G and beyond wireless networks. To that aim, we have identified the essential enabling technologies that will make 6G and Beyond-6G systems a success. We hope that by specifying both the working subtleties and open problems connected with each one, readers will get a complete knowledge of the next frontier in advanced wireless networks and be inspired to contribute to achieving the envisioned ubiquitous wireless future.

REFERENCES

1. N. Rajatheva, I. Atzeni, E. Bjornson, A. Bourdoux, S. Buzzi, J. B. Dore, S. Erkucuk, M. Fuentes, K. Guan, Y. Hu, X. Huang, J. Hulkkonen, J. M. Jornet, M. Katz, R. Nilsson, E. Panayirci, K. Rabie, N. Rajapaksha, M. Salehi, H. Sarieddeen, T. Svensson, O. Tervo, A. Tolli, Q. Wu, and W. Xu, "White Paper on Broadband Connectivity in 6G," 2020, arXiv:2004.14247.
2. S. Mohapatra, S. Mohanty, and S. Mohanty, "Smart Healthcare: An Approach for Ubiquitous Healthcare Management Using IoT," in *Big Data Analytics for Intelligent Healthcare Management*, Elsevier, 2019, pp. 175–196.
3. G. Bresson, Z. Alsayed, L. Yu, and S. Glaser, "Simultaneous Localization and Mapping: A Survey of Current Trends in Autonomous Driving," *IEEE Transactions on Intelligent Vehicles*, vol. 2, no. 3, pp. 194–220, September 2017.
4. R. Shakeri, M. A. Al-Garadi, A. Badawy, A. Mohamed, T. Khattab, A. K. Al-Ali, K. A. Harras, and M. Guizani, "Design Challenges of Multi-UAV Systems in Cyber-Physical Applications: A Comprehensive Survey and Future Directions," *IEEE Communications Surveys & Tutorials*, vol. 21, no. 4, pp. 3340–3385, 2019.
5. P. Fan, J. Zhao, and C.-L. I, "5G High Mobility Wireless Communications: Challenges and Solutions," *China Communications*, vol. 13, no. Supplement 2, pp. 1–13, 2016.

6. I. Akyildiz, J. Jornet, and C. Han, "TeraNets: Ultra-Broadband Communication Networks in the Terahertz Band," *IEEE Wireless Communications*, vol. 21, no.4, pp. 130–135, August 2014.
7. FCC, "FCC Take Steps to Open Spectrum Horizons for New Services and Technologies," March 2019. [Online]. Available: https://docs.fcc.gov/public/attachments/DOC-356588A1.pdf
8. S. Abadal, C. Han, and J. M. Jornet, "Wave Propagation and Channel Modeling in Chip-Scale Wireless Communications: A Survey from Millimeter-Wave to Terahertz and Optics," *IEEE Access*, vol. 8, pp. 278–293, 2019.
9. T. Nagatsuma and G. Carpintero, "Recent Progress and the Prospect Of Photonics-enabled Terahertz Communications Research," *IEICE Transactions on Electronics*, vol. 98, no.12, pp. 1060–1070, 2015.
10. T. Ishibashi, Y. Muramoto, T. Yoshimatsu, and H. Ito, "Uni Traveling-Carrier Photodiodes for Terahertz Applications," *IEEE Journal of Selected Topics in Quantum Electronics*, vol. 20, no. 6, pp. 79–88, November 2014.
11. H. Ito and T. Ishibashi, "Photonic Terahertz-Wave Generation Using Slot-antenna-Integrated Uni-traveling -Carrier Photodiodes," *IEEE Journal of Selected Topics in Quantum Electronics*, vol. 23, no. 4, pp. 1–7, July 2017.
12. J. L. Kloosterman, D. J. Hayton, Y. Ren, T.-Y. Kao, J. Hovenier, J.-R. Gao, T. M. K lapwijk, Q. Hu, C. K. Walker, and J. L. Reno, "Hot Electron Bolometer Heterodyne Receiver with a 4.7-THz Quantum Cascade Laser as a Local Oscillator," *Applied Physics Letters*, vol. 102, no. 1, p. 011123, 2013.
13. J. M. Jornet and I. F. Akyildiz, "Channel Modeling and Capacity Analysis for Electromagnetic Wireless Nanonetworks in the Terahertz Band," *IEEE transactions on Wireless Communications*, vol. 10, no. 10, pp. 3211–3221, October 2011.
14. S. Priebe and T. Kurner, "Stochastic Modeling of the Indoor Radio Channels," *IEEE Transactions on Wireless Communications*, vol. 12, no. 9, pp. 4445–4455, 2013.
15. D. He, K. Guan, A. Fricke, B. Ai, R. He, Z. Zhong, A. Kasamatsu, I. Hosaka, and T. Kürner, "Stochastic Channel Modeling for Kiosk Applications in the Terahertz Band," *IEEE Transactions on Terahertz Science and Technology*, vol. 7, no. 5, pp. 502–513, 2017.
16. C. Han, X. Zhang, and X. Wang, "On Medium Access Control Schemes for Wireless Networks in the Millimeter-Wave and Terahertz Bands," *Nano Communication Networks*, vol. 19, pp. 67–80, 2019.
17. R. R. Choudhury and N. H. Vaidya, "Deafness: A Mac Problem in Ad Hoc Networks when Using Directional Antennas," in Proceedings of the 12th IEEE International Conference on Network Protocols, 2004. ICNP 2004. IEEE, 2004, pp. 283–292.
18. IEEE 802.15. 3c, "Part 15.3: Wireless Medium Access Control (MAC) and Physical Layer (PHY) Specification for High-Rate Wireless Personal Area Networks (WPANs): Amendment 2: Millimeter-Wave Based Alter-native Physical Layer Extension," 2009.
19. X. Gao, L. Dai, Y. Zhang, T. Xie, X. Dai, and Z. Wang, "Fast Channel Tracking for Terahertz Beam Space Massive MIMO Systems," *IEEE Transactions on Vehicular Technology*, vol. 66, no. 7, pp. 5689–5696, 2016.
20. C. Han, W. Tong, and X.-W. Yao, "Madam: A Memory Assisted Angular-Division-Multiplexing Mac Protocol in Terahertz Communication Networks," *Nano Communication Networks*, vol. 13, pp. 51–59, 2017.
21. A. Tulumello, G. Belocchi, M. Bonola, S. Pontarelli, and G. Bianchi, "Pushing Services to the Edge Using a Stateful Programmable Data-Plane," in 2019 European Conference on Networks and Communications (EuCNC), June 2019.
22. I. F. Akyildiz, C. Han, and S. Nie, "Combating the Distance Problem in the Millimeter-Wave and Terahertz Frequency Bands," *IEEE Communications Magazine*, vol. 56, no. 6, pp. 102–108, June 2018.

23. J. Solé-Pareta, E. Alarcón, and A. Cabellos-Aparicio, "Computing and Communications for the Software-Defined Metamaterial Paradigm: A Con-text Analysis," *IEEE Access*, vol. 5, pp. 6225–6235, 2017.
24. Q. Wu and R. Zhang, "Intelligent Reflecting Surface-Enhanced Wireless Network: Joint Active and Passive Beamforming Design," in 2018 IEEE Global Communications Conference (GLOBECOM), 2018, pp. 1–6.
25. C. Huang, G. C. Alexandropoulos, A. Zappone, M. Debbah, and Yuen, "Energy-Efficient Multi-User Miso Communication Using Low Resolution Large, Intelligent Surfaces," in 2018 IEEE Globecom Workshops (GCWkshps), 2018, pp. 1–6.
26. L. Zhang, X. Q. Chen, S. Liu, Q. Zhang, J. Zhao, J. Y. Dai, G. D. Bai, X. Wan, Q. Cheng, G. Castaldi, et al., "Space-Time-coding Digital Metasurfaces," *Nature Communications*, vol. 9, no. 1, p. 4334, 2018.
27. N. Kato, Z. M. Fadlullah, B. Mao, F. Tang, O. Akashi, T. Inoue, and K. Mizutani, "The Deep Learning Vision for Heterogeneous Network Traffic Control: Proposal, Challenges, and Future Perspective," *IEEE Wireless Communications*, vol. 24, no. 3, pp. 146–153, June 2017.
28. N. Farsad and A.G. Goldsmith, "Neural Network Detection Of Data Sequences In Communication Systems," *IEEE Transactions on Signal Processing*, vol. 66, no. 21, pp. 5663–5678, 2018.
29. Y. Lv, Y. Duan, W. Kang, Z. Li, and F.-Y. Wang, "Traffic Flow Prediction with Big Data: A Deep Learning Approach," *IEEE Transactions on Intelligent Transportation Systems*, vol. 16, no. 2, pp. 865–873, 2014.
30. N. Taherkhani and S. Pierre, "Centralized and Localized Data Congestion Control Strategy for Vehicular Ad Hoc Networks Using a Machine Learning Clustering Algorithm," *IEEE Transactions on Intelligent transportation systems*, vol. 17, no. 11, pp. 3275–3285, 2016.
31. S. Nie, J. M. Jornet, and I. F. Akyildiz, "Deep-Learning-Based Resource Allocation for Multi-band Communications in CubeSat Networks," in 2019 IEEE International Conference on Communications Workshops (ICCWorkshops). IEEE, 2019, pp. 1–6.
32. G. Piao, "Machine & Deeplearning for Network Management: An Overview with Benchmarks," Blogpost, 2018.[Online]. Available: https://goo.gl/gp7gBb
33. R. Boutaba, M. A. Salahuddin, N. Limam, S. Ayoubi, N. Shahriar, F. Estrada-Solano, and O. M. Caicedo, "A Comprehensive Survey on Machine Learning for Networking: Evolution, Applications, and Research Opportunities," *Journal of Internet Services and Applications*, vol. 9, no. 1, June 2018.
34. P. Wang, S.-C. Lin, and M. Luo, "A Framework for QoS-Aware Traffic Classification Using Semi-supervised Machine Learning in SDNs," in 2016 IEEE International Conference on Services Computing (SCC), June 2016.
35. C. Gutterman, E. Grinshpun, S. Sharma, and G. Zussman, "RAN Re-source Usage Prediction for a 5G Slice Broker," in Proceedings of the Twentieth ACM International Symposium on Mobile Ad Hoc Networking and Computing, July 2019.
36. H. Mao, M. Alizadeh, I. Menache, and S. Kandula, "Resource Management with Deep Reinforcement Learning," in Proceedings of the 15th ACM Workshop on Hot Topics in Networks – HotNets '16, 2016.
37. D. Zeng, L. Gu, S. Pan, J. Cai, and S. Guo, "Resource Management at the Network Edge: A Deep Reinforcement Learning Approach," *IEEE Network*, vol. 33, no. 3, pp. 26–33, May 2019.
38. M. Dolati, S. B. Hassanpour, M. Ghaderi, and A. Khonsari, "DeepViNE: Virtual Network Embedding with Deep Reinforcement Learning," in IEEE INFOCOM 2019 – IEEE Conference on Computer Communications Workshops (INFOCOM WKSHPS), IEEE, Apr. 2019.

39. B. Li, W. Lu, S. Liu, and Z. Zhu, "Deep-Learning-Assisted Network Orchestration for On-Demand and Cost-Effective vNF Service Chaining in Inter-DC Elastic Optical Networks," *Journal of Optical Communications and Networking*, vol. 10, no. 10, p. D29, May 2018.
40. S. Ayoubi, N. Limam, M. A. Salahuddin, N. Shahriar, R. Boutaba, F. Estrada-Solano, and O. M. Caicedo, "Machine Learning for Cognitive Network Management," *IEEE Communications Magazine*, vol. 56, no. 1, pp. 158–165, January 2018.
41. P. Ahammad, B. Kennedy, P. Ganti, and H. Kolam, "QoE-Driven Unsupervised Image Categorization for Optimized Web Delivery," in Proceedings of the ACM International Conference Multimedia-MM'14, 2014.
42. M. Usama, J. Qadir, A. Raza, H. Arif, K. Lim Alvin Yau, Y. Elkhatib, A. Hussain, and A. Al-Fuqaha, "Unsupervised Machine Learning for networking: Techniques, Applications, and Research Challenges," *IEEE Access*, vol. 7, pp. 65579–65615, 2019.
43. H. Choi, M. Kim, G. Lee, and W. Kim, "Unsupervised Learning Approach for Network Intrusion Detection System Using Autoencoders," *The Journal of Supercomputing*, vol. 75, no. 9, pp. 5597–5621, March 2019.
44. J. Biamonte, P. Wittek, N. Pancotti, P. Rebentrost, N. Wiebe, and S. Lloyd, "Quantum Machine Learning," *Nature*, vol. 549, no. 7671, p. 195, 2017.
45. R. Li, Z. Zhao, X. Zhou, G. Ding, Y. Chen, Z. Wang, and H. Zhang, "Intelligent5g: When Cellular Networks Meet Artificial Intelligence," *IEEE Wireless Communications*, vol. 24, no. 5, pp. 175–183, 2017.
46. Akyildiz, A. Kak, E. Khorov, A. Krasilov, and A. Kureev, "ARBAT: A Flexible Network Architecture for QoE-Aware Communications in 5G Systems," *Computer Networks*, vol.147, pp.262–279, December 2018.
47. B. Pfaff, J. Pettit, T. Koponen, E. Jackson, A. Zhou, J. Rajahalme, J. Gross, A. Wang, J. Stringer, P. Shelar, K. Amidon, and M. Casado, "The Design and Implementation of Open v Switch," in 12th USENIX Symposium on Networked Systems Design and Implementation (NSDI'15), May 2015.
48. N. Nikaein, M. K. Marina, S. Manickam, A. Dawson, R. Knopp, and C. Bonnet, "Open Air Interface: A Flexible Platform for 5G Research," *ACM SIGCOMM Computer Communication Review*, vol. 44, no. 5, pp.33–38, October 2014.
49. I. Gomez-Miguelez, A. Garcia-Saavedra, P. D. Sutton, P. Serrano, C. Cano, and D. J. Leith, "Srs LTE: An Open-Source Platform for LTE Evolution and Experimentation," in Proceedings of the Tenth ACM International Workshop on Wireless Network Testbeds, Experimental Evaluation, and Characterization-WiNTECH'16, 2016.
50. J. Networks. What is Network Automation. [Online]. Available: https://www.juniper.net/us/en/products-services/what-is/network-automation/
51. N. McKeown, T. Anderson, H. Balakrishnan, G. Parulkar, L. Peterson, J. Rexford, S. Shenker, and J. Turner, "OpenFlow: Enabling Innovation in Campus Networks," *ACM SIGCOMM Computer Communication Review*, vol. 38, no. 2, pp. 69–74, 2008.
52. P. Bosshart, G. Varghese, D. Walker, D. Daly, G. Gibb, M. Izzard, N. McKeown, J. Rexford, C. Schlesinger, D. Talayco, and A. Vahdat, "P4," *ACMSIGCOMM Computer Communication Review*, vol. 44, no. 3, pp. 87–95, July 2014.
53. C. Sun, J. Bi, H. Chen, H. Hu, Z. Zheng, S. Zhu, and C. Wu, "SDPA: Toward a Stateful Data Plane in Software-Defined Networking," *IEEE/ACM Transactions on Networking*, vol. 25, no. 6, pp. 3294–3308, December 2017.
54. C. Cascone, R. Bifulco, S. Pontarelli, and A. Capone, "Relaxing State-Access Constraints in Stateful Programmable Data Planes," *ACM SIG-COMM Computer Communication Review*, vol. 48, no. 1, pp. 3–9, April 2018.
55. Technical Specification Group Services and System Aspects, "Telecommunication Management; Study on Management and Orchestration of Network Slicing for Next-Generation Network," 3rd Generation Partnership Project (3GPP), Tech. Rep. 28.801, 2018, Version 15.1.0.

56. I. Afolabi, T. Taleb, K. Samdanis, A. Ksentini, and H. Flinck, "Networks Living and Softwarization: A Survey on Principles, Enabling Technologies, and Solutions," *IEEE Communications Surveys & Tutorials*, vol. 20, no. 3, pp. 2429–2453, 2018.
57. Badmus, M. Matinmikko -Blue, J. S. Walia, and T. Taleb, "NetworkSlice Instantiation for 5G Micro-Operator Deployment Scenarios," in 2019 European Conference on Networks and Communications (EuCNC), IEEE, June 2019.
58. T. Szigeti, D. Zacks, M. Falkner, and S. Arena, *Cisco Digital Network Architecture: Intent-based Networking for the Enterprise*. Cisco Press, 2019.
59. M. Boucadair and C. Jacquenet, Eds., *Emerging Automation Techniques for the Future Internet*. Information Science Reference, 2019.
60. N. Feamster and J. Rexford, "Why (and How) Networks Should Run Themselves," in Proceedings of the Applied Networking Research Workshop on – ANRW 2018, 2018.
61. C. Kim, P. Bhide, E. Doe, H. Holbrook, A. Ghanwan, D. Daly, M. Hira, and B. Davie, "In-band Network Telemetry (INT)," *The P4 Language Consortium, Tech. Rep.*, 2016.
62. J. A. Marques, M. C. Luizelli, R. I. T. da Costa Filho, and L. P. Gaspary, "An Optimization-Based Approach for Efficient Network Monitoring Using In-band Network Telemetry," *Journal of Internet Services and Applications*, vol. 10, no. 1, June 2019.
63. M. Rais-Zadeh, J. T. Fox, D. D. Wentzloff, and Y. B. Gianchandani, "Reconfigurable Radios: A Possible Solution to Reduce Entry Costs in Wireless Phones," *Proceedings of the IEEE*, vol. 103, no. 3, pp. 438–451, 2015.
64. K. Tan, H. Liu, J. Zhang, Y. Zhang, J. Fang, and G. M. Voelker, "Sora: High-Performance Software Radio Using General-Purpose Multi-core Processors," *Communications of the ACM*, vol. 54, no. 1, pp. 99–107, 2011.
65. H. Zhu, S. Cheung, K. L. Chung, and T. I. Yuk, "Linear to Circular Polarization Conversion Using Metasurface," *IEEE Transactions on Antennas and Propagation*, vol. 61, no.9, pp. 4615–4623, 2013.
66. X. Liu, S. Yao, B. S. Cook, M. M. Tentzeris, and S. V. Georgakopoulos, "An Origami Reconfigurable Axial Mode Bifilar Helical Antenna," *IEEE transactions on Antennas and Propagation*, vol. 63, no. 12, pp. 5897–5903, 2015.
67. X. Liu, S. V. Georgakopoulos, and S. Rao, "A Design of an Origami Reconfigurable Qhawith a Foldable Reflector [Antenna Applications Corner]," *IEEE Antennas and Propagation Magazine*, vol. 59, no. 4, pp. 78–105, 2017.
68. J. M. Jornet and I. F. Akyildiz, "Graphene-Based Plasmonic Nano-Antenna for Terahertz Band Communication in Nanonetworks," *IEEE-JSAC, Special Issue on Emerging Technologies for Communications*, vol. 12, no. 12, pp. 685–694, December 2013.
69. Business Wire, "Astrocast Launches First Test Satellite of IoT Cube Sat Network," 2018.
70. S. Yao et al., "A Novel Reconfigurable Origami Spring Antenna," in Antennas and Propagation Society International Symposium (APSURSI), 2014, pp. 374–375.
71. Van Huynh, D. T. Hoang, X. Lu, D. Niyato, P. Wang, and D. I. Kim, "Ambient Backscatter Communications: A Contemporary Survey," *IEEE Communications Surveys & Tutorials*, vol. 20, no. 4, pp. 2889–2922, 2018.
72. I. F. Akyildiz, J. M. Jornet, and S. Nie, "A New CubeSat Design Wither Configurable Multi-band Radios for Dynamic Spectrum Satellite Communication Networks," *AdHoc Networks*, vol. 86, pp. 166–178, 2019. [Online]. Available: http://www.sciencedirect.com/science/article/pii/S1570870518309247
73. 3 GPP, "Unmanned Aerial Systems over 5G," November 2019. [Online]. Available: https://www.3gpp.org/technologies/keywords-acronyms/2090-unmanned-aerial-systems-over-5G
74. Liu, A. Parks, V. Talla, S. Gollakota, D. Wetherall, and J. R. Smith, "Ambient Backscatter: Wireless Communication out of Thin Air," in *ACMSIGCOMM Computer Communication Review*, vol. 43, no. 4. ACM, 2013, pp. 39–50.

75. H. Heidt, J. Puigsuari, A. S. Moore, S. Nakasuka, and R. J. Twiggs, "CubeSat: A New Generation of Picosatellite for Education and Industry Low-Cost Space Experimentation," in 14th Annual/USU Conference on Small Satellites, 2001.
76. I. F. Akyildiz and A. Kak, "The Internet of Space Things/CubeSats: Aubiquitous Cyber-Physical System for the Connected World," *Computer Networks*, vol. 150, pp. 134–149, 2019.
77. S. Madry, *Space Systems for Disaster Warning, Response, and Recovery*. Springer-Verlag GmbH, 2014.
78. Z. Pi and F. Khan, "An Introduction to Millimeter-Wave Mobile Broadband Systems," *IEEE Communications Magazine*, vol. 49, no. 6, pp. 101–107, June 2011.
79. I. F. Akyildiz, J. M. Jornet, and C. Han, "Terahertzband: Next Fontier for Wireless Communications," *Physical Communication*, vol. 12, pp. 16–32, September 2014.
80. S. Venkatesan, A. Lozano, and R. Valenzuela, "Network Mimo: Over-coming Intercell Interference in Indoor Wireless Systems," in Proc. Asilomar Conference on Signals, Systems, and Computers (ACSSC'07).Cite-seer, 2007, pp. 83–87.
81. B. Kiranmani, K. Gurucharan, and S. S. Kiran, "A Brief Insight on Game-Theoretic Approach for Wireless Communications," *Solid State Technology*, vol. mac_mac 63, no. 5, pp. 5174–5189, 2020. ISSN 0038-111X.
82. B. Kiranmai, V. Lavanya, and S. S. Kiran, "An Energy-Efficient Resource Allocation System Using OFDM DAS Model for LTE Applications," (eds.), *Communication Software and Networks, Lecture Notes in Networks and Systems* 134, 10.1007/978-981-15-5397-4_57 © Springer Nature Singapore Pte Ltd. 2021.
83. Guidotti, O. Kodheli, and A. Vanelli-Coralli, "Integration of 5G Technologies in LEO Mega Constellations," *IEEE 5G Tech Focus*, vol. 2, no. 1, March 2018.
84. K. Hosseini, W. Yu, and R. S. Adve, "Large-scale MIMO versus Network MIMO for Multicell Interference Mitigation," *IEEE Journal of Selected Topics in Signal Processing*, vol. 8, no. 5, pp. 930–941, 2014.
85. B. R. Devi, B. Kiranmai, K. Gurucharan, S S Kiran, and P. Srujana, A Coherent Hybrid Precoding for homogenizing Millimeter-Wave Multiple- in Multiple- out Systems for 5G Communication. Proceedings of the Fifth International Conference on Computing Methodologies and Communication (ICCMC 2021) DVD Part Number: CFP21K25-DVD: ISBN: 978-0-7381- 1204-6/21 © 202,. IEEE, pp. 207–211.
86. T. L. Marzetta, "Noncooperative Cellular Wireless with Unlimited Numbers of Base Station Antennas," *IEEE Transactions on Wireless Communications*, vol. 9, no. 11, pp. 3590–3600, 2010.
87. Z. Chen and E. Björnson, "Channel Hardening and Favorable Propagation in Massive Cell-Free MIMO with Stochastic Geometry," *IEEE Transactions on Communications*, vol. 66, no. 11, pp. 5205–5219, 2018.
88. H. Q. Ngo, E. G. Larsson and T. L. Marzetta, "Aspects of Favorable Propagation in Massive MIMO," in 2014 22nd European Signal Processing Conference (EUSIPCO), 2014, pp. 76–80.
89. I. F. Akyildiz, J. M. Jornet, and M. Pierobon, "Nanonetworks: A New Frontier in Communications," *Communications of the ACM*, vol. 54, no. 11, pp. 84–89, 2011.
90. I. F. Akyildiz and J. M. Jornet, "The Internet of Nano-Things," *IEEE Wireless Communications*, vol. 17, no. 6, pp. 58–63, 2010.
91. F. Akyildiz, M. Pierobon, and S. Balasubramaniam, "Moving Forward with Molecular Communication: From Theory to Human Health Applications [point of view]," *Proceedings of the IEEE*, vol. 107, no. 5, pp. 858–865, May 2019.
92. S. Pirandola, R. Laurenza, C. Ottaviani, and L. Banchi, "Fundamental Limits of Repeaters Quantum Communications," *Nature Communications*, vol. 8, p. 15043, 2017.

2 Comprehensive Method of Distributed Denial of Service Detection by Artificial Neural Network

K. Kumar, M. Khari, and A. Kumar
NSUT East Campus, Delhi, India

CONTENTS

2.1 INTRODUCTION

The DDoS can be demarcated as an intentional interruption to melt down system service and damage usual service. A DDoS attack occurs by imposing irrelevant services to the network or making the network busy, which behaves as an intrusive action made by an illegitimate handler. Hackers do not harm information reliably or permanently, but they work as a team and make it unavailable to legitimate users. The main objective of a Distributed Denial of Service (DDoS) strike is to affect

DOI: 10.1201/9781003230526-2

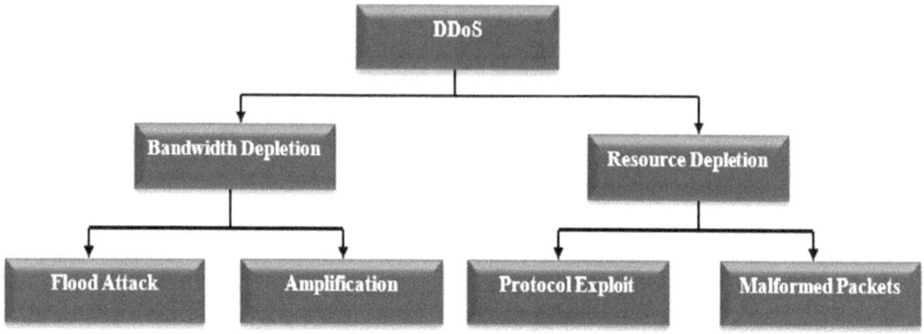

FIGURE 2.1 DDoS attack classification [3,4].

several systems by using the network with zombies and mark out botnets of networks [1,2]. The network of zombies is wished-for to target a specific object with a variety of information passing in the form of packets. The infected systems are slightly measured by either an invader or self-installed Trojans instructed to flood them with bogus information. The authors explain that the various classes of DDoS attacks are depicted as follows.

Figure 2.1 depicts the classification of the DDoS attacks, which are premediated attacks to the target system with unwelcome services that interrupt required services to the victim system. As the figure shows, the classification of different types of attacks that authors discuss as bandwidth-depletion attacks can delineate to flood the target with bogus information; flooding is the type of attack intended to clog an affected network or service by inundating it with a vast massive passage. An amplification attack is an attack that influences (DNS) resolvers to overcome a target with traffic, and a reduction attack is an outbreak that is scheduled for limiting the service to a legitimate user who is availing a malformed packet attack. In this attack, the packet comprises both identical sources with destination IP addresses, which complicates the operating system and instigates the target to break the system. This type of attack is known as a rules-based attack, which can deform packets that contain information and in which attackers damage a target server by congesting it with hefty irrelevant services. Various types of susceptibility can occur by DDoS attack [5,6]: susceptibility in networking protocol can permit the intruders to avoid mutual methods for sensing their movement, and UPnP has elevated safety fear in the context of security concerns over the years. The presence of UPnP-specific dynamic execution code controls the system after infecting it. The various ways UPnP attacks can be performed are XML-based exposure, utilizing XML files, and adapting port numbers. Sensational-based susceptibility occurs, divulging the inner code of machines to external networks. The application layer-based vulnerability involves HTTP GET and HTTP POST methods for requesting and sending information to the server. The attacker tries to exploit get or post requests to attack a web server or application. The last susceptibility is about exploitation of publicly accessible network time protocols, where an intruder abuses the server to devastate a target using UDP traffics and no flow control and congestion control in a network connection [7].

2.2 METHODS OF PERFORMING DDOS ATTACK

Various tools are currently in usage for performing DDoS attacks, such as agent-based DDoS tools, which are named trinoo and are highly utilized. The trinoo is a bandwidth-depletion attack tool that can craft synchronized UDP flood attacks by merging one or more logical addresses. [7] The shaft is another tool to perform a DDoS attack; it is an extension of the trinoo tool. It uses UDP communication between master and programmer. The famous IRC-based DDoS attack tools were made after the agent–handler attack tools. This has, as a consequence, made many IRC-based tools classier as they comprise some significant classes that can be found in numerous agent–handler attack tools. Knight is an IRC-based DDoS attack tool, one that is very trivial and influential. The Knight DDoS attack tool provides SYN attacks, UDP flood attacks, and an urgent pointer flooder by following the specific steps required to perform the DDoS attacks. The attacker chooses the programmer that will perform the attack. This equipment needs to have some susceptibly that the invader can practice penetrating to access them. An attacker must have abundant resources that will help them to construct influential attack streams. Initially, this process was done physically, but it was rapidly computerized by perusing tools [7]. The compromise in secondary approach permits this attack in which the attacker exploits loopholes in the system and weaknesses of the machines by inserting the malicious code. Moreover, attackers' efforts to shield the cipher from detection and deactivation. Another way of performing these DDoS attacks is known as self-propagating tools, such as the Ramen Worm [8] and Code Red [9], which rapidly computerized this stage. The master and operators of the programmer typically have no information about their participants who are participating in a DDoS attack. During support for performing a DDoS attack, each programmer package consumes only a fixed quantity of resources so that the masters of the process should have control of each participant to bring change to functioning. [10] Lasting but impactful attacks are known as communication attacks in which a mugger connects with managers to classify which vendors are up and controlling, and determines when to plan an attack or when to stimulate programmers to perform the attack [11].

2.3 PROBLEM STATEMENT

The DDoS can be outlined as a nonspecific intruder who wished to melt down existing connections and not deliver usual amenities. A DoS attack occurs by accessing a system resource that is expressly congested and tainted, similar to a nasty deed occupied by an alternate user. These spasms do not usually harm data unswervingly or everlastingly, but they cooperate with the accessibility of the resources. The main aim of a distributed denial-of-device (DDoS) bout is to target numerous systems through the network with zombies and outline botnets of networks. The network of zombies is intended to spasm a specific object or network with various types of data passing in the form of packets. The infected systems are slightly measured either by an invader or by self-installed Trojans instructed to flood it with bogus information. The problem in existing research reveals numerous complexities in how research needs to be improved to face intricacy. There is a process authors follow that researchers can evaluate. The research implication represents ways of organization of researching in a particular manner. In this organization of the paper, writers prepared a literature survey to analyze the current effort of authors in the context of information systems for security. The literature survey points to existing research findings and is named as an existing limitation. The limitation describes the problems in current work, which prompt us to do further research to improve a particular system. The statement of the problem needs to be considered using the present limitation. The identification of problem results gives inspiration about the purposes. The objective of a DDoS Detection System (DDDS) is to observe a system while an interloper stabs to invade the system. A signature-based DDDS explains outbreaks via a bench of malevolent signs during a rolling action that ties a sign in the bench; a notification is upturned. Numerous groups consume such devices because they are easy to use and comprehend, allowing managers to modify the known data and deliver accurate data about the ensued measures.

2.4 RESEARCH OBJECTIVE

The key objective is to propose a framework advanced or upgraded DDoS detection system. The secondary objective is to bring the framework into implementation by consuming specific parameters, such as the following.

- Propose a framework using an artificial neural network (ANN) for detecting the source of DDoS.
- Implement the DIDS for security to the spacious network using Python programming languages using a data set.
- The exactitude of classification must be improved.

The key objective of the chapter is to organize research in chronological order: Phase 1 defines the problem statement, Phase 3 states the objective, Phase 4 defines methodology, Phase 5 identifies related work, Phase 6 proposes a framework, Phase 7 describes the performance, Phase 8 discusses the research implication, Phase 9 the application work, and Phase 10 the conclusion; future scope is discussed in references.

2.5 RESEARCH METHODOLOGY

The research process is represented by research methodology, which depicts the methodical style to be obeyed when writing the script for leading the investigation in sequential order.

The Figure 2.2 represents ways of organization for researching in a particular manner. In this organization of the paper, writers prepared a literature survey to analyze the current effort of authors in the context of information systems for security. The literature survey points to existing research findings and is named as an existing limitation. The limitation describes the problems in current work, which prompts us to do further research to improve a particular system. The statement of a

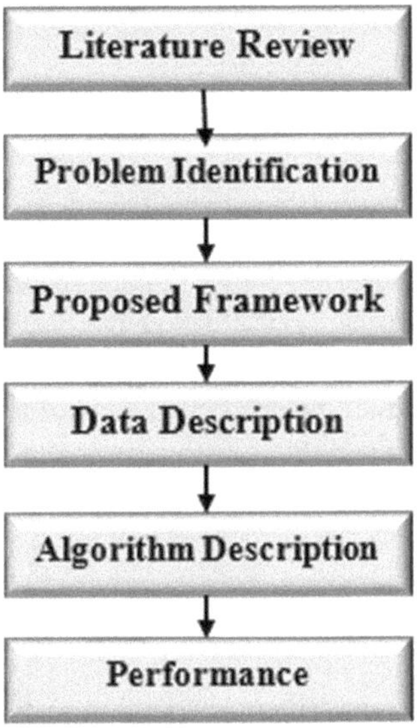

FIGURE 2.2 Research methodology.

problem needs to be considered using the present limitation. The identification of the problem results gives inspiration about the purposes. In this paper, the authors initially do a literature survey, unearth limitations, and recognize the specified drawback. The above outline is intended to attain the purpose successfully and travels to the model selection stage, where the suitable algorithm is selected for classification and the proposed model is arrayed using the selected algorithm. In this paper, writers consumed an artificial neural network (ANN) algorithm to classify the statistics. The algorithm's performance is inspected by measuring accuracy, recall, support, and f1 score.

2.6 LITERATURE REVIEW

The widespread usage of information technology in each domain has become significant to protect and strengthen data communication over the numerous platforms and systems consumed. Thus, data transfer over the internet must be secured using a leading technique.

The authors (Liao, H. J., et al. 2013) [12] suggest an extra sumptuous appraisal on IDSs. The authors abridged tables and figures to grip the public image. Additionally, the authors concisely present two well-known and exposed source tools for cramming IDSs. Another side highly significant virtualization technology is widely consumed in cloud platforms. The VM is a primary virtual module that directly associates operators, and consequently, authors observe a number of IDS problems in VMs. The authors (M. K. Elhadad et al. 2020) [13] propose a schema that can detect intrusions-based information passing through the network that depends on the source, such as WHO, UNICEF, and the DARPA as the information bases and epidemical material from the fact-checking websites. The results obtained by the author are 90.04% for Decision Tree, 90.09% for Naïve Bayes, 90.36% for

Linear Regression, 90.42% for KNN, 90.93% for Perceptron, 90.68% for Neural Networks.

The researchers (Grover, C. et al. 2019) [14] described the blockchain, and its dispersion looks to be changeable for different productions. The study's objective is to discover the blockchain technology dispersal in numerous businesses via a mixture of hypothetical works and e-community. Farouk, A., et al. 2020, elaborate on a combination of data, network, and blockchain technology in healthcare. Healthcare data have become an imperative element in this sector. There is a need to secure safe transmission of medical data across the organization. The big challenge is to collect information about patient diseases, which has become essential to restrict the pandemic in the world. Several are doing this, although it is impossible to obtain complete and temporal information at a given time. There is a need to share the data directly across the world to bring transparency between organizations. This can be done only via technology, which includes the combat of IoT networks and blockchain, which is decentralized in the system and can store the data in a distributed manner.

The writers (Linn, L. A., & Koo, M. B. 2016) [15] described blockchain technology as the benefit of this technology in various sectors such as healthcare to trace the date of disease, the government in terms of reducing the corruption, and health investigators to distribute a large amount of hereditary, intake, standard routine, ecological, and well-being data with security, privacy, and fortification. The collection, storing, and distributing of information place a methodical base for improving the curative study and fastidiousness of drugs, and aid in classifying for structuring a new means to cure and prevent disease by testing. The author (Anurag G et al., 2019) [16] paper suggested an approach to improve the detection of fake and real news spread in our society. The results are: 80% AUC and 75% F1 Score for K-Nearest Neighbor Classifier, 72% AUC and 75% F1 Score for Naïve Bayes, 85% AUC and 81% F1 Score for Random Forest, 79% AUC and 76% F1 Score for Support vector machine. Random Forest gives the best results from other algorithms. In this paper (Debar, H., 1992), [17] authors expressed that neural networks are used in an intrusion-detection system (IDS). The operator model writers made here is the balance of a statistical model because neural networks cannot passably knob all the accessible data.

The authors (M. Granik et al., 2017) [18] in this paper use the simple approach to identify the fake news from Facebook, a social media site using Naïve Bayes Classifier. The author proposed a solution for identifying fake news. The overall accuracy on the test set is 79%. In the paper, authors (Siaterlis, C., & Maglaris, B. 2004) [19] reveal the fusion of data to detect DoS attacks. The authors show DempsterShafer's Theory of Evidence as the measured and scientific basis for expanding a new DoS detection engine. The approach is used by authors of fusing data that associates many multiple signs created from modest rule-of-thumb to serve D-S inference engine and spot deluging attacks. The method has maximum benefits for sculpting the influence of signs in stating opinions in various premises, the aptitude to add the concepts of ambiguity and illiteracy in the structure, and the reckonable dimension of the certainty and likelihood in the approach of finding outcomes. The detection-engine prototype occurs via a set of trails that were operated with real-time operations and by using the tools of DDoS. The conclusion of this paper is where the data-fusion technique is consumed, which increases the detection rate and decreases the false alarm rate.

The sources (Joldzic, O. et al. 2016) [20] labeled a framework that screens system passage for DoS attacks and is skilled at averting them. The scheme is entirely apparent and can be implanted on any connection in the system. It presents negligible delay in transforming sachets as it is intended to be accessible via proficient managing tactics and a proper balancing approach system over numerous passage processors. The existing variant of the scheme can be used to comprise recognition variants of the systems for different network attacks, which can insecurely be ordered as DoS at tacks. Outbreaks tempted by unacceptable TCP order figures, SYN-floods, and past bouts can be recognized without any complex variants to the platforms and any extra processing space. Similar attacks could be detected without significant changes to the platform and additional memory or processing requirements. The recognition process utilized for these bouts is the network traffic redistribution and can work on the proposed platform. The proposed review in the study provides hypothetical, real answers but would be problematics to obtain in real life. Questionably, by hindering the terminus host after a distributed attack is noticed, the targets of the bouts are somewhat satisfied, but the structure is conserved from harms. However, segregating the cooperated hosts in a DDoS attack is a likely theme for future study.

The authors (Idhammad, M. et al. 2017) [21] described DoS tools for attacks that have overgrown and become progressively refined, perplexing the existing detection systems to increase their enactments constantly. The authors show how a sufferer ends DoS recognition tactics using the Artificial Neural Networks (ANN). The planned approach using a Feed-forward Neural Network (FNN) is improved to spot the DoS attack with the minor assets consuming. The suggested approach in the review involves three key stages: assortment of inward traffic, selection of significant traits for DDoS spotting by consuming a machine-learning branch as unsupervised learning and sorting or classification of inward traffic or regular traffic. Numerous trials were operated to evaluate the enactment of the suggested approach in the paper using datasets known as UNSW-NB15 and NSL-KDD. The acquired outcomes are adequate when likened to the advanced DoS detection methods.

The authors (Lyamin, N. et al., 2019) [22,23] explain data mining-based approaches that are usable in real-time recognition of radio congestion Denial-of-Service (DoS) attacks in 802.11p infrastructure. The process of detection entirely relies on 802.11p protocol processing. The approach is assessed for two congested frameworks, which include the random nature of active and nonactive. Hence, the proposed method is comparatively better than sigmoid. Review analysis describes existing research and research limitations that need to be improved. The analysis provides better reachability to the current research, encouraging more research to improve the existing research.

2.7 LITERATURE REVIEW ANALYSIS

Various researchers have surveyed the literature, which is essential to obtain the accuracy and method of each paper with a definite dataset.

TABLE 2.1
Analysis based on algorithm and accuracy

Author	Year	Algorithm	Accuracy (in %)
Liao, H. J., et al.	2013	DECISION TREE + NAÏVE BAYES + LINEAR REGRESSION + PERCEPTRON	91
Grover, C., et al.	2019	BLOCKCHAIN	75
Linn, L. A., & Koo, M. B	2016	KNN + NAÏVE BAYES + RANDOM FOREST	76
M. Granik, et al.	2017	NAÏVE BAYES	79
Siaterlis, C., & Maglaris, B.	2004	DEMPSTER SHAFER	69
Joldzic, O., et al.	2016	MACHINE LEARNING	70
Idhammad, M., et al.	2017	MACHINE LEARNING	80
Lyamin, N., et al	2019	DATA MINING	82

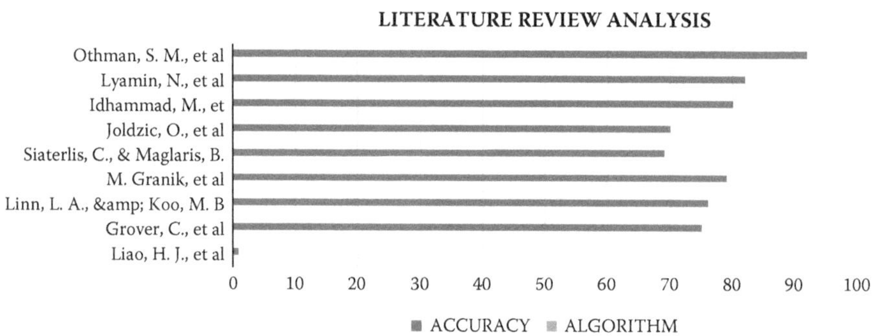

FIGURE 2.3 Existing research paper analysis.

Table 2.1 views which literature is analyzed by researchers and is required to obtain the statement concerning the accuracy and method consumed in each paper with a definite dataset (Figure 2.3).

The above graph reveals the performance of the existing method in existing papers to know the statement concerning the accuracy and method consumed in each paper with a definite dataset.

2.8 EXISTING LITERATURE GAP

The author studied various research papers published in the same domain where authors are working, and authors have found some gaps in the existing system, which are discussed as follows.

- The quality of data needs to be utilized, data having traversed cycle more than six months to fulfill requirement and data modeling correctly.
- The number of malicious packets must be less with imperative features.
- The unbalance classes in several samples need to be balanced.
- The detection system needs to occur at the significant location of the whole network as at outside and inside the dematerialized zone of the network.

2.9 PROPOSED FRAMEWORK

An explorer plans the process architecture to mention how an artificial neural network (ANN) concept is trailed for surfacing the DDDS. Figure 2 validates the projected DDDS agenda to function with processes to notice the attack where the DDoS dataset is collected for detection. The algorithm via the machine-learned outcomes is true or false (Figure 2.4).

The whole process consists of different phases via which detection of DDoS by an artificial neural network is possible. The first phase consists of collecting data that is needed to be collected from the Kaggle site. The collected data need to be sensed to get an insight about significant existing features in data that are useful to predict data with high accuracy, and it is significant for reducing the training time of

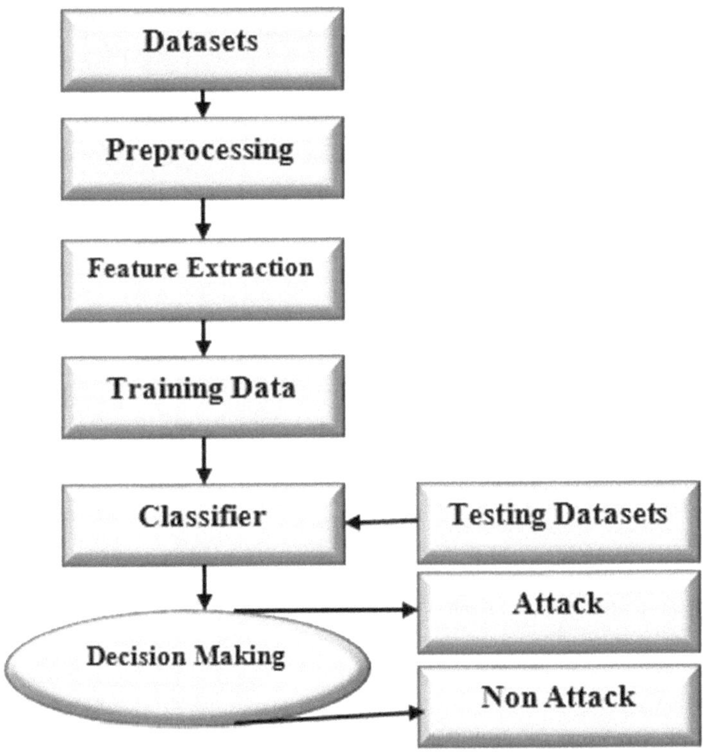

FIGURE 2.4 Flow diagram of DDoS detection.

the classifier. The next phase is about data preparation, which helps transform the data into the usual pattern to make data suitable to train the classifier. The following phase shows a need to process the data after interpretation or sensing the data to reduce the size, clean the data, and fill the null value with a constant or specific number.

The subsequent phase is to select the impactful features with intense impact to produce a highly accurate result. The second-to-last phase is to prepare the data for training, which can be attained by segmentation of the entire data set, so that obtained information segregates into the given condition and non-given condition, in the case of a labeled dataset. The final stage is to train thee model to perform the classification and predict the outcomes.

By the contemporary mechanism of the machine-learning concept, which is critical functionalities of the detection system and entails a Signature Dataset (SDS) and a machine-learning algorithm (MLA), the prime purpose is to fetch DDoS data, which are labeled and exist in storage. SDS is needed to craft traits choice tactics. Lastly, the furthermost precise MLA selects, trains, and imports into the classification system. The phase is about collecting the data named DDoS and Network concerned. The second step is to sample the data for preprocessing by sampling, cleaning, collection, integration, transforming, and pattern evaluation by performing data engineering and modeling. The control or observed process, a bundle of instances, is compulsory for drilling the process in the model. The known data is expressed as the signature database. The example of the record has a usual of features related to a marquee or a course. This stage is to classify features in traffic that are adept at extricating the standard net system from DoS attacks. The fourth stage is about selecting a suitable algorithm needed to perform the classification. The process involves selecting significant traits by consuming a different algorithm, such as AdaBoost or an extra classifier, to produce a specific impactful feature that produces the outcomes. In this toil, the collection choice of variables was made in two phases. The primary stage is consumed for recursive Feature Elimination with Cross-Validation (RFECV) was utilized with the concept of machine learning. The result is about to classify data by the detection system in a measured network setting using numerous rates. In the trials, raw data of the CIC- DoS, CICIDS2017, and CSE-CIC-IDS2018 datasets, and the raw data amassed in the modified testbed trails were consumed. DIDS detection systems have gotten expected accuracy and low false-positive rates. The final step is about detection by using the classic technique of observing the data passing through a network.

2.10 FRAMEWORK DEPLOYMENT

A step-by-step framework is deployed by providing a depth implementation using the specific hyperparameters, which are following:

- **Data-Collection**
  Defining the problem and assembling a dataset: Data collection aims to implement tools to monitor and analyze the data passing through the network and system. For instance, it acts as the malicious data come to the

firewall and other security devices. After storing the data into the various databases, storage needs to be collected and stored in the data warehouse.

- **Data-Preparation**
  Preparing your data: The data preparation is consumed to prepare the data using cleaned data, filling the missing value, balancing the imbalanced classes, and finally, transforming the data into an appropriate form that can suit a classifier.
- **Model Selection**
  This step is to choose the model or algorithm for selection based on performance by the surveyed paper and compare the performance of numerous algorithm in the context of accuracy.
- **Train-model**
  Developing a model that does better than a baseline: The phase is most significant for modeling and utilizing the training data to train the model.
- **Evaluate-model**
  Choosing a measure of success: Here is the need to measure the training time consumed. Deciding on an evaluation protocol: Here is the need to measure the number of epoch consumed.
- **Parameter tuning**
  Scaling up: developing a model that overfits. The phase is used to observe the overfitting and underfitting of the model. It was regularizing your model and tuning your parameters. It is used to bring the model into a normal state to perform with all hyperparameters.
- **Predict**
  Prediction is determined by performing the classification: The phrase is used to predict the outcome for detection purposes.

2.11 DATA DESCRIPTION

Distributed Denial of Service (DDoS) is a threat to system service security that aims to fatiguge the object nets with malevolent transportation. Though several numerical processes have been intended for DDoS outbreak detection, manipulating a real-time detector with less pressure needs to be an important factor [24,25]. Another side is about determining the estimation of the novel recognition process and methods, which depends on the presence of stylish datasets. CICDDoS2019 comprises the kind and the maximum up-to-date common DDoS outbreaks, which look like real-world data (PCAPs). It encompasses the consequences of the system traffic study using CICFlowMeter-V3 with considered movement built on the timestamp, source, and destination IPs, source and destination ports, protocols, and attack (CSV files) [24]. Creating accurately related passages of data has been a top priority in making a dataset. The authors have consumed the B-Profile system to outline the intangible action of human interactions and generate realistic, gentle passage of data in the planned testbed (Figure 2.5).

The above figure shows how events are occurring through the various modules and producing information about attacks. In this way, data is collected and integrated with a label to know the features of attacks or non-attacks.

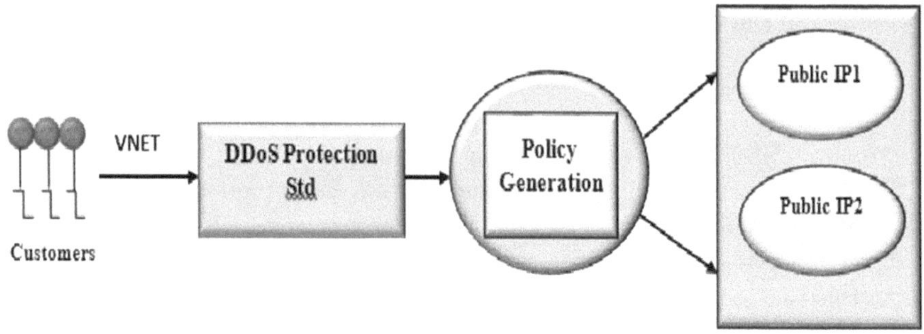

FIGURE 2.5 DDoS features [26].

2.12 ARTIFICIAL NEURAL NETWORK

The authors explained here that the algorithm consumed in the chapter is artificial neural networks (ANNs), which come under the category of neural networks (NNs), or processing systems elusively stirred by the philosophy of neurons in biology found in mammals [27–29] (Figure 2.6).

The above depiction reveals a neural network consisting of many connections or links, called processing units or artificial neurons similar to biotic neurons existing in the human brain. The artificial neuron consists of associated links similar to synapses in the neuron in the brain of a living being. It is called processing units because it captured signals, determined them, processed them, acted according to signal, and passed them to other neurons to make certain actions. The artificial neural works based on mathematics, which takes numerical values, a real number in nature. The neurons connect with other neurons to form a bundle of mathematical connections known as neural networks. The artificial neural network consists of

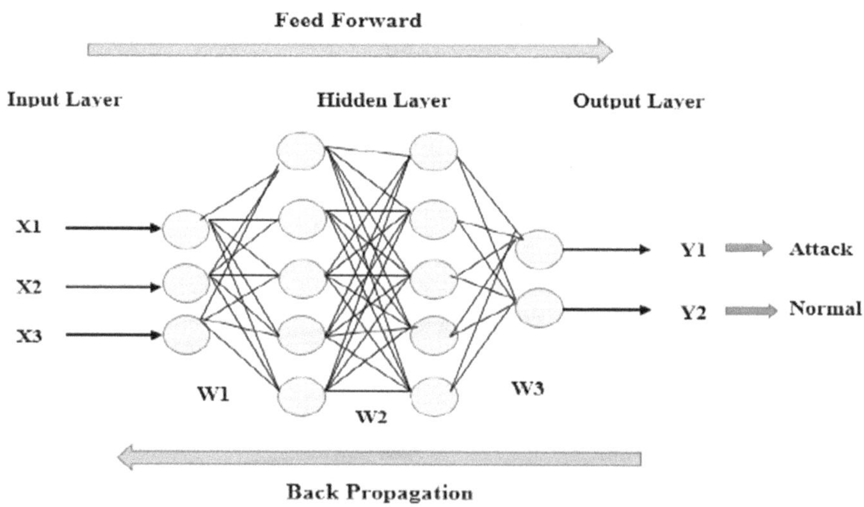

FIGURE 2.6 Artificial neural network [27].

three layers: the first layer has the function to take data and pass it to hidden layers responsible for processing data to provide the approximate result as expected. The third layer is known as the output layer, which is used to produce outcomes [28–30]. The ANNs are composed of a processing unit in the neural network, which is theoretically resulting from biotic neurons. Every single processing has data inputs and yields a single outcome that can transfer to numerous processing units. The entered data can be the items standards of an instance of peripheral data and outcomes of other processing units. The outcome of the last processing unit is to achieve a target to detect a mark of the entity in the picture [31–33].

For unearthing the outcomes of the processing units, initially, one must take the partial sum of all the inputs, biased by the weights of the links from the input data to the processing unit. Extra effort is needed, which is known as a bias to this addition. This partial sum is a function for activation. The sum is transferred to the nonlinear activation function to yield the outcomes. The initial inputs are external data, such as images and documents. The outcomes achieve the mission, likewise knowing an entity in appearance in the image [34–36]. The system entails a network in which each connection delivers the processing result of one processing unit as a feedback input, as in pipelining to another processing unit. The network links are assigned a weight that depicts their imperativeness. The single processing units may have numerous input and output links [37]. The main function calculates the data to processing units from its previous processing units and their network as a partial sum. A terminology that is named bias can be included in the outcomes of propagation function. Hence, the algorithm discussed above is absolute and fine to apply to the data set, CICIDS, to perform the classification for detection. The performance of the given algorithm is much better in comparison to others in the context of accuracy.

2.13 PERFORMANCE

The consequence signifies the enactment of the algorithm in the trial and solidifies the outcome, which displays the accuracy, precision, f1-score, and support.

Table 2.2: Keep the effect of accuracy, precision, recall, and f1-score by artificial neural network algorithms (ANN).

Figure 2.7 displays the analysis using a chart that shows the accuracy, precision, and recall value for artificial neural networks (ANN).

TABLE 2.2
Performance of (ANN) classifier

	"Precision"	"Recall"	"F1*Score"	"Support"
"0"	"0.9321"	"0.9321"	"0.9331"	"8285"
"1"	"0.9312"	"0.9311"	"0.9341"	"8985"
"ACCURACY"			"0.9367"	"15462"
"MICRO AVG"	"0.9313"	"0.9315"	"0.9314"	"15462"
"WEIGHTED AVG"	"0.9312"	"0.9314"	"0.9315"	"15462"

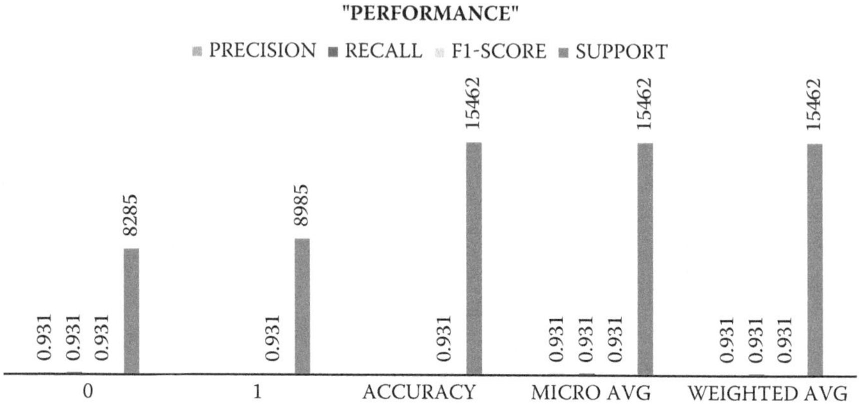

FIGURE 2.7 Performance analysis.

2.14 RESEARCH IMPLICATION

The research implication shows how research needs to be improved if it is to face implication or intricacy. The authors followed a process that researchers can evaluate. The research implication represents ways of organization of researching in a particular manner. In this organization of the paper, writers prepared a literature survey to analyze the current effort of authors in the context of information systems for security. The literature survey points to existing research findings and is named as an existing limitation. The limitation describes the problems in current work, which prompt us to do further research to improve a particular system. The statement of a problem must be considered.

By considering the present limitation, the identification of problem results gives inspiration about the purposes. In the chapter, the authors explained how security in information is playing a highly significant role to produce innocuous and secure communication by rapid improvement to information technology. Various kinds of contravention can be created by DDoS, such as slowing down the network performance (opening files or accessing websites), making a particular website unavailable, and creating an inability to access any website; this interruption will irritate clients, cause them to leave, and create failure deployment of service due to which system is restricted, leading to a loss of money. The threat of DDoS is continuously increasing over the significant continuous service provider. Therefore, there is a need to develop a robust DDoS detection system. Existing systems based on DDoS are complex and ineffective in producing a result because they perform everyday activities with undermining approaches. Therefore, the invader can choke the network for the absence of essential resources and service to users. In this chapter, the authors proposed a DDDS framework by using an artificial neural network mechanism. The authors utilized the CIDDS dataset for determining the correctness, precision, and recall of the system, and, as a result, decided parameters for classification. The authors collected the CIDDS data by using web-mining algorithms to estimate the performance of numerous algorithms for selecting the optimized algorithm. The selected algorithm is ANN, known for a feed-forward neural network that is effective and can train on a massive amount of data to produce the optimal result in accuracy, precision, recall, and f1-score.

2.15 OPEN CHALLENGES IN DDOS DETECTION SYSTEM

The research boundaries specify the issues that explorers face to prompt the framework into execution and control the areas of analysis strayed by authors. The absence of quality research papercraft is a problem for scholars who lack knowledge and authentication for doing the research. The extensive research environment can make likelihood in study to remove restrictions and recompense of dominant learning from related work led by scholars. The research was prepared in a particular field by recurrent methods that sacked arrivals of giving outcomes initially.

2.16 CONCLUSION

The security in information is playing a highly significant role to produce innocuous and secure communication by rapid improvement to information technology. DDoS can create various kinds of contravention, such as slowing down network performance (opening files or accessing websites) and making a particular website unavailable. An inability to access any website or encountering interruptions irritate clients, cause them to leave, and create failure deployment of service due to which the system is restricted, leading to loss of money. The threat of DDoS is continuously increasing over the significant continuous service provider. Therefore, there is a need to develop a robust DDoS detection system. Existing systems based on DDoS are complex and ineffective in producing a result because they perform everyday activities with undermining approaches. Therefore, the invader can choke the network due to denying essential resources and interrupting service. In this chapter, the authors proposed a DDDS framework by using an artificial neural network mechanism. The authors utilized the CIDDS dataset to determine the system's correctness, precision, and recall, and then decided parameters for classification. The authors collected the CIDDS data by using web-mining algorithms to estimate the performance of numerous algorithms for selecting the optimized algorithm. The selected algorithm is ANN, a feed-forward neural network that is effective and can train on vast amounts of data to produce the optimal result in accuracy, precision, recall, and f1-score. The result obtained by experimenting in terms of accuracy is 93%, which is effective compared to existing models.

REFERENCES

1. A. A. Aburomman and M. B. I. Reaz, "A survey of intrusion detection systems based on ensemble and hybrid classifiers," *Computers & Security*, vol. 65, pp. 135–152, 2017.
2. K. Al Jelled, M. Algida, and M. S. Dismuke, "Big data analysis and distributed deep learning for next-generation botnet detection system optimization," *Journal of Big Data*, vol. 6, no. 1, p. 88, 2019.

3. M. Almeida, M. Lazuli, S. Kovacs, and M. Alkasassbeh, "Evaluation of machine learning algorithms for a botnet detection system," In 15th International Symposium on Intelligent Systems and Informatics (SISY), 2017, pp. 000277–000282.
4. I. A. Basheer and M. Hajmeer, "Artificial neural networks: Fundamentals, computing, design, and application," *Journal of Microbiological Methods*, vol. 43, no. 1, pp. 3–31, 2000.
5. U. Banerjee, A. Vashishtha, and M. Saxena, "Evaluation of the capabilities of wireshark as a tool for botnet detection," *International Journal of Computer Applications*, vol. 6, no. 7, pp. 1–5, 2010.
6. I. M. Bapiyev, B. H. Aitchanov, I. A. Tereikovskyi, L. A. Tereikovska, and A. A. Korchenko, "Deep neural networks in cyber-attack detection systems," *International Journal of Civil Engineering and Technology (IJCIET)*, vol. 8, no. 11, pp. 1086–1092, 2017.
7. A. Bijalwan, "Botnet forensic analysis using machine learning," *Security and Communication Networks*, vol. 2020, pp. 1–9, 2020.
8. W. Bulrajoul, A. James, and S. Shaikh, "A new architecture for network botnet detection and prevention," *IEEE Access*, vol. 7, pp. 18558–18573, 2019.
9. T. Chandak, S. Shukla, and R. Wadhvani, "An analysis of A feature reduced botnet detection system using ANN classifier by Akashdeep et al., expert systems with applications," *Expert Systems with Applications*, vol. 130, pp. 79–83, 2019.
10. P. R. Chandre, P. N. Mahalle, and G. R. Shinde, "Machine learning-based novel approach for botnet detection and prevention system: A tool based verification," Global Conference on Wireless Computing and Networking (GCWCN), 2018, pp. 135–140.
11. R. Chen, W. Niu, X. Zhang, Z. Zhuo, and F. Lv, "An effective conversation-based botnet detection method," *Mathematical Problems in Engineering*, vol. 2017, pp. 1–9, 2017.
12. H. J. Liao, C. H. R. Lin, Y. C. Lin, and K. Y. Tung, "Intrusion detection system: A comprehensive review," *Journal of Network and Computer Applications*, vol. 36, no. 1, pp. 16–24, 2013.
13. M. K. Elhadad, K. F. Li, and F. Gebali, "Detecting misleading information on COVID-19," *IEEE Access*, vol. 8, pp. 165201–165215, 2020.
14. P. Grover, A. K. Kar, and M. Janssen, "Diffusion of blockchain technology," *Journal of Enterprise Information Management*, vol. 8, no. 2, pp. 1–11, 2019.
15. L. A. Linn and M. B. Koo, "Blockchain for health data and its potential use in health it and healthcare-related research," In *ONC/NIST Use of Blockchain for Healthcare and Research Workshop*, Gaithersburg, Maryland, United States. ONC/NIST, 2016, pp. 1–10.
16. A. Jain and A. Kasbe, "Fake news detection," In the Proceedings of International Students' Conference on Electrical, Electronics and Computer Science (SCEECS), IEEE, pp. 1–5.
17. H. Debar, M. Becker, and D. Siboni, "A neural network component for an intrusion detection system," In Proceedings 1992 IEEE Computer Society Symposium on Research in Security and Privacy, 1992, pp. 240–240.
18. M. Granik and V. Mesyura, "Fake news detection using Naive Bayes classifier," In the proceedings of IEEE First Ukraine Conference on Electrical and Computer Engineering (UKRCON), Kiev, IEEE, pp. 900–903.
19. C. Siaterlis and B. Maglaris, "Towards multisensor data fusion for DoS detection," In Proceedings of the 2004 ACM symposium on Applied computing, 2004, pp. 439–446.
20. O. Joldzic, Z. Djuric, and P. Vuletic, "A transparent and scalable anomaly-based DoS detection method," *Computer Networks*, vol. 104, pp. 27–42, 2016.
21. M. Idhammad, K. Afdel, and M. Belouch, "Dos detection method based on artificial neural networks," *International Journal of Advanced Computer Science and Applications*, vol. 8, no. 4, pp. 465–471, 2017.

22. N. Lyamin, D. Kleyko, Q. Delooz, and A. Vinel, "Real-time jamming DoS detection in safety-critical V2V C-ITS using data mining," *IEEE Communications Letters*, vol. 23, no. 3, pp. 442–445, 2019.
23. S. RoselinMary, M. Maheshwari, and M. Thamaraiselvan, "Early detection of DOS attacks in VANET using Attacked Packet Detection Algorithm (APDA)," In 2013 international conference on information communication and embedded systems (ICICES), 2013, pp. 237–240.
24. P. P. Lee, T. Bu, and T. Woo, "On the detection of signaling DoS attacks on 3G wireless networks," In IEEE INFOCOM 2007-26th IEEE International Conference on Computer Communications, 2007, pp. 1289–1297.
25. W. Wei, F. Chen, Y. Xia, and G. Jin, "A rank correlation-based detection against distributed reflection DoS attacks," *IEEE Communications Letters*, vol. 17, no. 1, pp. 173–175, 2013.
26. Z. Wu, L. Zhang, and M. Yue, "Low-rate DoS attacks detection based on network multifractal," *IEEE Transactions on Dependable and Secure Computing*, vol. 13, no. 5, pp. 559–567, 2015.
27. G. Diaz, M. Sen, K. T. Yang, and R. L. McClain, "Simulation of heat exchanger performance by artificial neural networks," *Hvac & R Research*, vol. 5, no. 3, pp. 195–208, 1999.
28. Y. Ding and Y. Zhai, "Botnet detection system for NSL-KDD dataset using convolutional neural networks," In Proceedings of the 2018 2nd International Conference on Computer Science and Artificial Intelligence, 2018, pp. 81–85.
29. L. Feng, H. Wang, Q. Han, Q. Zhao, and L. Song, "Modeling peer-to-peer botnet on scale-free network," *Abstract and Applied Analysis*, 2014, Hindawi. 2014, pp. 1–8.
30. S. Ganapathy, P. Yogesh, and A. Kannan, "Intelligent agent-based botnet detection system using enhanced multiclass SVM," *Computational Intelligence and Neuroscience*, pp. 1–10, 2012.
31. N. Singh, M. Dayal, R. S. Raw, and S. Kumar, "SQL injection: Types, methodology, attack queries and prevention," In 3rd International Conference on Computing for Sustainable Global Development (INDIACom), 2016, pp. 2872–2876.
32. M. D. Ambedkar, N. S. Ambedkar, and R. S. Raw, "A comprehensive inspection of cross-site scripting attack," 2016 International Conference on Computing, Communication and Automation (ICCCA), vol. 8, no. 2, 2016, pp. 1–6.
33. R. Kamal, R. S. Raw, N. G. Saxena, and S. K. Kaushik, "Implementation of security & challenges on vehicular cloud networks," In Communication and Computing Systems: Proceedings of the International Conference on Communication and Computing Systems (ICCCS), Gurgaon, India, 2017, pp. 9–11.
34. R. S. Raw, "The amalgamation of blockchain with smart and connected vehicles: Requirements, attacks, and possible solution," 2nd International Conference on Advances in Computing, Communication Control and Networking (ICACCCN), vol. 11, no. 2, 2020, pp. 14–17.
35. A. K. Yadav, R. K. Bharti, and R. S. Raw, "Security solution to prevent data leakage over multitenant cloud infrastructure," *International Journal of Pure and Applied Mathematics*, vol. 118, no. 7, pp. 269–276, 2018.
36. R. S. Raw, M. Kumar, and N. Singh, "Software-defined vehicular adhoc network: A theoretical approach," Cloud-based big data analytics in vehicular ad-hoc networks. *IGI Global*, 2021, pp. 141–164.
37. A. Aliyu, A. H. Abdullah, O. Kaiwartya, Y. Cao, M. J. Usman, S. Kumar, D. K. Lobiyal, and R. S. Raw, "Cloud computing in VANETs: Architecture, taxonomy, and challenges," *IETE Technical Review*, vol. 35, no. 5, pp. 523–547, 2018.

3 Energy-Efficient Massive MIMO for Future Generation Wireless Communication Systems

Jagtar Singh
Department of Electronics and Communication Engineering, N.C. College of Engineering, Israna, Panipat, Haryana, India

Anuj Singal and Deepak Kedia
Department of Electronics and Communication Engineering, Guru Jambheshwar University of Science & Technology, Hisar, Haryana, India

CONTENTS

3.1 INTRODUCTION

To meet the requirements for higher data rates in future wireless communication systems, the area throughput (in bit/s/km^2) is required to be improved by a factor of 1000 through the next few years [1,2]. The step toward increasing the throughput will lead to higher power consumption at the transmitters. This is because existing cellular systems are developed on central infrastructure, which is rigid and powered by electric grids or, in some cases, by high-power generators. At the same time, a more significant consumption of power by the cellular industry and a corresponding decrease in energy efficiency (EE) have become major economic and societal concerns [3]. It is forecasted that an increase in energy efficiency of 1000 times the

DOI: 10.1201/9781003230526-3

current level is needed to meet higher power consumption [4]. Cellular networks' significantly increased power consumption has led to severe environmental and economic issues in the past decades [5]. The requirement of higher power consumption creates a financial burden on the cellular industry [6,7]. Therefore, the aim is to design new technologies that are competent to fulfil the growing requirements for greater spectral efficiency without increasing power consumption at the base stations.

Massive MIMO is one of the 5G technologies that can satisfy the requirements of higher data rates and low power consumptions at the transmitter. The Massive MIMO technology provides all the benefits of a conventional MU-MIMO communication system [8]. In this technology, the base stations contain a considerable number of transmitting antennas compared to conventional systems. These large numbers of antennas are serving a small number of antennas, for example, 100 antennas are serving 10 user terminals in a particular cell. The system takes advantage of enhanced spatial multiplexing gain achieved by the significant array gains. A Massive MIMO communication scenario is shown in Figure 3.1.

The EE is calculated by the number of bits transmitted per Joule in a wireless communication system, and computed as:

$$Energy\ Efficiency = \frac{\sum_{k=1}^{K} R_k}{P_{Total}} \tag{3.1}$$

Where R_k represents the rate of transmission of user k and P_{total} is the overall power consumption.

Equation (3.1) represents the energy-efficiency metric and is measured in bit/Joule. The energy efficiency of communication networks has been investigated in the past years [9,10]. Most of the previous works have analyzed the energy efficiency of single-cell systems and ignored the interference from other neighboring

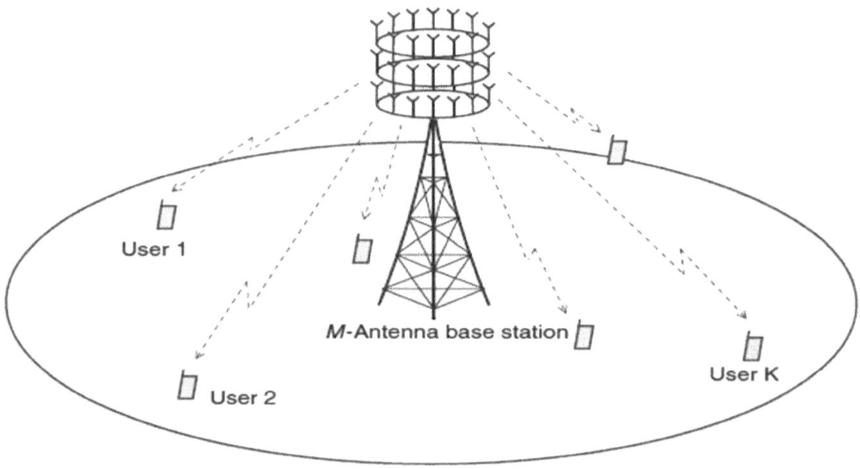

FIGURE 3.1 Illustration of massive-MIMO system.

cells of the network [11–16]. Authors in [12] assumed the uplink power allocation of the MU-MIMO system and proved the energy maximization is achieved when some specific users are powered off. The transmission during uplink environment was studied in [13], where energy efficiency was related to antennas at transmitter and rates of users. In the model [17], the overall consumption of power is calculated as the summation of power radiated and a fixed factor relating to the circuit power consumption [18]. The model considered in [17] produces misleading results, as the antennas increase; this model provides unbounded energy efficiency and data rates, which is not feasible in practical cases. Achieving infinity energy efficiency is impossible in this model as it does not consider the power consumed by the analog circuits and digital signal processing as the transmitter antennas and users increase. This explains that its contribution is low for conventional MU-MIMO systems and affects more in case of a Massive MU-MIMO network [19–22]. The work in [23] uplinks spectral efficiency expression is derived with an optimal number of users and antennas and ignores the channel estimation overhead.

This chapter aims to collaboratively design the downlink and uplink of the Massive MIMO system to maximize energy efficiency (EE). We investigate the effect of various antennas and users and transmit power selection to cover a particular area with maximum energy efficiency. We show how BS antennas, users and power transmitted affect the energy efficiency with different precoding techniques employed at the base station. We investigate the effect of precoding and combining methods like ZF, MMSE, and MRC/MR transmission on energy efficiency [24]. The proposed model scales power linearly with varying M and users K compared to existing literature models. We focus on the ZF precoding scheme for single-cell network and utilize a proposed system for computing energy efficiency optimal values of parameters. Our simulation indicates that an energy-efficient system is achieved using 100–200 antennas and serving users of the same order of magnitude. We also observe that the system becomes more energy-efficient with ZF precoding than MRT and MMSE precoding methods.

The rest of the chapter is arranged as follows. System model is presented in Section 3.2. Section 3.3 discusses the process of channel estimation and various precoding schemes. Energy-efficiency analysis is discussed in Section 3.4. The simulation results are presented in Section 3.5. In Section 3.6, we present the conclusion of this work.

3.2 SYSTEM MODEL

We assume uplink and downlink of a Massive multi-user MIMO communication system. The transmitter is provided with M number of antennas and K number of users. The flat-fading channel is assumed and h_k is the channel vector between the kth user and the base station. We assume block flat-fading channels with coherence time T_c and BW B_c. Therefore, the channel is assumed to be constant in the coherence interval, $\tau_c = B_c T_c$ as shown in Figure 3.2.

The base station is operating according to the TDD principle, as shown in Figure 3.2. The user terminals transmit the uplink pilots toward the BS at the start of the coherence block. The BS utilizes these pilots to estimate the channels of user

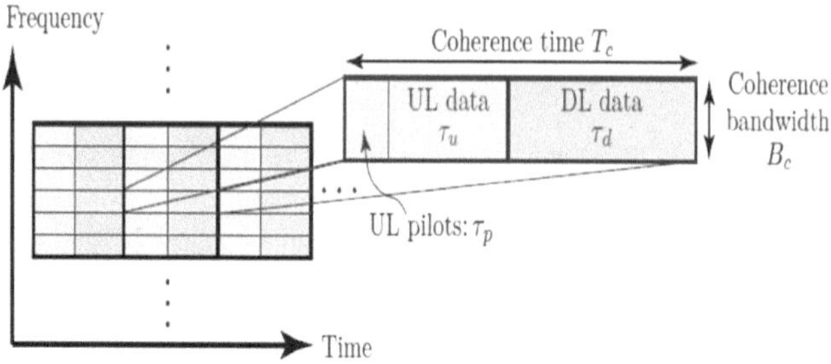

FIGURE 3.2 Time-division duplex (TDD) protocol.

terminals. Since the TDD scheme utilizes reciprocity, the BS uses uplink channel estimates for transmission in the downlink.

We assume a multi-user MIMO system that comprises a BS and K number of user's equipment (UEs). The BS has M number of antenna elements providing service to K single-antenna users in the cell. The multi-user MIMO system is presented in Figure 3.1. Consider $h_{k,n}$ expresses the channel coefficient among the kth UE and n^{th} antenna element of the BS. The channel coefficient $_{HKn}$ is the multiplication factor responsible for small-scale fading and an amplitude factor representing large-scale fading and geometric attenuation. The channel coefficient $_{HKn}$ can be represented by Equation (3.2)

$$h_{k,n} = g_{k,n}\sqrt{d_k} \tag{3.2}$$

Where $g_{k,n}$ denotes the complex minor-scaling fading components and d_k represents large-scale fading components, shadow fading, and path loss.

We consider that the array of base station antennas is compact adequately such that the coefficients of considerable scaling fading are the same for the various antenna elements at the same base station. The small-scale fading coefficients are considered distinct for various UEs or various antenna elements. Therefore, the matrix H amongst UEs and transmitter can be written as:

$$H = \begin{bmatrix} h_{11} & h_{21} & . & . & h_{K,1} \\ h_{12} & h_{22} & . & . & h_{K,2} \\ \vdots & \vdots & \vdots & \vdots & \vdots \\ h_{1,Mt} & h_{2Mt} & . & . & h_{KM_t} \end{bmatrix} = GD^{1/2} \tag{3.3}$$

where

$$G = \begin{bmatrix} g_{11} & g_{21} & . & . & g_{K,1} \\ g_{12} & g_{22} & . & . & g_{K,2} \\ \vdots & \vdots & \vdots & \vdots & \vdots \\ g_{1Mt} & g_{2Mt} & . & . & g_{KM_t} \end{bmatrix}, \quad D = \begin{bmatrix} d_1 & & \\ & \ddots & \\ & & d_k \end{bmatrix}$$

The kth column of G represents the small-scale fading among the kth UE and the M_t antennas, and the $D^{1/2}$ is the diagonal matrix whose k^{th} component represents the coefficient of large-scale fading.

The multi-user MIMO system has a vast array having greater number of antenna elements in comparison to served UEs. Then, under the favorable propagation condition, the columns of the channel matrix H are asymptotical orthogonal, such that

$$\left(\frac{H^H H}{M_t}\right)_{M_t >> K} = D^{1/2}\left(\frac{G^H G}{M_t}\right)_{M_t >> K} D^{1/2} \approx D \tag{3.4}$$

Based on these theoretical concepts, we discuss the uplink and downlink transmission in the following subsection.

3.3 CHANNEL ESTIMATION AND PROCESSING SCHEMES

The base station requires CSI to design MU precoding in the downlink transmission and detect data transmitted by the UEs. The CSI is acquired through the pilots, which UEs transmit during uplink transmission. Every UE is allocated an orthogonal pilot and transmits this pilot to the base station. The base station then estimates the channel based upon the received pilots.

For realistic environments, the CSI estimation has to be performed at the BS in a particular cell. The estimation of the channel depends on the duplexing schemes utilized. Two types of duplexing schemes are utilized, i.e., time-division duplexing (TDD) and frequency-division duplexing (FDD). In the FDD scheme, the downlink and uplink utilize different frequency spectrums, and thus, the CSI for downlink and uplink transmissions are distinct. The channel estimation with the FDD scheme comprises two segments. In the first segment, the BS transmits pilots to all the UEs in the cell. In the second segment, UEs estimate CSI and provide feedback to the BS over a dedicated feedback channel. Therefore, the required time for the transmission of downlink pilots is proportional to BS antennas. Hence, in the FDD scheme, the estimation overhead rises with the growth in the number of BS antenna elements in the LS-MU-MIMO system and the channel estimation scheme for downlink becomes impractical. Consider that over Ω successive subcarriers, the frequency response of the channel remains unchanging. There is a requirement of at least M_t/Ω time slots for orthogonal pilot transmission. In practice, the coherence time is limited, when numbers of antennas grow large, the time needed to transmit the pilot sequences can exceed the channel coherence time. Therefore, the estimated downlink CSI can become outdated.

The TDD scheme can solve the problem identified in the FDD scheme. A TDD transmission scheme is depicted in Figure 3.2, which utilizes the channel-reciprocity principle where only uplink CSI is required to be estimated [25]. The frequency-time resources are parted into W Hz and T seconds structures, as shown in Figure 3.2.

The downlink pilots and feedback of CSI are not required, unlike the FDD scheme. The base station can process uplink and downlink signals utilizing uplink channel estimates due to the reciprocity in the TTD system. The benefit of TDD is that the required overhead due to channel estimation is proportionate to the number of user terminals in the cell.

The base station antenna elements are situated sufficiently and separated such that the channel elements among users and the base station antenna elements are not correlated. The channel vector $h_k = [h_{k,1}, h_{k,2}, \dots\dots\dots\dots\dots\dots h_{k,M_t}]^T \in^{M_t x1}$. We have assumed a Rayleigh-fading channel model, valid for large and small arrays [26]. Linear processing is utilized for downlink precoding and uplink detection. The CSI is estimated by utilizing uplink pilots radiated by the cell users.

The MRT, MMSE, and ZF precoding techniques are utilized at BS for downlink transmission. The MRT precoding can be represented by the conjugate of channel matrix H and denoted by the Equation (3.5) below [21].

$$W_{MRT} = H^H \tag{3.5}$$

The ZF precoding matrix can be denoted by the Equation (3.6) as below [21]

$$W_{ZF} = (H^H H)^{-1} H^H \tag{3.6}$$

The MMSE method is expected to minimize the error between the received and transmitted symbol, which is due to interferences and noise distortions. The MMSE precoding scheme can be represented by the Equation (3.7)

$$W_{MMSE} = \left(H^H H + \frac{1}{SNR} I\right)^{-1} H^H \tag{3.7}$$

Similarly, the MRC, ZF, and MMSE receive combining matrices and can be represented by the Equation (3.8), Equation (3.9), and Equation (3.10), respectively.

$$G_{MRC} = H^H \tag{3.8}$$

$$G_{ZF} = (H^H H)^{-1} H^H \tag{3.9}$$

$$G_{MMSE} = \left(H^H H + \frac{1}{SNR} I\right)^{-1} H^H \tag{3.10}$$

The assumption of W = G reduces the complexity of computation, but it is not necessary.

3.4 ENERGY EFFICIENCY ANALYSIS

The energy efficiency of wireless systems is quantified in bit/J and equal to the ratio of mean sum rate and power consumption at BS. The consumed power of traditional macro base stations is proportional to the power radiated [18]. However, this condition does not hold in the case of an extensive multi-user MIMO system [19]. Therefore, there is a requirement of a more realistic and exhaustive model. The circuit power consumption results from the analog filters utilized for digital-signal processing, RF and baseband processing, and power consumption by power amplifiers. Motivated by the power-consumption model in [18,27,28], we propose an improved model that indicates power scaling with BS antennas and users.

3.4.1 Uplink Transmission

With the proper choice of processing methods, the uplink rate of user k can be [29]

$$R_k^{UL} = \xi^{UL}\left(1 - \frac{\tau^{UL}K}{\tau_c \xi^{UL}}\right)\tilde{R}_k^{UL} \tag{3.11}$$

ξ^{UL} denotes the fragment of uplink transmission. The objective is to achieve the equal overall rate $\tilde{R}_k^{UL} = \tilde{R}$ for every UE. This rate objective is achieved by employing the scheme in [30] if the power assignment vector p^{UL} is so as:

$$p^{UL} = \sigma^2 (D^{UL})\, 1_K \tag{3.12}$$

The (k, l)th coefficient of $D^{UL} \in \mathbb{C}^{K\times K}$is [31].

$$[D^{UL}]_{k,l} = \begin{cases} \dfrac{|g_k^H h_k|^2}{\left(2^{\frac{\tilde{R}}{B}} - 1\right) \| g_k \|^2} & when \quad k = l \\ -\dfrac{|g_k^H h_k|^2}{\| g_k \|^2} & when \quad k \neq l \end{cases} \tag{3.13}$$

Equation (3.12) can be utilized to compute uplink power consumption, as follows:

$$P_{UL}^{Tx} = \frac{B\xi^{UL}}{\eta^{UL}} E\left[1_K^T P_{UL}\right] \tag{3.14}$$

$$P_{UL}^{Tx} = \frac{B\xi^{UL}}{\eta^{UL}} E\left[1_K^T \,(D^{UL})^{-1} 1_K\right] \tag{3.15}$$

Where η^{UL} represents power efficiency of UEs.

3.4.2 Downlink Transmission

Assume that a kth user is allocated precoding vector w_k and power of P_k^{DL}. Considering CSI of perfect nature, the achievable rate of the user k in downlink with precoding method can be [32]

$$R_k^{DL} = \xi^{DL}\left(1 - \frac{\tau^{DL}K}{\tau_c\xi^{DL}}\right)\tilde{R}_k^{DL} \tag{3.16}$$

The mean power of power amplifier in downlink can be represented as:

$$P_{DL}^{Tx} = \frac{B\xi^{DL}}{\eta^{DL}} \sum_{k=1}^{K} E\,[P_k^{DL}] \tag{3.17}$$

Here η^{DL} denotes power amplifier efficiency at the base station. Similar to uplink case, downlink power is:

$$p^{DL} = \sigma^2 (D^{DL})\, 1_K \tag{3.18}$$

Here the (k, l) component $D^{DL} \in \mathbb{C}^{K\times K}$ is

$$[D^{DL}]_{k,l} = \begin{cases} \dfrac{|h_k^H w_k|^2}{\left(2^{\frac{R}{B}} - 1\right) \| h_k \|^2} & for \quad k = l \\ -\dfrac{|h_k^H w_l|^2}{\| w_l \|^2} & for \quad k \neq l \end{cases} \tag{3.19}$$

The downlink power by utilizing Equation (3.18) can be:

$$P_{DL}^{Tx} = \sigma^2 \frac{B\xi^{DL}}{\eta^{DL}} E\,\{1_K^T (D^{DL})^{-1} 1_K\} \tag{3.20}$$

We observe that $(D^{UL})^T = D^{DL}$ when the identical processing technique is employed for transmit precoding and receive combining.

The system's energy efficiency can be evaluated by the expression that is the fraction of the mean sum rate to the overall consumption of power. The overall energy efficiency can be determined by assuming downlink and uplink transmission. Mathematically, the EE can be represented by [33,34]

$$Energy\ \ Efficinecy = \frac{\sum_{k=1}^{K} (E\,\{R_k^{UL}\} + E\,\{R_k^{DL}\})}{P^{CP} + P_{DL}^{Tx} + P_{UL}^{Tx}} \tag{3.21}$$

Where P^{CP} is the circuit power consumption.

The optimum EE setup for Massive MIMO can be obtained by maximizing Equation (3.21) and expressed by Equation (3.22) as below [33]

$$\max\ Energy\ \ Efficinecy = \frac{\sum_{k=1}^{K} (E\,\{R_k^{UL}\} + E\,\{R_k^{DL}\})}{P^{CP}(Mt,\ NrK,\ R_k) + P_{DL}^{Tx} + P_{UL}^{Tx}} \tag{3.22}$$

On the basis of earlier works [35,36], the model for circuit power consumption can be implemented, which assume the consumption of power due to load-dependent backhaul (P^{BH}), transceiver chains (P^{TC}), channel estimation process (P^{CE}), channel decoding and coding (P^{CD}), processing (P^{P}) at the base station. The circuit power consumption can be modeled as:

$$P^{CP} = P^{fix} + P^{TC} + P^{CE} + P^{BH} + P^{CD} + P^{P} \tag{3.23}$$

The terms in Equation (3.23) relies on the essential parameters (M_t, N_rK, R_k). The consumption of power caused by transceiver components is:

$$P^{TC} = M_t P_{bs} + P_{lo} + K\ P_U \tag{3.24}$$

Here P_{bs} represents the required power to operate the circuit components like filters, converters, and mixers related with every base-station antenna element. P_U denotes the power required by every one of the circuit components like amplifiers, filters, oscillators, and mixers of user equipments. The term P_{lo} represents the local oscillator power consumption.

The estimation of the channel is carried out with uplink pilot sequences once per coherence block, and B/τ_c is the coherence block per second. During transmission in the uplink, the base station acquires pilot sequences of $M_t \times K\tau^{UL}$ and estimates the every user channel by associating the respective pilot length of $\tau^{UL}K$ [29]. This process needs uplink power consumption P_{UL}^{CE} as represented by Equation (3.25) [37]

$$P_{UL}^{CE} = \frac{B}{\tau_c}\frac{2M_t K^2 \tau^{UL}}{L_{BS}} \quad Watt \tag{3.25}$$

Here L_{BS} denote the arithmetical complex process per joule calculated in flops/W.

Similarly, the power consumption P_{DL}^{CE} in downlink is represented by Equation (3.26).

$$P_{DL}^{CE} = \frac{B}{\tau_c}\frac{4K^2 \tau^{UL}}{L_U} \quad Watt \tag{3.26}$$

Thus, the consumption of overall power for the channel-estimation operation can be obtained by adding Equations (3.25) and (3.26) and represented by Equation (3.27).

$$P^{CE} = \frac{B}{\tau_c}\frac{2M_t K^2 \tau^{UL}}{L_{BS}} + \frac{B}{\tau_c}\frac{4K^2 \tau^{UL}}{L_U} \quad Watt \tag{3.27}$$

The backhaul power can be calculated by Equation (3.28) [28]

$$P^{BH} = \sum_{k=1}^{K} (E\{R_k^{DL} + R_k^{UL}\}) P^{BT} \quad Watt \tag{3.28}$$

Here P^{BT} represents the traffic power due to backhaul.

The base station employs coding and modulation to K symbols, while transmission in downlink and reverse process in the uplink. The power consumed due to such process is proportionate to the number of bits and determined as below. [38]

$$P^{CD} = \sum_{k=1}^{K} \left(E\left\{R_k^{DL} + R_k^{UL}\right\}\right) \left(P^{COD} + P^{DEC}\right) \; Watt \tag{3.29}$$

Where P^{DEC} and P^{COD} represents the required powers due to decoding and coding.

The combining operation processes the signal received at the BS, and the precoding operation generates transmitted signal vector. The power consumption required for these actions can be represented by Equation (3.30) [38]

$$P^{P} = B\left(1 - \frac{\left(\tau^{DL} + \tau^{UL}\right)K}{\tau_c}\right)\frac{2KM}{L_{BS}} + P^{C} \; Watt \tag{3.30}$$

If MRT/MRC processing method is employed, and this operation needs power consumption $P^{C}_{MRC/MRT}$ as represented by Equation (3.31) [38].

$$P^{C}_{MRC/MRT} = \frac{B}{\tau_c}\frac{3M_t K}{L_{BS}} \; Watt \tag{3.31}$$

The consumption of power required for ZF process is given by [38]:

$$P^{C}_{ZF} = \left(\frac{K^3}{3L_{BS}} + \frac{3M_t K^2 + KM_t}{L_{BS}}\right)\frac{B}{\tau_c} \; Watt \tag{3.32}$$

The MMSE process needs greater complexity because Equation (3.12) is a constant point expression that requires iterations until convergence. To make this process more straightforward, the number of iterations is fixed to a prespecified number Q. Thus, this operation requires the power P^{C}_{MMSE} represented by Equation (3.33)

$$P^{C}_{MMSE} = QP^{C}_{ZF} \; Watt \tag{3.33}$$

By utilizing the expressions considered above, the derivations of EE with various processing methods are performed. First, the overall EE by considering ZF method from Equation (3.22) can be represented as below:

$$\max \quad (Energy \;\; Efficinecy)^{ZF} = \frac{K\left(1 - \frac{\left(\tau^{DL} + \tau^{UL}\right)K}{\tau_c}\right)\tilde{R}}{P^{CP}_{ZF}\left(Mt, \;\; NrK, \;\; R_k\right) + P^{Tx}_{DL} + P^{Tx}_{UL}} \tag{3.34}$$

where P_{ZF}^{CP} can be written as:

$$P_{ZF}^{CP} = P_{ZF}^{P} + P^{CD} + P^{BH} + P^{CE} + P^{TC} + P^{fix} \quad (3.35)$$

$$P_{ZF}^{CP} = P_{Fix} + [M_t P_{bs} + P_{Osc} + K_D P_U] + \left[\frac{B}{\tau_c}\frac{2M_t K^2 \tau^{UL}}{L_{BS}} + \frac{B}{\tau_c}\frac{4K^2\tau^{UL}}{L_U}\right]$$
$$+ \left[\sum_{k=1}^{K} (E\{R_k^{DL} + R_k^{UL}\})P^{BT}\right] + \left[\sum_{k=1}^{K} (E\{R_k^{DL} + R_k^{UL}\})\ (P^{COD} + P^{DEC})\right]$$
$$+ \left[B\left(1 - \frac{(\tau^{DL} + \tau^{UL})K}{\tau_c}\right)\frac{2KM}{L_{BS}} + P^C\right] \quad (3.36)$$

Equation (3.36) can be simplified and represented as below:

$$P_{ZF}^{CP} = \sum_{j=0}^{3} K^j C_{0i} + M_t \sum_{j=0}^{2} B_{0,j} K^j + \left(1 - \frac{(\tau^{DL} + \tau^{UL})K}{\tau_c}\right) AK\tilde{R} \quad (3.37)$$

Here, ZF precoding coefficients in Equation (3.37) are given by:

$C_{0,1} = P_U$, $C_{0,0} = P_{Fix} + P_{Osc}$, $C_{0,3} = \frac{B}{3\tau_c L_{BS}}$, $C_{0,2} = \frac{4B\tau^{DL}}{\tau_c L_U}$, $B_{0,0} = P_{bs}$, $B_{0,1} = \frac{B}{L_{BS}}\left(2 + \frac{1}{\tau_c}\right)$, $B_{0,2} = \frac{B}{\tau_c L_{BS}}(3 - 2\tau^{DL})$, $A = P_{BT} + P_{DEC} + P_{COD}$

Substituting Equation (3.37) in Equation (3.34), we obtain

$$(Energy\ \ Efficinecy)^{ZF} = \frac{K\left(1 - \frac{(\tau^{DL} + \tau^{UL})K}{\tau_c}\right)\tilde{R}}{\sum_{j=0}^{3} K^j C_{0i} + M_t \sum_{j=0}^{2} K^j B_{0,j} + \left(1 - \frac{K(\tau^{DL} + \tau^{UL})}{\tau_c}\right)AK\tilde{R} + P_{DL}^{Tx} + P_{UL}^{Tx}} \quad (3.38)$$

By following a similar process, the derivation for EE with MRT and MMSE methods can be carried out and mentioned below:

$$(Energy\ \ Efficinecy)^{MRT} = \frac{K\tilde{R}\left(1 - \frac{K(\tau^{DL} + \tau^{UL})}{\tau_c}\right)}{P_{MRT}^{CP}(M_t,\ \ KN_r,\ \ R_k) + P_{DL}^{Tx} + P_{UL}^{Tx}} \quad (3.39)$$

$$(Energy\ \ Efficinecy)^{MMSE} = \frac{K\tilde{R}\left(1 - \frac{K(\tau^{DL} + \tau^{UL})}{\tau_c}\right)}{P_{MMSE}^{CP}(M_t,\ \ KN_r,\ \ R_k) + P_{DL}^{Tx} + P_{UL}^{Tx}} \quad (3.40)$$

Therefore, the energy efficiency with different precoding schemes and multiple antenna UEs can be computed with the derived expressions represented by

Equations (3.38), (3.39), and (3.40). These energy-efficiency expressions are utilized for simulation work in the following section.

3.5 SIMULATION RESULTS

In this portion, various simulations are accomplished to validate the numerical and mathematical processing methods described in previous portions. Table 3.1 shows the different simulation parameters used in this work. We evaluate the performance of the Massive MIMO system by employing Monte-Carlo simulations, which considers ZF, MMSE, and MRT precoding techniques at the BS. To obtain smooth numerical results, we perform 300 Monte-Carlo realizations.

We present the simulation results by assuming perfect and imperfect CSI. The MATLAB Monte Carlo simulations are utilized to simulate the results. First, we assume a single-cell scenario with a radius of 250 m and an operational frequency is

TABLE 3.1
Simulation parameter values

Parameter Name	Value
Cell radius	250 m
Number of UEs	K = 15
Samples per coherence interval	$\tau_c = 200$
Number of antennas at base station	M = 200
Coherence BW	$B_c = 200$ kHz
Coherence interval	$T_c = 1$ ms
Number of Monte Carlo simulations	300
SNR	5 dB
Noise power	–96 dBm
BS transmit power	20 dBm
Channel model	Rayleigh fading
Fraction of downlink transmission	$\xi^{DL} = 0.6$
BS power amplifier efficiency	$\eta^{DL} = 0.39$
Fraction of uplink transmission	$\xi^{UL} = 0.4$
UEs power amplifier efficiency	$\eta^{UL} = 0.3$
User transmit power	20 dBm
Required power for backhaul traffic	$P_{BT} = 0.25$ W/(Gbit/s)
BS computational efficiency	$L_{BS} = 12.8$ Gflops/W
Required power for UEs circuit elements	$P_U = 0.1$ W
UEs computational efficiency	$L_U = 5$ Gflops/W
Required power for BS circuit elements	$P_{BS} = 1$ W
Local oscillator power consumption	$P_{lo} = 2$ W
Fix power consumption	$P^{fix} = 18$ W

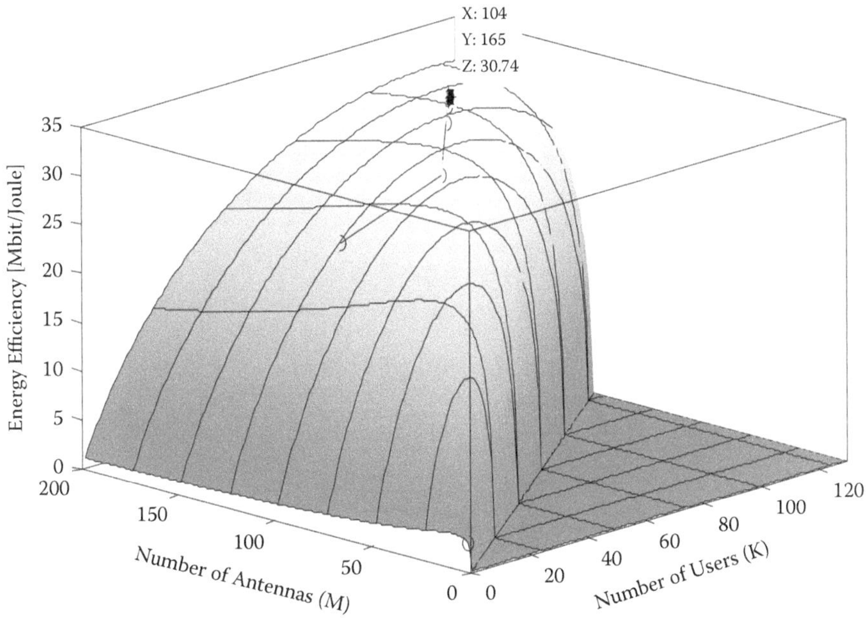

FIGURE 3.3 EE with ZF scheme in the single-cell environment.

2 GHz. Figure 3.3 depicts the energy-efficiency values by considering perfect CSI, varying BS antennas M and users K and ZR precoding. In ZF precoding, the number of base-station antenna elements satisfies M > K + 1. This figure depicts that there are optimal energy-efficiency points at K = 104 and M = 165. The optimum points are a sizeable MU-MIMO system. The surface in Figure 3.3 is smooth and concave; hence, there are different system parameter values those result in optimum energy-efficiency values.

Figure 3.4 depicts the possible energy-efficiency values with MMSE precoding. The global optimum points were obtained at M = 144, K = 89 and EE = 30.59 Mbit/J.

Figure 3.5 shows the simulation with MRC/MRT precoding scheme. The optimum points achieved are M = 72, K = 69 and EE = 9.996 Mbit/J. Figure 3.6 considers ZF precoding with imperfect CSI. Monte Carlo simulations obtain the results. The optimum points achieved are K = 110, M = 185and Energy Efficiency = 25.88 Mbit/J.

MRC/MRT precoding scheme provides optimum energy points smaller than ZF and MMSE schemes. Still, this is called a Massive MIMO system because there are several antenna elements in this setup. However, power saving is not enough to provide compensation for lower rates. To obtain similar data rates as zero-forcing processing, MRC/MRC needs M > K, which increases circuit power and does not enhance the system's EE.

Figure 3.7 shows the plot of maximum energy vs. the number of antennas at BS. We consider MMSE, MRT, and ZF schemes in the simulation. The MMSE and ZF perform similarly at high SNRs, and MRT shows a lower energy efficiency than MMSE and ZF processing schemes. It is observed that three-times difference

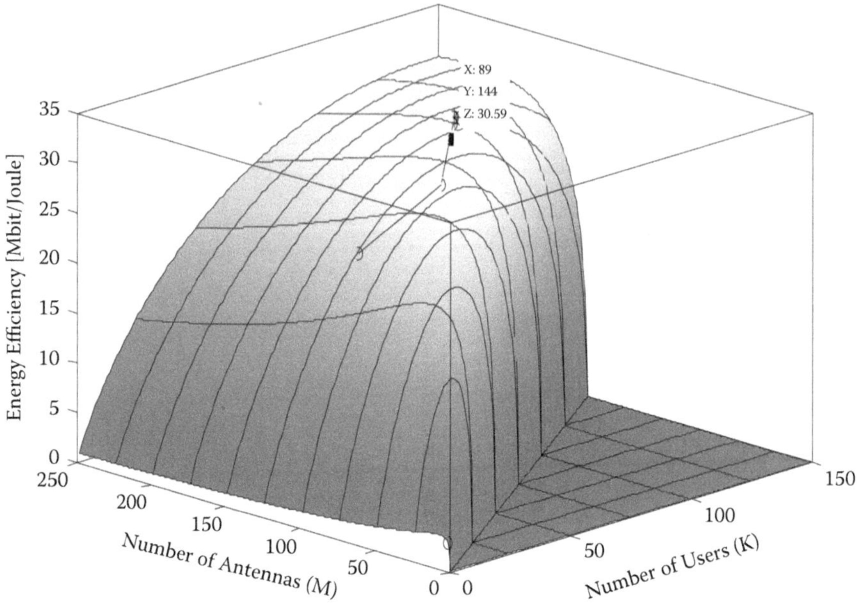

FIGURE 3.4 EE with MMSE scheme in single-cell environment.

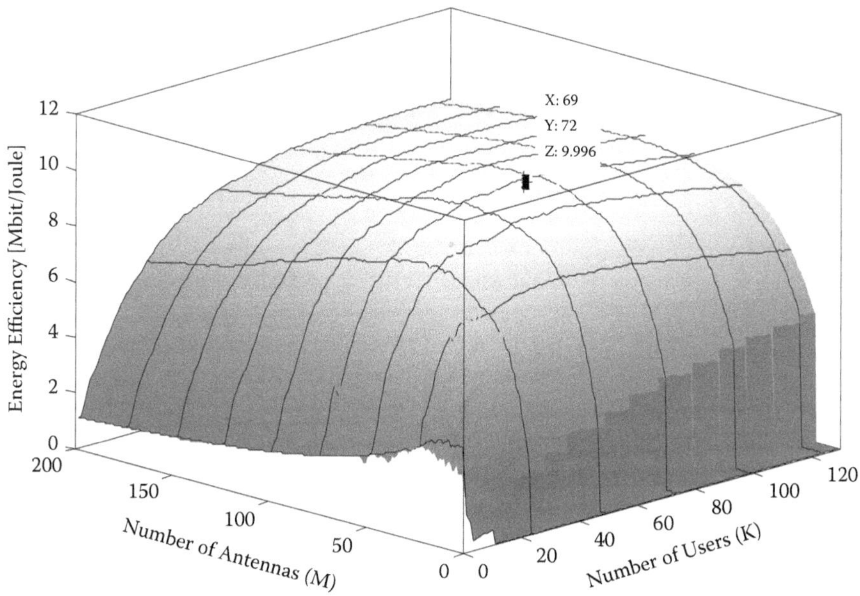

FIGURE 3.5 EE with MRT/MRC method in single-cell environment.

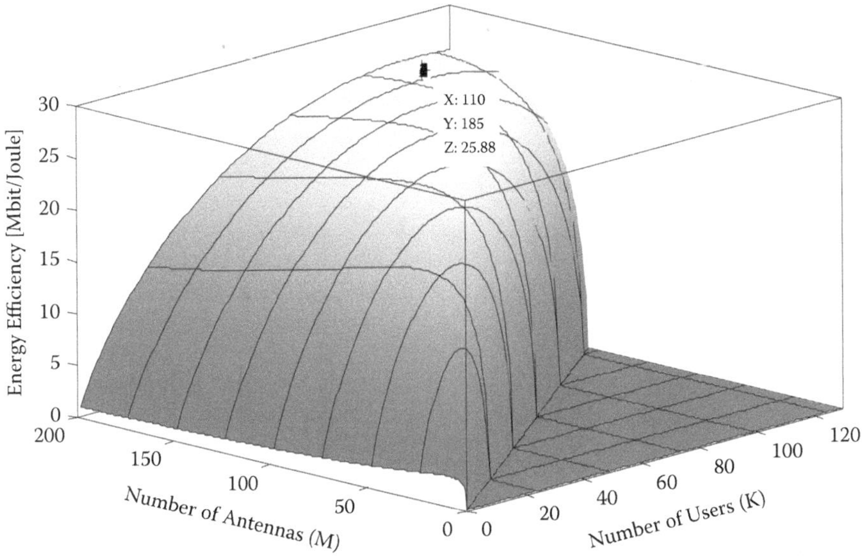

FIGURE 3.6 EE with ZF method in the single-cell environment having imperfect CSI.

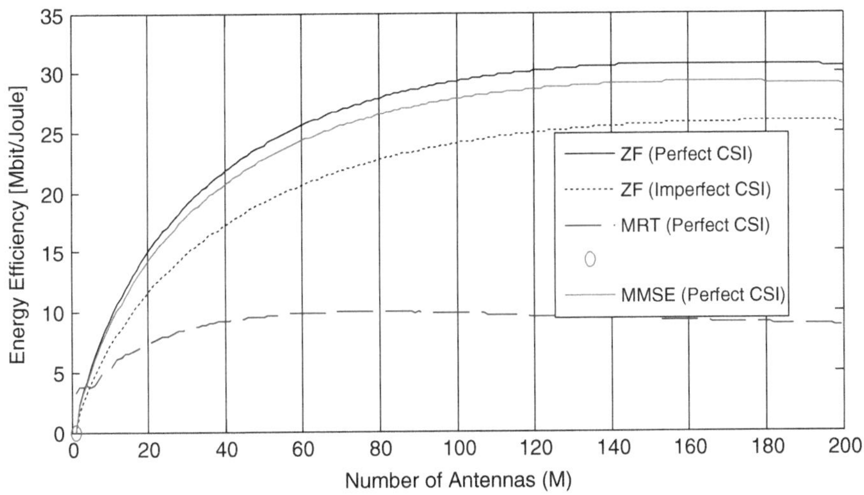

FIGURE 3.7 Energy efficiency vs number of base-station antenna elements and various processing techniques in the single-cell environment.

between optimum energy efficiency occurs between MMSE/ZF and MRT under perfect CSI. Therefore, proper selection of precoding schemes provides maximum spectral and energy efficiency in the MU-MIMO system.

Figure 3.8 depicts the average power of PA that provides the highest energy efficiency for various BS antennas. This figure represents that the power transmitted

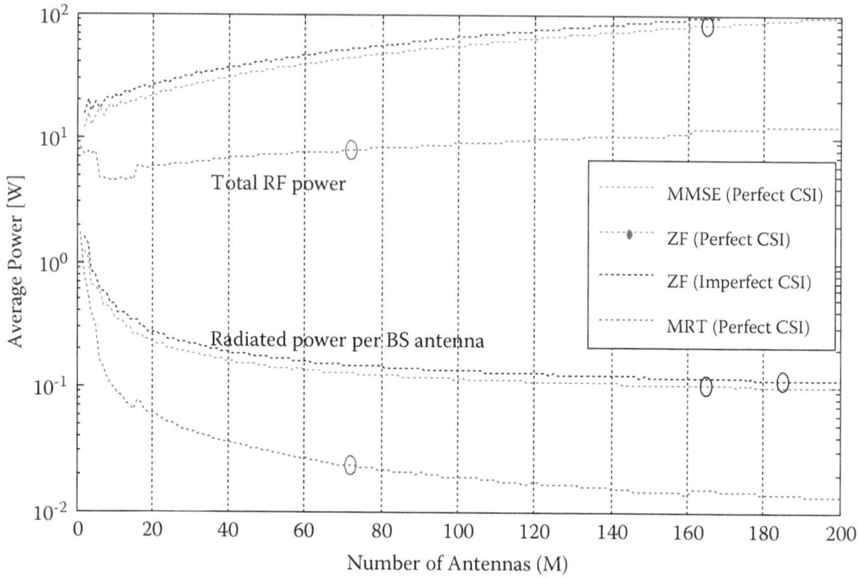

FIGURE 3.8 Overall power of PA at the energy efficiency-maximizing solution for several base-station antenna elements in the single-cell environment. The power radiated per base station antenna is also displayed.

per base-station antenna element decreases with M. The power transmitted in downlink with MMSE and ZF precoding is approximately 100 mW per antenna; however, power reduces to 23 mW per antenna employing MRT scheme because it yields more significant interference and therefore forces the network interference-limited at smaller power levels. The power is much lower than the traditional macro-BS, which works at approximately 40×10^3. Similar levels of power are noticed for the users in the uplink.

Figure 3.9 depicts the area throughput that maximizes the EE for various values of antennas M. It is noticed from Figure 3.7 that there is a three times enhancement in optimum energy efficiency for MMSE and ZF compared to MRC/MRT scheme. Figure 3.9 depicts that a simultaneous eight-times enhancement occurs in area throughput. Most of the performance gain is achieved under the consideration of imperfect CSI, which indicates that a massive MU-MIMO with a suitable precoding scheme can yield higher area throughput and energy efficiency. However, it is not suitable to employ an MRT/MRC scheme with a more significant number of BS antennas since it dramatically limits throughput and energy efficiency.

Now, we consider multicell massive MIMO system as shown in Figure 3.10. Every cell of size 500 × 500 square with uniformly distributed user terminals. As observed from the single-cell scenario, we only consider the ZF scheme here as the ZF is the optimum processing scheme in the single-cell case. The simulation results are compared by the utilization of different pilot reuse factors. Here, the cells are split into four clusters, as depicted in Figure 3.10. The pilot reuse factors of 1, 2, 4, i.e., f = 1, 2, 4, are considered.

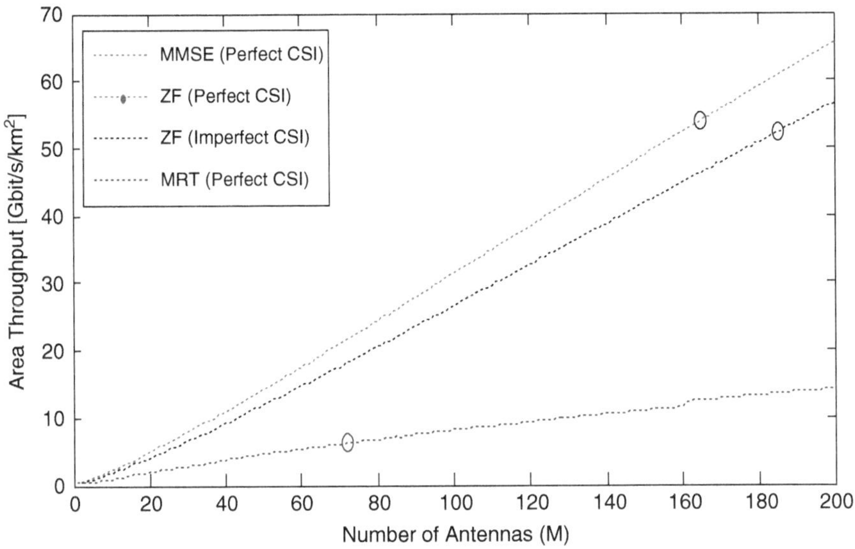

FIGURE 3.9 Area throughput at the energy efficiency-maximizing points for various base-station antenna elements in the single-cell environment.

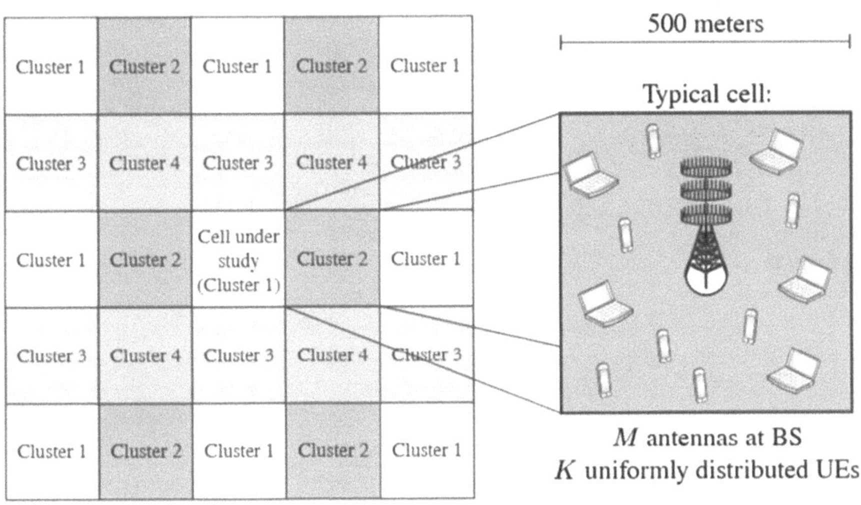

FIGURE 3.10 A multi-cell environment with various pilot reuse factors.

The maximum energy efficiency for different BS antennas is depicted in Figure 3.11, while Figure 3.12 depicts the corresponding power amplifier (PA) power. Figure 3.13 shows the area throughput with a different number of antenna elements at BS. The simulated figures are similar to the single-cell scenario but having smaller energy efficiency and throughput values. The lower values of energy efficiency are due to the increase in interference from the neighboring cells. The main observation is that the maximum throughput and energy efficiency are

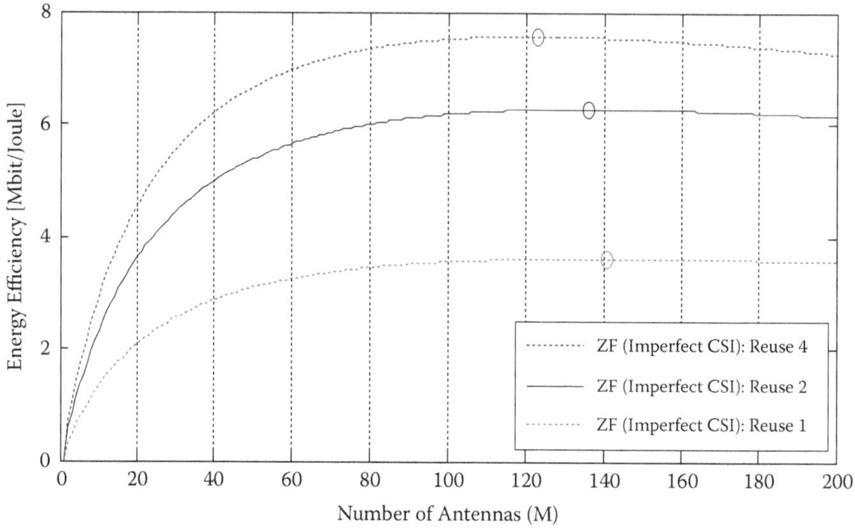

FIGURE 3.11 EE in the multicell environment for different pilot reuse factors and number of base-station antenna elements.

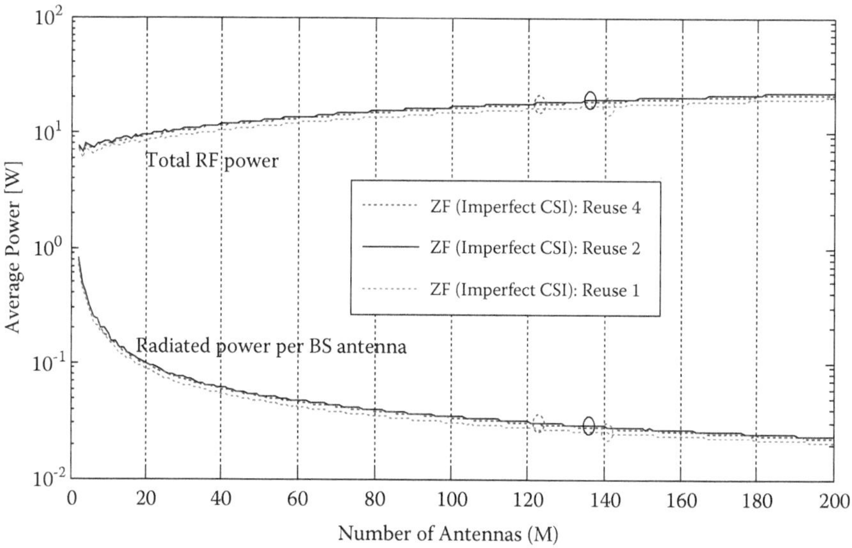

FIGURE 3.12 Overall power of PA at the energy efficiency-maximizing points in the multicell environment for various BS antenna elements. The power radiated per base-station antenna is also displayed.

obtained with f = 4. This illustrates the requirement of actively reducing pilot contamination in a massive multicell system.

Figure 3.14 represents the set of EE values with various values of BS antennas and users. Figure 3.14 assumes a pilot-reuse factor of value four because it provides

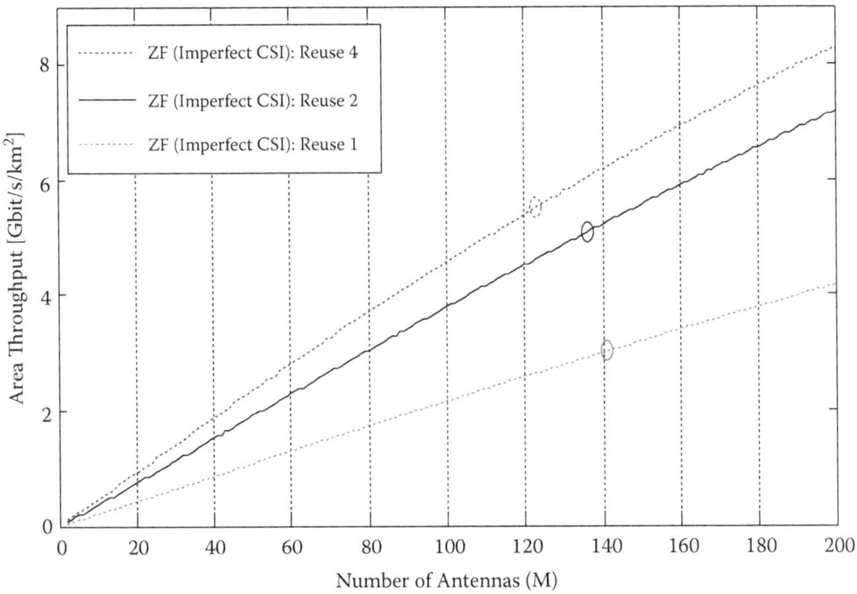

FIGURE 3.13 Area throughput at the energy efficiency-maximizing points in the multicell environment with various BS antennas.

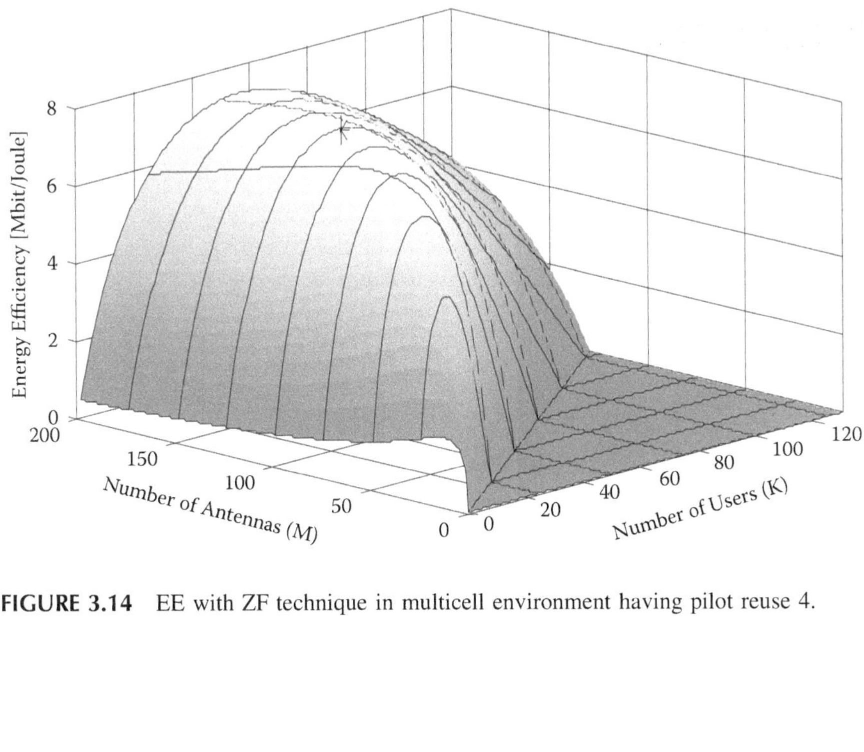

FIGURE 3.14 EE with ZF technique in multicell environment having pilot reuse 4.

the maximum value of energy efficiency. The curve having a shape similar to the single-cell scenario is noticed, but the obtained value of energy efficiency is more minor. This is due to the increase in interference that forces the cells to forfeit some freedom.

We observe that pilot overhead is almost similar to the single-cell case, although the pilot reuse factor allows fewer users. It is concluded that a massive MIMO system provides the maximum energy efficiency with ZF precoding and combining scheme.

3.6 CONCLUSIONS

This chapter focused on enhancing the EE of Massive MIMO systems considering various precoding schemes and a varying number of antennas and users in the cell. We analyzed how to choose several antennas, users, and precoding schemes to maximize the energy efficiency of Massive MIMO communication systems. The energy efficiency (bit/J) is a function of antennas at the base station and users. Our simulation results show that the optimum energy-efficiency solution employs 200 BS antennas to serve many user terminals. We investigated single-cell, as well as multicell, scenarios of Massive MIMO systems. In a single-cell system, ZF performs best and shows the highest energy efficiency compared to the MRT scheme. In the multicell case, the cell encounters more intercell interference and energy efficiency with precoding schemes, which reduces and in the multicell case; ZF precoding also performs better in terms of energy efficiency. We observe from simulation results those systems that are energy efficient are not preferable in small SNR scenarios but beneficial to operate in a scenario where interference elimination schemes are employed instead of interference-ignoring schemes such as MRC/MRT. The emitted power from antennas decreases with antennas at the transmitter. The simulation results showed that the ZF method is more energy-efficient than MMSE and MR methods. The proposed Massive MIMO system can achieve the requirements of next generation wireless systems. This shows that large MU-MIMO can be implemented by utilizing small power transceiver components on the base station compared to traditional high-power equipment. Proposal of energy-efficient systems reduces the cost of deployment of massive systems in future generation wireless communication systems.

REFERENCES

1. J. G. Andrews, S. Buzzi, W. Choi, S. V. Hanly, A. Lozano, A. C. K. Soong, and J. C. Zhang, "What will 5G be?," *IEEE J. Sel. Areas Commun.*, vol. 32, no. 6, pp. 1065–1082, 2014.
2. B. Naga, J. Li, D. Malladi, R. Gilmore, D. Brenner, A. Damnjanovic, R. T. Sukhavasi, C. Patel, and S. Geirhofer, "Network densification: the dominant theme for wireless evolution into 5G," *IEEE Communications Magazine*, vol. 52, no. 2, pp. 82–89, 2014.
3. G. Auer et al., D2.3: Energy efficiency analysis of the reference systems, areas of improvements and target breakdown. INFSO-ICT-247733 EAR H, ver. 2.0, 2012. [Online]. Available at: http://www.ict-earth.eu/

4. J.M.H. Elmirghani, T. Klein, K. Hinton, T. E. H. El-Gorashi, A. Q. Lawey, and X. Dong "GreenTouch GreenMeter core network power consumption models and results." In 2014 IEEE Online Conference on Green Communications (OnlineGreenComm), pp. 1–8, IEEE, 2014.
5. K. N. R. S. V. Prasad, E. Hossain, and V. K. Bhargava, "Energy efficiency in massive MIMO-Based 5G networks: Opportunities and challenges," *IEEE Wireless Communications*, vol. 24, pp. 86–94, June 2017.
6. C. Han, T. Harrold, S. Armour, I. Krikidis, S. Video, P. M. Grant, H. Haas, J. S. Thompson, I. Ku, C.-X. Wang, et al., "Green radio: Radio techniques to enable energy-efficient wireless networks," *IEEE Communications Magazine*, vol. 49, no. 6, pp. 46–54, 2011.
7. C. X. Wang, F. Haider, X. Gao, X. H. You, Y. Yang, D. Yuan, H. Aggoune, H. Haas, S. Fletcher, and E. Hepsaydir, "Cellular architecture and key technologies for 5G wireless communication networks," *IEEECommunicationsMagazine*, vol. 52, no. 2, pp. 122–130, 2014.
8. E. G. Larsson, O. Edfors, F. Tufvesson, and T. L. Marzetta. "Massive Mimo for next-generation wireless systems," *IEEE Communications Magazine*, vol. 52, no. 2, pp. 186–195, February 2014.
9. D. Feng, C. Jiang, G. Lim, L. J. Cimini, G. Feng, and G. Y. Li, "A survey of energy-efficient wireless communications," *IEEE Commun. Surveys Tutorials*, vol. 15, no. 1, pp. 167–178, 2013.
10. G. Y. Li, Z. Xu, C. Xiong, C. Yang, S. Zhang, Y. Chen, and S. Xu, "Energy-efficient wireless communications: Tutorial, survey, and open issues," *IEEE Wireless Commun.*, vol. 18, no. 6, pp. 28–35, 2011.
11. E. Bj¨ornson, M. Kountouris, and M. Debbah, "Massive MIMO and small cells: Improving energy efficiency by optimal soft-cell coordination," in Proc. Int. Conf. Telecommun. (ICT), 2013.
12. G. Miao, "Energy-efficient uplink multi-user MIMO," *IEEE Trans. Wireless Commun.*, vol. 12, no. 5, pp. 2302–2313, 2013.
13. Y. Hu, B. Ji, Y. Huang, F. Yu, and L. Yang, "Energy-efficiency resource allocation of extensive multi-user MIMO systems," *Wireless Network*, vol. 20, 2014.
14. D. Ha, K. Lee, and J. Kang, "Energy efficiency analysis with circuit power consumption in massive MIMO systems," in Proc. IEEE Int. Symp. Personal, Indoor and Mobile Radio Commun. (PIMRC), 2013.
15. H. Yang and T. L. Marzetta, "Total energy efficiency of large cellular scale antenna system multiple access mobile networks," in Proc. IEEE Online Green Comm, 2013.
16. S. Mohammed, "Impact of transceiver power consumption on the energy efficiency of the zero-forcing detector in massive MIMO systems," *IEEE Trans. Commun.*, vol. 62, no. 11, pp. 3874–3890, 2014.
17. A. Fehske, G. Fettweis, J. Malmodin, and G. Bicz'ok, "The global footprint of mobile communications: The ecological and economic perspective," *IEEE Trans. Commun.*, vol. 49, no. 8, pp. 55–62, 2011.
18. K. N. R. S. V. Prasad, E. Hossain, and V. K. Bhargava. "Energy efficiency in massive Mimo-based 5g networks: Opportunities and challenges," *IEEE Wireless Communications*, vol. 24, no. 3, 86–94, June 2017.
19. E. Bj¨ornson, J. Hoydis, M. Kountouris, and M. Debbah, "Massive MIMO systems with non-ideal hardware: Energy efficiency, estimation, and capacity limits," *IEEE Trans. Inf. Theory*, vol. 60, no. 11, pp. 7112–7139, 2014.
20. T. Marzetta, "Noncooperative cellular wireless with unlimited numbers of base station antennas," *IEEE Trans. Wireless Commun.*, vol. 9, no. 11, pp. 3590–3600, 2010.
21. F. Rusek, D. Persson, B. Lau, E. Larsson, T. Marzetta, O. Edfors, and F. Tufvesson, "Scaling up MIMO: Opportunities and challenges with vast arrays," *IEEE Signal Process. Mag.*, vol. 30, no. 1, pp. 40–60, 2013.

22. H. Ngo, E. Larsson, and T. Marzetta, "Energy and spectral efficiency of extensive multi-user MIMO systems," *IEEE Trans. Commun.*, vol. 61, no. 4, pp. 1436–1449, 2013.
23. S. K. Mohammed, "Impact of transceiver power consumption on the energy efficiency of zero-forcing detector in massive MIMO systems," *IEEE Transactions on Communications*, vol. 62, no. 11, pp. 3874–3890, Nov. 2014.
24. A. Al-Rimawi , L. Ibrahim, and W. Ajib, "Achievable rate of multi-cell downlink massive MIMO systems with D2D underly," in 2020 IEEE 91st Vehicular Technology Conference (VTC2020-Spring), Antwerp, Belgium, pp. 1–5, May 2020.
25. E. Björnson, E. G. Larsson, and M. Debbah, "Massive MIMO for maximal spectral efficiency: How many users and pilots should be allocated?," *IEEE Transactions on Wireless Communications*, vol. 15, no. 2, pp. 1293–1308, February 2016.
26. X. Gao, O. Edfors, F. Rusek, and F. Tufvesson, "Massive MIMO in real propagation environments," *IEEE Trans. Wireless Commun.*, 2014. submitted. [Online]. Available at: http://arxiv.org/abs/1403.3376
27. S. Tombaz, A. V¨astberg, and J. Zander, "Energy- and cost-efficient ultrahigh- capacity wireless access," *IEEE Wireless Commun. Mag.*, vol. 18, no. 5, pp. 18–24, 2011.
28. E. Björnson, L. Sanguinetti, H. Wymeersch, J. Hoydis, and T. L. Marzetta, "Massive MIMO is a reality – What is next? Five promising research directions for antenna arrays," *Digital Signal Processing*, vol. 94, pp. 3–20, November 2019.
29. Hoydis J., Brink S. ten, and Debbah M., "Massive MIMO in the UL/DL of cellular networks: How many antennas do we need?," *IEEE Journal on Selected Areas in Communications*, vol. 31, no. 2, pp. 160–171, February 2013.
30. S. U. Pillai, T. Suel, and S. Cha, "The Perron-Frobenius theorem: some of its applications," in *IEEE Signal Processing Magazine*, vol. 22, no. 2, pp. 62–75, March 2005.
31. A. Wiesel, Y. C. Eldar, and S. Shamai, "Linear precoding via conic optimization for fixed MIMO receivers," *IEEE Transactions on Signal Processing*, vol. 54, no. 1, pp. 161–176, January 2006.
32. E. Björnson, E. Jorswieck, "Optimal resource allocation in coordinated multi-cell systems," *Optimal Resource Allocation in Coordinated Multi-Cell Systems, now*, vol. 9, no. 2–3, pp. 113–381, 2013.
33. E. Björnson, L. Sanguinetti, J. Hoydis, and M. Debbah, "Designing multi-user MIMO for energy efficiency: When is massive MIMO the answer?," in IEEE Wireless Communications and Networking Conference (WCNC), Istanbul, pp. 242–247, April 2014.
34. G. Auer, O. Blume, V. Giannini, I. Godor, M. Imran, Y. Jading, E. Katranaras, M. Olsson, D. Sabella, P. Skillermark, and W. Wajda, "D2. 3: Energy efficiency analysis of the reference systems, areas of improvements and target breakdown," *Earth*, vol. 20, no. 10, November 2010. [Online]. Available at: http://www.ict-earth.eu/
35. S. Tombaz, K. W. Sung, and J. Zander, "Impact of densification on energy efficiency in wireless access networks," in IEEE Globecom Workshops, Anaheim, CA, pp. 57–62, December 2012.
36. D. Ha, K. Lee, and J. Kang, "Energy efficiency analysis with circuit power consumption in massive MIMO systems," in IEEE 24th Annual International Symposium on Personal, Indoor, and Mobile Radio Communications (PIMRC), London, pp. 938–942, September 2013.
37. S. Boyd and L. Vandenberghe, "Numerical linear algebra background." 2015 [Online]. Available at: ww.ee.ucla.edu/ee236b/lectures/num-lin-alg.pdf
38. A. Mezghani and J. A. Nossek, "Power efficiency in communication systems from a circuit perspective," in IEEE International Symposium of Circuits and Systems (ISCAS), Rio de Janeiro, pp. 1896–1899, May 2011.

4 IoT Based One-to-One Security Web Application for Smart Home

Harsha Vardan Maddiboyina
R & D Department, Planet Sigma Embedded Systems Pvt. Ltd., Hyderabad, India

V. A. Sankar Ponnapalli
Department of Electronics and Communication Engineering, Sreyas Institute of Engineering and Technology, Hyderabad, India

A. Naresh Kumar
Department of Electrical and Electronics Engineering, Institute of Aeronautical Engineering, Hyderabad, India

CONTENTS

4.1 INTRODUCTION

Internet of Things (IoT) is an innovative idea to build communication between devices through the internet. Here, the internet is a network of networks, whereas things can be any devices such as home appliances, vehicles, and machinery [1–7]. The IoT concept was implemented in different fields, such as health, medical, agriculture, transportation, home automation, etc. [8–10]. For smart homes, the user can control home appliances, such as fans, lights, etc., remotely from a developed web application through PC/mobile phones [11–13]. To develop the IoT applications for any field, the developer should focus on establishing communication infrastructure, speed of communication, data-transmission reliability, and cyber security [14,15]. An intelligent and

DOI: 10.1201/9781003230526-4

secured software application for IoT-based smart homes has been discussed in [16]. In this method, an android mobile application was developed to control home appliances from anywhere in the world. This system uses an advanced message-queuing protocol as a communication between the software, host, and server. The drawback behind this method is that the innovative home mobile application will be supported only by android mobiles and not PC/laptops. A point-n-press is an intelligent universal remote-control system for home appliances, as explained in [17]. In this scheme, the home appliances can be controlled using an intellectual universal remote, including an infrared (IR) sensor. The home appliances can be controlled only for a short range of communication.

Ancestral homes to smart homes using IoT and mobile application has been addressed in [18]. In this technique, a mobile application has been designed to control home appliances. Here, the base station and satellite station both communicate by using the radio-frequency module. There is no security for the developed mobile application. The design of the IoT innovative home system has been presented in [19]. In this work, an intelligent home application has been demonstrated to control home appliances remotely. The developed application can be supported by mobiles, PCs, and laptops and even controlled by an IR remote. An Internet of Things-based prototype for intelligent application control was discussed in [20]. In this paper, the IoT frame with automatic intelligent sensing-based application control was developed. It contains four sections: data acquisition, communication, processing, and performing section. Initially, the data-acquisition section acquires the analog data from the respected sensors and broadcasts the data to NodeMCU to convert analog data into digital form. The NodeMCU will be in the communication section. After converting the data into digital form, the communication section broadcasts the digital data to the processing section, i.e., to the raspberry pi. Raspberry pi converts the digital data into the Message Queuing Telemetry Transport (MQTT) frame format and sends back the MQTT data from the processing section to the communication section. Now the communication section broadcasts the MQTT frame format data to the performance section, i.e., application output.

Enhanced smart home automation systems based on the Internet of Things has been discussed in [21]. In this paper, a secure IoT-based smart home automation with diminution of security issues was discussed. A real-time broker cloud was used for processing the data between the source (user) and destination (device) at a very high speed by performing the cryptographic operation. This method saves the data-encryption process time. This method receives the data from the sensors and backs the data to the user through a developed web application. Here the devices act based on the current sensor values instead of user requests. So, in this method, the user cannot take charge of the devices and only needs to monitor the devices' status based on the current sensor values. A review of connectivity challenges in IoT smart homes has been discussed in [22]. In this paper, network connectivity challenges using different wireless protocols for smart homes have been discussed. Wi-Fi, Zigbee, Bluetooth, Z-wave, Thread, etc. are some of the IoT wireless communication protocol devices. Interoperability, self-management, maintainability, signaling, bandwidth, and power consumption are network connectivity challenges for smart homes. Interoperability is the primary concern for IoT devices and smart homes because good communication to be established between the devices and the user without losing any data during communication. In the self-management process, smart devices should scrutinize their own operating performance and notify the user about their status before getting shut down. While designing the communication network for the smart home, the designed network should have less maintenance in terms of cost. In an IoT network, the communication between devices and the user should be consistent; bidirectional signaling is essential for accumulating and routing the data between the devices. As the number of IoT devices increases, the amount of data and bandwidth increases simultaneously. So, to avoid the load on the network, a lightweight network protocol should be established. An efficient IoT network needs minimal battery drain and low power consumption in such communication. Generally, Zigbee-based networks utilize only 25% of power energy when compared with the Wi-Fi network. Whereas the Z-wave network consumes a significantly smaller amount of power when compared with the Wi-Fi network, the data exchange rate is prolonged compared with Wi-Fi.

The design of a secure IoT platform for the smart home system was discussed in [23]. In this paper, Elliptic-Curve Diffie-Hellman's (ECDH) asymmetric cryptography-based protocol was developed to communicate data between devices and local gateway. This method reduces the computational and storage requirements on the devices and stores the data into a local server while preserving confidentiality, privacy, and authentication. The cost for the hardware section gets high because, every time, the data is storing in a separate server. A novel smart-energy theft system (SETS) for IoT-based smart homes has been discussed [24].

In this paper, an energy-exposure system called SETS based on machine learning and a statistical model was developed. This method contains three levels of decision-making modules. The first level is the prediction model, which employs a multi-model forecasting system. This forecasting system is used to find the level of power consumption. The second level is the primary decision-making model that exploits a simple moving average (SAM) technique for filtering anomalous. In the final level, the detection of energy theft mechanisms was developed. However, the above methods suffer from some deficiencies. According to the literature survey, there are few research gaps between the security of an application, the compatibility, and the range of communication. Meanwhile, security-based web applications play a vital role in controlling home appliances. This work elucidates the usage of home appliances and devices to communicate smartly with the user using the Internet of Things with high security-based authentication levels [25–27].

The paper is structured as follows: Section 4.1 is devoted to the introduction and the related works. Section 4.2 focuses on the specifications of the tools, and Section 4.3 explains the details of the proposed method. Section 4.4 presents a prototype demonstration of the proposed method's performance and the obtained results. Finally, the conclusion remarks about the proposed work are presented in Section 4.5.

4.2 TOOLS SPECIFICATIONS

NodeMCU is a low-cost IoT device that is integrated with ESP8266 WiFi SOC. It has developed by Espressif Systems in 2013. It has been successfully employed in

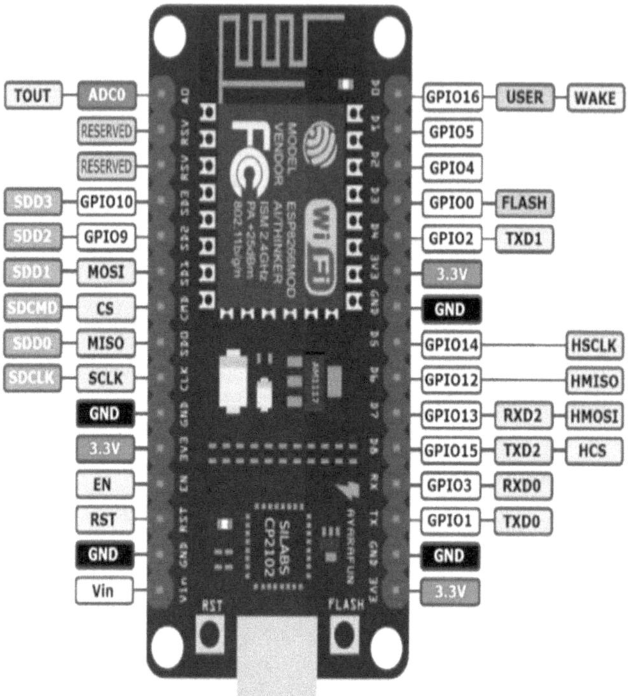

FIGURE 4.1 Pin layout of nodeMCU.

various fields, like internet smoke alarms, security alarms, octopod, VR tracker, serial port monitor, incubator controller, IoT home automation, and ESP lamp. The pin layout of NodeMCU is depicted in Figure 4.1.

A relay is a switch used to control high-voltage devices such as fans, TVs, refrigerators, ACs, etc., with low-voltage devices such as microcontrollers. Wi-Fi is a technology used to provide network connectivity with the Wi-Fi modules to access the internet. Port forwarding is a method in Wi-Fi technology used to connect a specific device or computer within a private local area network. HyperText Markup Language (HTML) is a standard markup language that is used for developing web pages. There are different editors/tools available to write the HTML program. Even we can write the HTML program in notepad and save the program with an '.html' extension. Arduino IDE is an open-source platform used to write the programs for different microcontrollers, such as Arduino Uno, Arduino Mega, Arduino Mini, NodeMCU, etc. The Arduino IDE supports different types of programming languages such as C, C++, and Java. Arduino IDE contains majorly two functions such as 'void setup' and 'void loop.' Here 'void setup' function executes for one time, whereas the 'void loop' function executes continuously.

The step-by-step procedure for programming in Arduino IDE is as follows:

- Open Arduino IDE.
- Now go to File → New.

- Now write the program in Arduino IDE and save the program.
- Now compile the program.
- For a successful compilation, it will show the message as 'Done Compiling.'
- Now connect the target board/microcontroller through a USB cable to the pc/laptop in which the Arduino IDE was installed.
- Now go to Tools → select the respected board and com port.
- Now upload the program into the target board.
- For successful uploading, it will show the message as 'Done Uploading.'

4.3 PROPOSED METHOD

In this section, a web application is developed to control home appliances through the internet. The web application should pass several security checks before it connects to home appliances. The block diagram of the project is shown in Figure 4.2. The proposed web application is compatible with any device such as mobiles/tablets/PCs/laptops. Nine different users can access the web application at a time. Here, NodeMCU is connected with eight different home appliances, which are present in four rooms with a relay as an intermediate. For instance: In Room-1, fan, and light are connected to GPIO16 and GPIO5 pins of NodeMCU and in Room-2, fan and light are connected to GPIO4 and GPIO2 pins of NodeMCU and in Room-3, fan and light are connected to GPIO14 and GPIO12 pins of NodeMCU and in Room-4, fan and light are connected to GPIO13 and GPIO15 pins of NodeMCU. For a relay, the input voltage is given as 5 V and common ground is connected. The above GPIO pins are connected to the input of the relay; the output of the relay should be connected to the

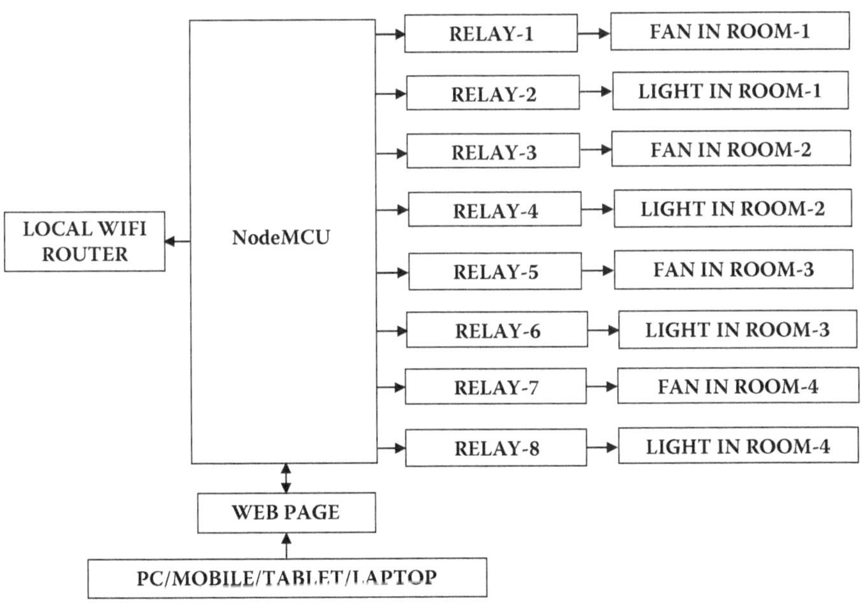

FIGURE 4.2 Block diagram of the project.

Syntax Code for Device Initialization:

```
#include <ESP8266WiFi.h>

#include <EEPROM.h>



const char* Username = "**********";

const char* Password = "**********";

WiFiServer server(80);

int p=0;

void setup()

{

        pinMode(GPIO_NUMBER,MODE);

        EEPROM.begin(4096);

        Serial.begin(9600);

        WiFi.begin(Username,Password);

        server.begin();

}
```

FIGURE 4.3 Syntax code for device initialization.

respected home appliances. The syntax code for device initialization is shown in Figure 4.3. In this syntax, the header files 'ESP8266WiFi.h' and 'EEPROM.h' are used to run the functions related to the 'wifi module' and internal 'EEPROM' of NodeMCU, respectively. 'Username' and 'Password' are the two-character pointer variables used to store the values of 'Username' and 'Password' of the Wi-Fi, respectively. WiFiServer is a structure with a variable 'server' initialized with a port value '80'. Here, 'setup' is a one-time initialization function. 'pinMode' is used to initialize with two parameters: NodeMCU's GPIO number and type of mode (input or output), respectively. 'EEPROM.begin' is an EEPROM initialization function, which is loaded with the size of the internal EEPROM in bytes.

Syntax Code for HTML Page:

```
void loop(){

    if(p==1){

        client.println("<!DOCTYPE HTML>");

        client.println("<html>");

        client.println("<div>");

        client.println("<center><p><b>successfully

        created..........</b></p><center>");

        client.println("<center><br><br>");

        client.println("<a href=\"/log\"\"><button>

        Goto Login</button></a><br    /><center>");

        client.println("</div>");

        client.println("</html>");

    }

}
```

FIGURE 4.4 Syntax code for HTML page.

'WiFi.begin' is a Wi-Fi initialization function loaded with a Wi-Fi parameter such as 'Username' and 'Password' respectively. 'server. Begin' is a server-initialization function used to start up the server. The syntax code for the HTML page is shown in Figure 4.4. In this syntax, 'void loop' is an infinity execution loop that executes the primary function of the program. 'client.println' is a function used to transmit the data serially and to execute the HTML code; the respected code should be loaded into the 'client.println' function. The syntax code for linking two HTML pages is shown in Figure 4.5. In this syntax, the respected HTML code acts as a bridge between two HTML pages. After uploading the program into

Syntax Code for Linking two HTML Pages:

```
void loop() {

        if (request.indexOf("/log") != -1) {

                client.println("<!DOCTYPE html>");

                client.println("<html>");

                client.println("<head>");

                client.println("<script
                src=\"https://ajax.googleapis.com/

                ajax/libs/jquery/3.3.1/jquery.min.js\"></script>");

                client.println("<script>");

                client.println("$(document).ready(function(){");

                client.println("$(\"#div\").empty();");

                client.println("});");

                client.println("</script>");

                client.println("</head>");

                client.println("</html>");

                p=1;

        }

}
```

FIGURE 4.5 Syntax code for linking two HTML pages.

NodeMCU, open the serial terminal of Arduino IDE. Now an IP address of the wifi router will be printed on the serial terminal.

Note down the IP address of the Wi-Fi router. Now connect the NodeMCU to a mobile adaptor for the power supply. Open a web browser and type the IP address in the URL of the web browser. Now a 'Login/Registration Page' will appear. The HTML code is written for developing web pages. The Login/Registration page will

FIGURE 4.6 Login/registration page.

have four options: Login, New Registration, Forgot Password, and Forgot Username, as shown in Figure 4.6. For a new user, go to 'New Registration' and fill the form. In this, everyone should create a 'Username' and 'Password' with conditions as shown in the below figure, and everyone should enter a unique 'Key' and a unique 'Device Number which was allocated along with the device. 'Key' and 'Device Number' are different from device to device. Here, the programmer creates 'Key and Device Number.' Finally, everyone should write the user's 'First Name' and 'Last Name.' The Registration Page is shown in Figure 4.7. For successful registration, a message 'successfully created' will be appeared. If anyone wants to keep the username as a password, it shows the message as 'username and password are the same, and please give username and password differently.' If anyone enters the username in the registration form already in use, it shows the message as 'username already in use. If anyone wants to log in to the homepage, go to 'Login' and fill in the username and password with valid credentials, as shown in Figure 4.8. For a successful login, a message will appear with your registered name. Now click on the 'Goto Home' button to enter into a home page. There will be three buttons on the home page: Connect to my home, Change Password and Logout. The Home Page is shown in Figure 4.9. By clicking on the 'connect to my home' button, the current page will be redirected to another page that contains four-room buttons such as Room-1, Room-2, Room-3, Room-4, and one back button as shown in Figure 4.10. By clicking on the Room-1 button, one can on/off the home appliances such as fan/light, which are present in Room-1 as shown in Figure 4.11.

If anyone wants to change the password, they access a button called 'Change Password' on the home page. By clicking on that button, the home page will be redirected to the change password page. For a valid credential, it will show the message as 'password successfully changed.' After completing their requirements, one can log out of the page by clicking the 'Logout' button on the home page. A successful logout shows the message as 'successfully log out, close this window tab for security purposes.' If anyone forgot their password, go to 'Login/Registration Page' and click on the 'Forgot Password' button. Now the page will be redirected to forgot password step-1 page. Here one should enter their username. After giving the valid credentials, the page will be redirected to the forgot password step-2 page.

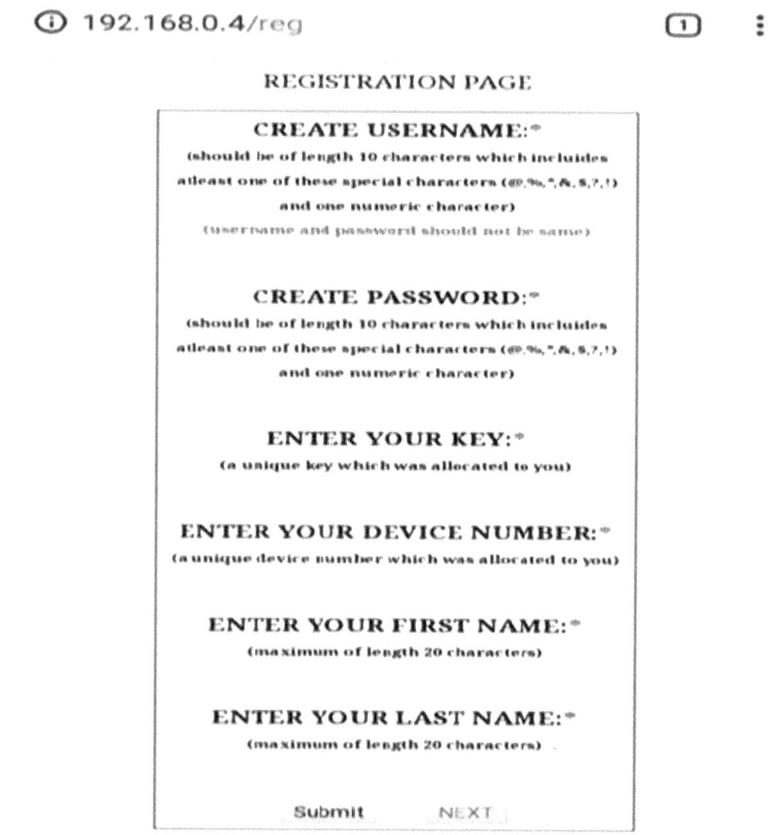

FIGURE 4.7 Registration page.

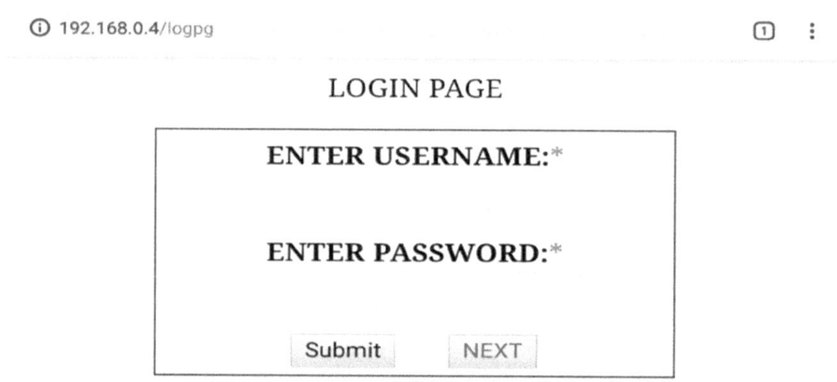

FIGURE 4.8 Login page.

FIGURE 4.9 Home page.

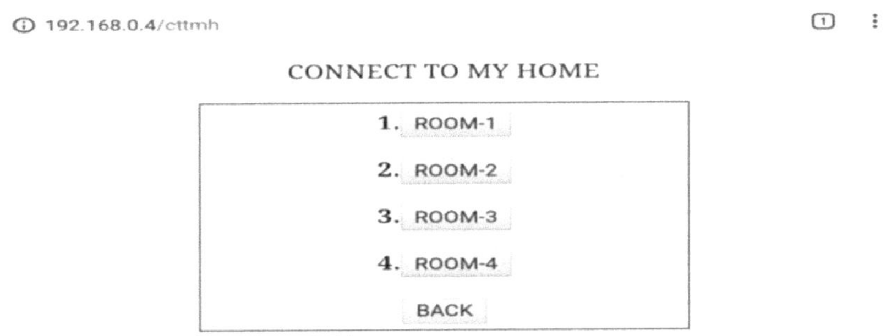

FIGURE 4.10 Connect to my home.

FIGURE 4.11 Room-1.

Here one should enter their unique key. After giving the valid credentials, the page will be redirected to the forgot password step-3 page. Here one should enter their unique device number. After giving the valid credentials, the page will be redirected to the forgot password step-4 page, the final step in this process. Here one should create their new password. For a valid credential, it will show the message as 'password successfully changed.' If anyone forgot their username, go to 'Login/Registration Page' and click on the 'Forgot Username' button. Now the page will be redirected to the forgot username step-1 page. Here one should enter their unique key and device number. After giving the valid credentials, the page will be redirected to the forgot username step-2 page, the final step in this process. Here one should enter their first name and last name. For a valid credential, it will show the message as 'your username is.' For an invalid credential, it will show the message as 'invalid credentials.'

4.4 RESULTS AND DISCUSSION

The flow chart of the proposed work is shown in Figure 4.12. This section shows the practical outputs discussed earlier. A bulb is connected to the NodeMCU for a specified pin, discussed in the proposed method. After creating an account successfully, log into the web application with the desired details. From here, we can control the home appliances smartly. The previous state (ON/OFF condition) will

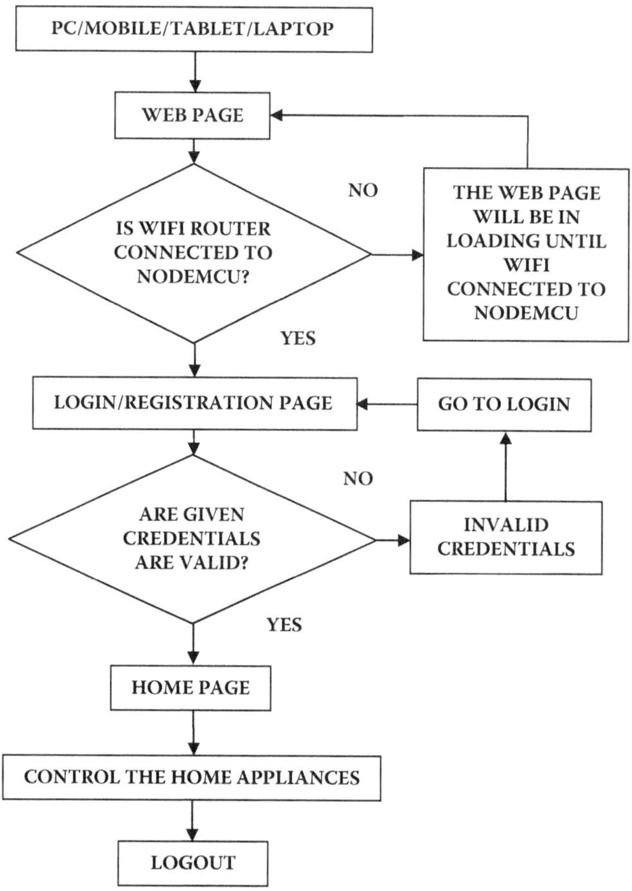

FIGURE 4.12 Flow chart of the proposed work.

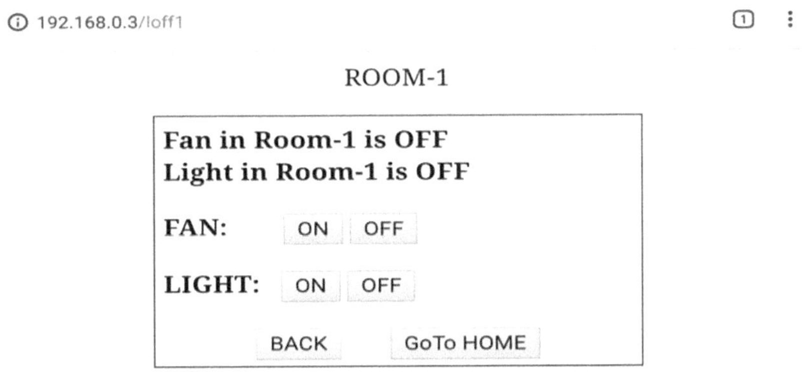

FIGURE 4.13 Current state of the light in Room-1 on a web page is OFF.

be carried until we manually update the state of the appliances. The results of the proposed work for different cases are discussed below.

Case 1. The current state of the light in Room-1 on the web page is 'OFF' as shown in Figure 4.13. The light in Room-1 connected to NodeMCU and relay is in the 'OFF' state, as shown in Figure 4.14.

Case 2. The current state of the light in Room-1 on the web page is 'ON' as shown in Figure 4.15. The light in Room-1 connected to NodeMCU and relay is in 'ON' state, as shown in Figure 4.16.

FIGURE 4.14 Light in Room-1 connected to nodeMCU and relay is in OFF state.

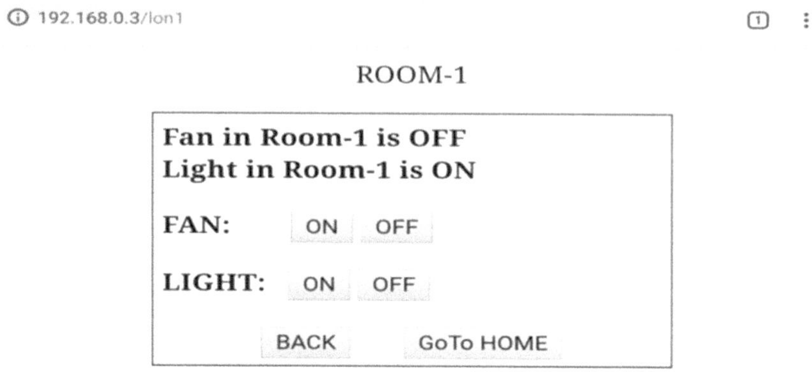

FIGURE 4.15 Current state of the light in Room-1 on a web page is ON.

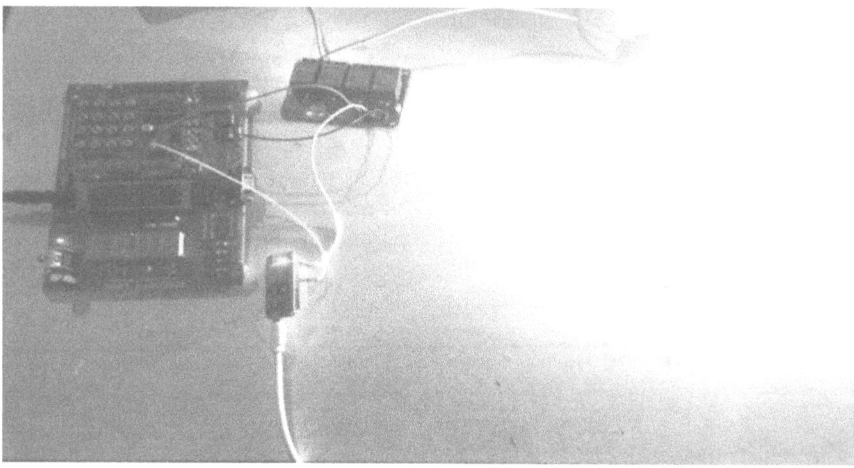

FIGURE 4.16 Light in Room-1 connected to NodeMCU and relay is in ON state.

CONCLUSION

In this work, a security-based web application has been developed to control the home appliances using NodeMCU. Home appliances can be controlled through a web page from anywhere in the world. The developed web application should pass several security checks before it connects to the home. So only the authorized person can access or control their home smartly. At the programming level, the header files and functions in Arduino IDE have been beneficial for developing the web application. The proposed web application for the controlling of home appliances has been demonstrated using a realistic test system.

REFERENCES

1. P. P. Gaikwad, J. P. Gabhane, and S. S. Golait, "A Survey based on Smart Homes System Using Internet-of-Things," International Conference on Computation of Power, Energy, Information and Communication (ICCPEIC), pp. 330–335, Chennai, India, 2015.
2. W. Ejaz, M. Naeem, A. Shahid, A. Anpalagan, and M. Jo, "Efficient Energy Management for the Internet of Things in Smart Cities," *IEEE Communications Magazine*, vol. 55, pp. 84–91, 2017.
3. P. A. Catherwood, D. Steele, M. Little, S. Mccomb, and J. Mclaughlin, "A Community-Based IoT Personalized Wireless Healthcare Solution Trial," *IEEE Journal of Translational Engineering in Health and Medicine*, vol. 6, pp. 1–13, 2018.
4. A. Zanella, N. Bui, A. Castellani, L. Vangelista, and M. Zorzi, "Internet of Things for Smart Cities," *IEEE Internet of Things Journal*, vol. 1, pp. 22–32, 2014.
5. Wang Lu, Yang Hailiang, Qi Xiaoke, Xu Jun, and Wu Kaishun, "iCast Fine-Grained Wireless Video Streaming Over the Internet of Intelligent Vehicles," *IEEE Internet of Things Journal*, vol. 6, pp. 111–123, 2018.
6. A. A. Ahmed and M. M. AL-Shaboti, "Implementation of Internet of Things (IoT) Based on IPv6 over Wireless Sensor Networks," *International Journal of Sensors, Wireless Communication and Control*, vol. 7, no. 2, pp. 129–137, 2017.
7. G. Verma and V. Sharma, "Efficient RF Energy Harvesting Circuit Design for WSN and IoT Application," *International Journal of Sensors, Wireless Communication and Control*, vol. 8, no. 1, pp. 37–46, 2018.
8. S. Vatsikas, G. Kalogridis, T. Lewis, and M. Sooriyabandara, "The Experience of Using the IES Cities Citizen-Centric IoT Platform," *IEEE Communications Magazine*, vol. 55, pp. 40–47, 2017.
9. P. M. Jacob and P. Mani, "Software Architecture Pattern Selection Model for the Internet of Things Based Systems," *IET Software*, vol. 12, pp. 390–396, 2018.
10. A. Li, X. Ye, and H. Ning, "Thing Relation Modeling in the Internet of Things," *IEEE Access*, vol. 5, pp. 17117–17125, 2017.
11. P. A. Dhobi and N. Tevar, "IoT-Based Home Appliances Control," 2017 International Conference on Computing Methodologies and Communication (ICCMC), pp. 648–651, Erode, India, 2017.
12. R. K. Deore, V. R. Sonawane, and P. H. Satpute, "Internet of Thing Based Home Appliances Control," 2015 International Conference on Computational Intelligence and Communication Networks (CICN), pp. 898–902, Jabalpur, India, 2015.
13. T. Malche and P. Maheshwary, "Internet of Things (IoT) for Building Intelligent Home System," 2017 International Conference on I-SMAC (IoT in Social, Mobile, Analytics and Cloud) (I-SMAC), pp. 65–70, Palladam, India, 2017.
14. W. Li, T. Logenthiran, V. Phan, and W. L. Woo, "Implemented IoT-Based Self-Learning Home Management System (SHMS) for Singapore," *IEEE Internet of Things Journal*, vol. 5, pp. 2212–2219, 2018.
15. C. Lin, D. Deng, and L. Lu, "Many-Objective Sensor Selection in IoT Systems," *IEEE Wireless Communications*, vol. 24, pp. 40–47, 2017.
16. T. Adiono, B. A. Manangkalangi, R. Muttaqin, S. Harimurti, and W. Adijarto, "Intelligent and Secured Software Application for IoT Based Smart Home," 2017 IEEE 6th Global Conference on Consumer Electronics (GCCE), pp. 1–2, Nagoya, Japan, 2017.
17. K. Lee, W. Teng, and T. Hou, "Point-n-Press: An Intelligent Universal Remote Control System for Home Appliances," *IEEE Transactions on Automation Science and Engineering*, vol. 13, pp. 1308–1317, 2016.

18. V. Govindraj, M. Sathiyanarayanan, and B. Abubakar, "Customary Homes to Smart Homes Using Internet of Things (IoT) and Mobile Application," 2017 International Conference On Smart Technologies For Smart Nation (SmartTechCon), pp. 1059–1063, Bangalore, India, 2017.
19. A. Khan, A. Al-Zahrani, S. Al-Harbi, S. Al-Nashri, and I. A. Khan, "Design of an IoT Innovative Home System," 2018 15th Learning and Technology Conference (L&T), pp. 1–5, Jeddah, Saudi Arabia, 2018.
20. S. Aniruddha, S. Shubham, and M. R. Mahboob, "An Internet of Things Based Prototype for Smart Application Control," 2017 International Conference on Computing, Communication and Automation (ICCCA), pp. 1358–1363, Greater Noida, India, 2017.
21. C. Tushar and J. Prashanth Kumar, "Enhanced Smart Home Automation System Based on Internet of Things," 2019 Third International conference on I-SMAC (IoT in Social, Mobile, Analytics and Cloud) (I-SMAC), pp. 709–713, Palladam, India, 2019.
22. S. Sujin Issac Samuel, "A Review of Connectivity Challenges in IoT-Smart Home," 2016 3rd MEC International Conference on Big Data and Smart City (ICBDSC), pp. 1–4, Muscat, Oman, 2016.
23. Yoanes B. Arvandy, "Design of Secure IoT Platform for Smart Home System," 2018 5th International Conference on Information Technology, Computer, and Electrical Engineering (ICITACEE), pp. 114–119, Semarang, Indonesia, 2018.
24. Li Weixian, P. Van-Tung, and W. Wai Lok, "A Novel Smart Energy Theft System (SETS) for IoT Based Smart Home," *IEEE Internet of Things Journal*, vol. 6, no. 3, pp. 5531–5539, 2019.
25. H. V. Maddiboyina and V. S. Ponnapalli, "Fuzzy Logic Based VANETS: A Review on Smart Transportation System," 2019 International Conference on Computer Communication and Informatics (ICCCI), pp. 1–4, IEEE, India, 2019, January.
26. S. Soumya and V. A. Sankar Ponnapalli, "A Survey—Vanets and Protocols. In: A. Kumar, M. Paprzycki and V. Gunjan (eds.), *ICDSMLA 2019. Lecture Notes in Electrical Engineering* 2020, vol 601. Springer, Singapore.
27. H. V. Maddiboyina and V. S. Ponnapalli, "Mamdani Fuzzy-based Vehicular Grouping at the Intersection of Roads for the Smart Transportation System," *International Journal of Intelligent Engineering Informatics*, vol. 7, no. 2–3, pp. 263–285, 2019.

5 Use of Multi-Channels and MAC Layer Issues in Mobile Ad Hoc Networks: A Detailed Review

Rohit Kumar and Gaurav Bathla
Department of CSE, Chandigarh University, Punjab, India

CONTENTS

5.1 INTRODUCTION

QoS given by a system characterizes its ease of use. QoS differs from one system to another, and various applications have prerequisites for QoS. A few real-time applications require higher QoS than other applications do. The factors like throughput, delay, jitter, PDR, reaction time, and so on lead QoS. Enhancement in QoS governs more significant expense as new infrastructure might be set up. The QoS of existing systems can likewise be expanded by upgrading the current system and consolidating some significant changes like multiplexing and so forth.

With improvement in transmitters/receiver hardware and sensory hardware resources, multichannel implementation is foreseen in ad-hoc networks, and many new systems have 868/916 MHz multichannel transceivers, e.g., Berkeley's third-generation system mica2 mote. One other technology, known as "Rockwell's ad hoc networks nodes," provides 40-channel capabilities. The node has to switch to any given channel using a controller from the ISM band [1–4].

DOI: 10.1201/9781003230526-5

One new standard, called Zigbee, provides 11 communication frequencies at 868/915 MHz and 16 similar channels on the frequency at 2.4 GHz [2]. Two other standards, MICA and Telo, use CC2420 radios providing many operational frequencies. In quantitative terms, 16 non-overlapping channels are provided with a guard band of 5 MHz [2]. So, it is mandatory to design and develop new MC-MAC protocols to avail complete advantages of parallel-transmission capabilities to improve communication and other functioning parameters of MANETs.

Rigorous work has been done in developing data-link layer MAC protocols for wire-free ad-hoc networks with mobility support. Different aspects of these MAC protocols can be studied in [5–16]. However, as far as multichannel MAC protocols are concerned, minimal literature is available. This critical aspect motivated us to provide a detailed survey on MC-MAC protocols in the domain of ad-hoc networks.

Many factors affect the design and development of MC-MAC protocols for ad-hoc or MANETS. Some of them are: (a) the number of communication channels can be practically used (affected by the interference), (b) the channel assignment (CA) method, and (c) rendezvous point. The rendezvous point is the place where two-node decides upon a channel for communication. Other parameters are discussed in detail in later sections.

5.1.1 Types of Channels for MANET

Historically, MANETs has used single channels for communication. However, this history does not necessarily limit MANETs from using multiple channels. Figure 5.1, shown below, presents the types of channels available for use in modern-day MANETs. Though an ad-hoc system uses one channel while exchanging information, this single channel is used or shared by all the competing hosts in the network. As the name specifies, a multichannel contrast system operates on different channels for information exchange, and nodes can utilize the same. Different channels in MANETs are underneath.

Single-Channel Approach: Single channels are the de-facto standard for communication in ad-hoc networks. In a single-channel approach, the same channel is shared among all the participating nodes, and nodes will communicate to access

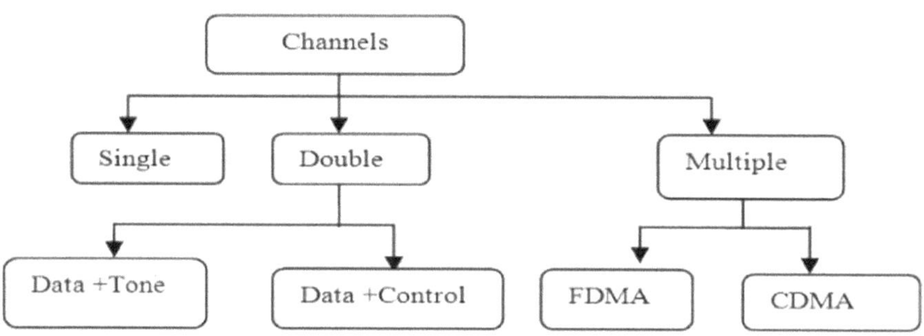

FIGURE 5.1 Possible channel types in MANET.

the channel/medium. Channel allocation and energy conservation are the prime objectives of single-channel-based mobile ad-hoc networks.

Double Channel Approach: The double channel approach can be implemented using two primary techniques: (a) data channel with tone single and (b) data channel with a control channel. In the first approach, a unique single channel called tone is used to control access to the medium, and, in other approaches, a separate channel is used for control purposes. The control channel serves synchronization and channel negotiation. After the negotiation is over, communication is initiated among the channel between selected nodes.

Multiple Channels Approach: In the case of multiple channels, more than one channels are available for data transmission. These techniques divide the available spectrum into multiple channels by using different techniques for enhanced spectrum utilization. The practical techniques to do so in theory and practice are FDMA and CDMA. The FDMA technique divides the bandwidth into several smaller bands on which nodes can send data concurrently/simultaneously without any collision. The CDMA on different sides uses only a single communication channel or single carrier in amalgamation with a group of different codes. The transmitting node XORs the data packet with a specific code before transmitting it. Upon receiving the data packet, 'the destination XORs it again with the same code to obtain the original packet. This communication is just random noise for the other receivers in the transmission range using other different codes. In this way, CDMA allows collision-free simultaneous multiple transmissions, thus, achieving the parallel transmission objective.

5.1.2 Challenges in Single Channel Ad Hoc Networks

Single Channel Access Protocols operate by sharing a single data channel among multiple nodes. Nodes need to compete with each other to access the channel. The capacity of the channel restricts the capacity of such networks. Such networks cannot handle the high-density rise in a particular geographical area. Some of the problems associated with single-channel ad-hoc networks are:

When to Selected: This problem is concerned with the problem of channel selections, i.e., when to select the channel for initiating communication. Waiting time and back-off times determine when and how the channel selection can be made.

Collision: Despite careful channel selection, the possibility of collisions cannot be ruled out in wireless networks due to their unreliable nature. The collisions occur, and methods must be put in place to detect them to carry out retransmissions.

Energy Efficiency: It is the primary design goal of single-channel ad-hoc networks. All the sources that are a vital contributor to energy waste need to be controlled efficiently. Routing protocols are explicitly tailored to minimize energy consumption.

Hidden/Exposed Node Problem: This problem is prevalent in wireless networks and is widely studied. The existence and use of multiple radio devices raise the problem of hidden and exposed node problems. RTS and CTS are used to resolve this issue, but they have not proved to be very effective. Some mechanism must be devised to ensure effective utilization of resources & resolve this problem.

Channel Allocation and Medium Access: Channel allocation is the biggest challenge that wireless ad-hoc networks tackle. As the medium is shared among all the participating nodes, a decision must be made about who will access the medium. Care needs to be taken to make the minor energy consumption during this process.

QoS and Security: Quality of service has become an important design goal for all wireless networks. Provisioning of QoS is a daunting task in ad-hoc networks due to their changing topology, mobility, and limited battery resources. Provisioning security is the other objective of these networks. Designing security algorithms for ad-hoc networks is challenging as energy efficiency needs to be taken care of.

This paper has been divided into four sections, excluding the introductory section. Section 5.2 provides classifications and other information specific to multichannel networks. Section 5.3 discusses and analyzes different protocols. Section 5.4 describes a comparison among different protocols, and last, Section 5.5 presents the conclusions.

5.2 CLASSIFICATIONS OF MAJOR MULTICHANNEL MEDIUM ACCESS CONTROL PROTOCOLS

The MC-MAC layer protocols can be categorized into different classes and groups based on how these protocols execute. Techniques must be chosen to categorize according to communication transceivers, type of rendezvous, initiator among communicators, synchronization method for communication, and the control channel. Figures 5.2 and 5.3 present the channel-assignment strategies for ad-hoc networks and include methods that allow parallel data transmission over a single channel. The fixed and semi-dynamic channel assignment methods do not allow fully parallel multiple transmissions over a single channel. However, the techniques falling under the "Dynamic" category provide multiple parallel communications over a single channel. Here are details of these approaches falling under the fully dynamic category:

Dedicated Control Channel: Every gadget utilizes two or more virtual communication channels in reserved control channel MAC protocols. One is controlling, and the rest are for information exchange (see Figure 5.3(a)). One handset of every gadget, for example, node, is consistently tuned to the system's control

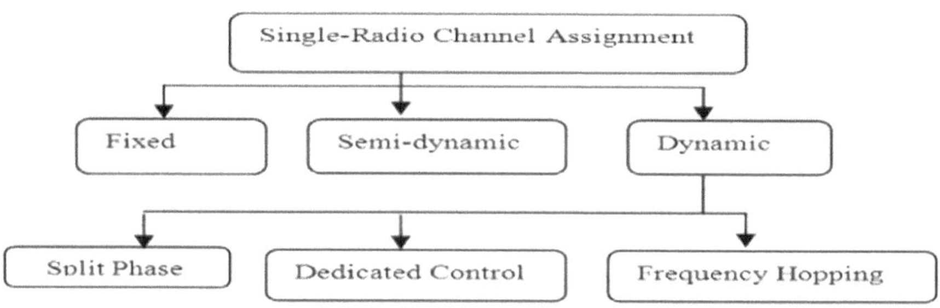

FIGURE 5.2 Different channel assignments policies.

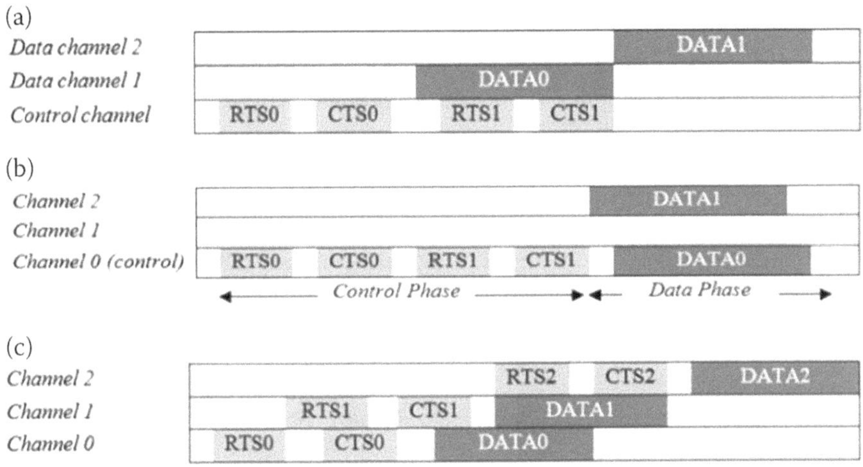

FIGURE 5.3 Methods of dynamic channel assignment in single radio (a) Dedicated channel for control approach (b) Split phase approach (c) Frequency hopping approach.

channel to give synchronization and allow channel access. Other leftover channels are utilized for information interchanges [1].

Common Hopping: Channel hopping is an enhancement over the protocols reserving a channel since they can work with only one channel and all the rest of the channels are accessible for information exchange. A node switches over to another channel after a stipulated (settled early) time in these processes. Figure 5.3(c) presents the life systems of the standard, hoping mechanism. For starting an information exchange, one communicator first detects the channel if it is occupied, then defers & switches over to the next hop for the possibility of transmission. However, while detecting, assuming the channel is observed to be free, it sends an RTS to the receiver, and if the receiver is additionally accessible, it reacts with the CTS message. After this exchange, the node stays on this channel until communication session is finished. After the exchange is over, nodes join the regular hopping plan once more [8,17].

Split Phase: This plan is like the committed control-channel approach; however, it needs only one transceiver to use the different access channels. In this approach, the communicating devices shift back and forth between the data and control phase. The node changes to control the channel on fixed periods and makes resulting channel reservations. The consultations made during 1st phase are executed during the 2nd phase. Figure 5.3(b) presents the working of this method. This methodology has a critical inconvenience of exchanging overheads and synchronization [9].

Parallel Rendezvous: Such protocols do not rely on one control channel for channel dealings and synchronization. Instead, each channel freely controls data trade and performs a connection setup. This technique eliminates the channel exchanging time impressively as each channel can play out the entirety of its activities freely. This strategy likewise eliminates the bottleneck under a single control channel and can handle high-density rise viably.

5.2.1 Challenges in Designing and Implementing MC-MAC Protocols

Except for common challenges of ad-hoc networks like energy efficacy and high scalability, there are complex challenges to tackle while developing multichannel MAC solutions. In a multichannel regimen, the transceiver can switch to multiple available channels to enhance performance. Both the receiver and sender must be on the same channel to carry out information exchange. So, it is challenging to arrange two nodes to come to a common channel. Synchronization and coordination remain challenging problems. The following paragraphs present the fundamental problems in designing and implementing MC MAC protocols:

The MC Hidden Terminal Problem: This problem exists in single-channel and multichannel networks but becomes more complex in a multichannel context. The problem is illustrated through Figure 5.4, wherein a node like C is busy receiving or transmitting data and then a nearby node like A starts channel reservation and making negotiation connection establishment on a given domain. As host C is live on its channel, it cannot determine that the communication channel has been selected by a neighbor node and may wrongly choose the same channel while starting the following data exchange. To bind this problem, some protocols assign different frequencies in the two-hop range, i.e., within two, i.e., two interference hops, to get rid of this problem.

Deafness or Missing Receiver Problem: A problem may occur when a node tries to connect to a busy node on another channel. This may also be stated as "when a given host desires to send a data packet to a recipient/destination host whose transceiver is tuned to a different channel, the deafness or missing receiver problem may occur." This situation is represented in Figure 5.5, where host A tries to send data to node B and transmit the RTS packet. In this case, as B is busy being occupied in other communication & will originate this problem.

Support for Broadcast: A new problem called broadcast support may occur due to the missing receiver (MR) problem. In the case of one-channel networks, the connected hosts can listen to the transmission under broadcast mode (Figure 5.6(a)). Broadcasting becomes complex in the MC domain as the nodes may be on different radio channels when broadcast occurs (Figure 5.6(b)). In this case, some of the nodes may miss the broadcast completely. Broadcast is considered very important for ad-hoc networks, e.g., in reactive-routing protocols, broadcasting is an essential step.

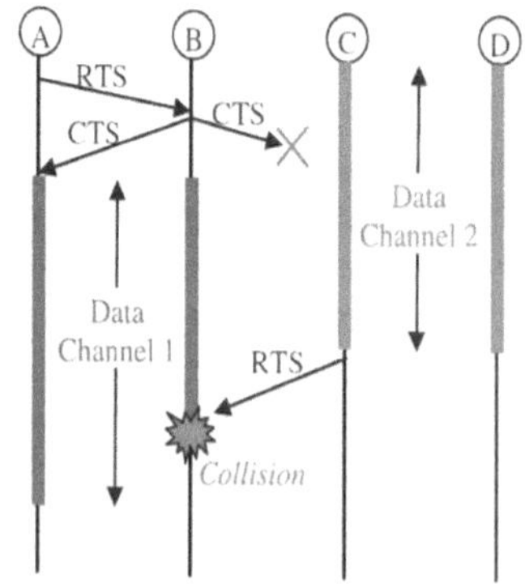

FIGURE 5.4 MC hidden node problem.

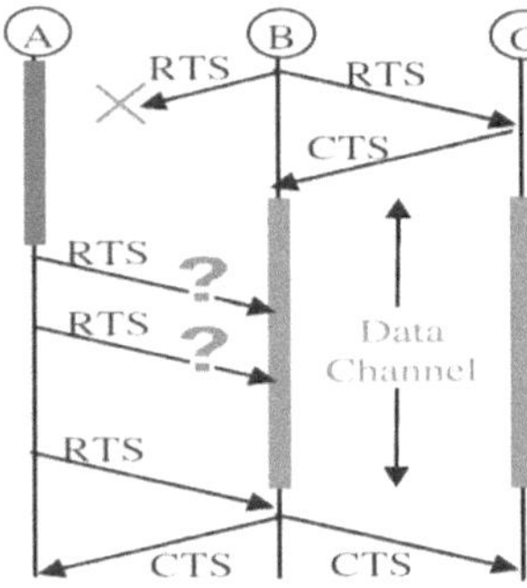

FIGURE 5.5 Representation of missing receiver problem.

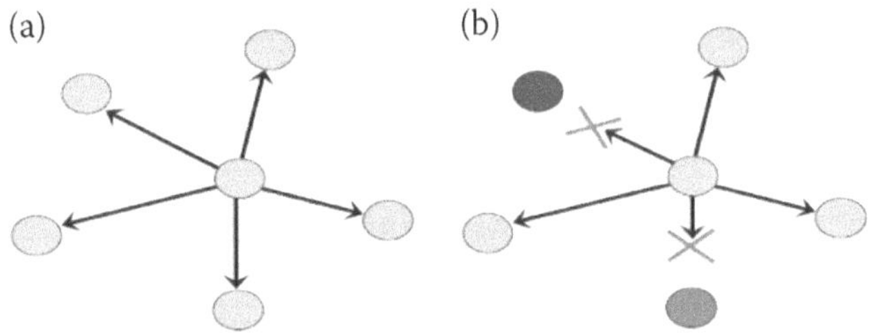

FIGURE 5.6 The problem of broadcast support (a) Broadcasting in single-channel (b) Broadcasting in multi-channels.

Switching Delay: A node needs to switch to different channels depending on availability and other requirements in a multichannel environment. Nearly 200 microseconds get consumed while switching from a specific frequency to another frequency. While the swapping is in progress, the actual data transfer cannot occur, wasting valuable bandwidth.

Method of Channel Assignment: When multiple channels are available, selecting an appropriate one is challenging. Each channel may have its own limitations in terms of bandwidth and average traffic, etc. To allow channel access with minimum interference is commonly called a two-hop coloring problem in literature and falls in the NP-complete problem set.

Inter-Channel Interference: The existence of multichannel gives rise to inter-channel interference and must be tackled proactively. If not resolved, this problem will let collisions occur and ensure a loss of packets. So nonadjacent channels must be chosen by nodes. If one sees the IEEE 802.11 a/b standard where one channel can interfere with at least four channels, the same is shown in Figure 5.7.

The Problem of Synchronization: Providing synchronization is a big problem and must be solved efficiently when dynamic channel allocation is used. When the channel access is dynamic, thorough coordination is required between sender and receiver to carry out channel switching as somehow, they have to agree on the same channel.

The Problem of Network Joining: A new node joining a network may play a spoilsport, and it may lead to channel reorganization, and the new node may need to go through or test all the available channels to find one to use. So, protocols must ensure the smooth joining of a new node in the network.

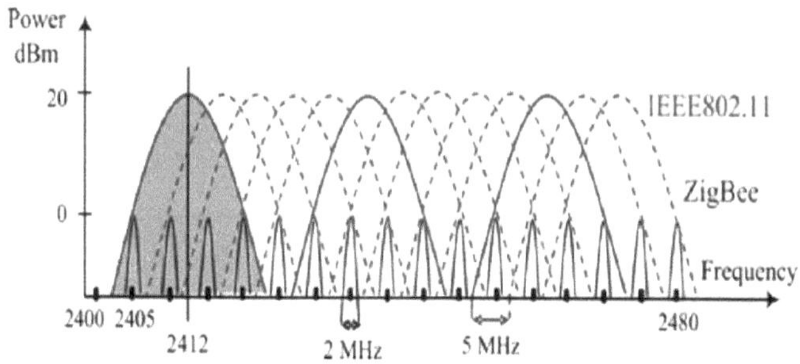

FIGURE 5.7 The 2.4 GigaHertz ISM IEEE-802.11 and IEEE-802.15.4 band.

Network Partitioning Problem: The nodes communicate over different radio frequencies, so figuring out network partitioning is difficult. Innovative methods must be put in place to detect and diagnose network partitioning problems.

5.3 ANALYSIS OF EXISTING MC MAC PROTOCOLS

Many protocols under the multichannel category have been designed and implemented. In the following part, all major multiple channel protocols have been discussed in detail.

1. Duc Ngoc Minh Dang et al. [1] have presented a novel method to implement a multichannel protocol using IEEE 802.11 b standard. The 802.11 MAC has been crafted for single-channel and uses the standard CSMA/CA method for communication. The overall throughput is restricted by the given single channel and can cause starvation and other problems with Omnidirectional antenna. This paper has resulted in developing the MMAC-DA protocol, which stands for MC Protocol with Directional Antennas. It makes use of the energy-saving mechanism provided by IEEE 802.11 and utilizes multiple channels resources with further integration of directional antennas.

Working Method and Setup: The authors have assumed that there are in total N non-overlapping communication channels. The half-duplex transceiver is assumed to be possessed by all the nodes. Synchronization among the nodes is mandatory, and it can be done by using GPS or by using another standard method called TFS (Timing Synchronization Mechanism) of the IEEE 802.11 standard. The time structure of the IEEE 802.11 power-saving mechanism (PSM) has been assumed wherein time is divided or slotted. Next, every time interval is slotted into ATIM (Ad Hoc Traffic Indication) message and data window.

Before selecting any channel or initiating data-transfer nodes, settle it in the given ATIM window. During ATIM, window handshaking is done and control packets are exchanged. While on the ATIM window, nodes may pick any channel among multiple available channels for data transfer in the following data window. The critical attribute of this protocol is that the data transmission can take place during the ATIM window, which was not possible earlier. Figure 5.8 shows the

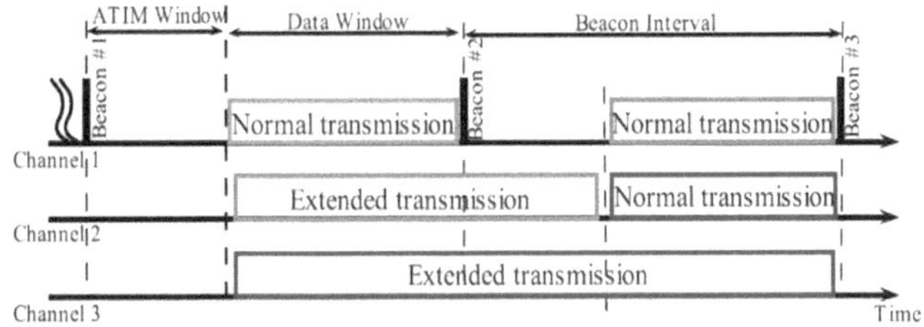

FIGURE 5.8 Illustration of transmission modes.

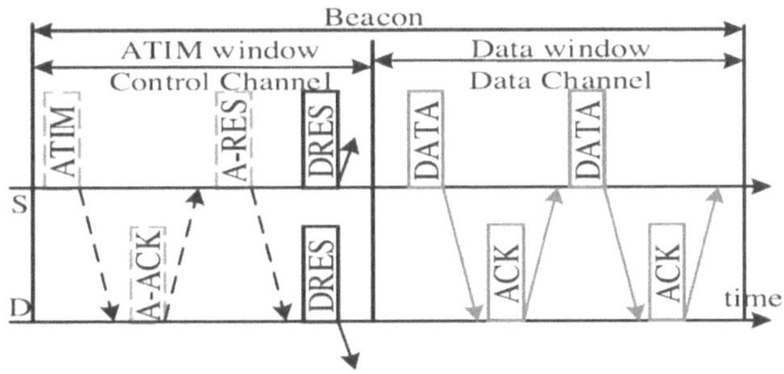

FIGURE 5.9 Basic logic of MMAC-DA.

channel and timing structure of the MMAC-DA protocol. The actual message exchanges are presented in Figure 5.9. During the active ATIM window, the nodes perform the regular three-way handshake in the omnidirectional mode for gaining access to a data channel.

In Figure 5.9, the starting node S sends the ATIM message, which has the accessible channels information in its context. Upon receiving such a message, the receiver finds out the beam direction in the direction of the node and then selects the channel based on direction calculated and availability of its own and channels available at the source. The destination D replies with ATIM-ACK or A-ACK to node S. The A-ACK, where A stands for AITM, includes the beam-direction information and the selected-channel information. After the receipt of A-ACK, node S confirms with A-RES (AITM reservation) message. The last DRES (Directional Reservation) message is sent by the participating nodes and other nodes for information updates. After DRES, the data communication is initiated over the agreed data channel without any interference.

2. Jaeseon Hwang et al. [17] have presented an interference-resistance protocol called RcMAC. The RcMAC stands for "Receiver Centric MC MAC protocols for Wireless Networks." With the availability of many channels, the participant nodes must regularly switch among available channels to capitalize on increased network performance. However, this switching is very costly and can severely affect performance. So, in the proposed system, work has been done to reduce the switching delay considerably. The protocol is receiver centric, and the receiver makes critical decisions

in channel selections and switching. The protocol also enables the originator node to change independently and asynchronously to channels where the desired receiver lives without explicit negotiations.

Receiver-Centric Multichannel MAC Basic Operation: Data communication using RcMAC consists of two phases: the channel selection phase over the control channel and, after that, occupancy of the data channel for transferring the information. Figure 5.10 presents the basic functioning of RcMAC. Both the receiver and sender first have to pick a channel for initiating the transfer. When both the receiver and sender are on the control channel during the channel-selection phase, the channel selection is made by RTS/CTS/CFM messages handshake. However, when the receiver and sender are not on the same channel, then RTS/NCTS/CFM handshake is done where the NCTS message is called neighbor CTS. Similarly, CFM broadcasting can be used for channel selection.

The purpose behind using the RTS/NCTS/CFM handshake is to gather channel information for its desired receiver. It helps the sender switch to a channel where the receiver resides without explicit negotiations, i.e., without switching to the control channel. After the channel selection is made, regular data transfer takes place on the data channel using the send and acknowledge paradigm, and multiple data packets can be transmitted to reduce channel switching. For transferring a data packet, the standard method of CSMA with the binary exponential back-off method has been used in DCF (Distributed Coordination Function) configuration.

Receiver-Oriented Channel Switching: As mentioned above, receiver-centric switching is used when the source and destinations are not on the same channel. Figure 5.11 shows the working method and method exchange process of receiver-centric switching with two associated methods, as shown in Figure 5.11(a), (b), respectively. When the sender does not have any information, it transmits an RTS message on the control channel and later obtains NCTS from the neighbors, which provides the channel information, e.g., channels two, as shown in Figure 5.11(a). After it, it broadcasts the CFM message messages and then makes a transition to channel 2. However, if the originator knows the channel of the destination, then it simply transmits the CFM and immediately makes the transition to the channel.

3. Yuhan Moon et al. [18] proposed a CDMA-based cooperative protocol called CCM-MAC, where CCM-MAC stands for "Cooperative CDMA-based MC-MAC protocol for MANETs." In the CCM-MAC method, a node equipped with one half-duplex receiver-transmitter obtains channel information from cooperating neighboring nodes, significantly improving performance. Besides performance gain, CCM-MAC use remedies many multichannel ad-hoc network problems like CDMA's hidden/exposed terminal problems.

Channel Negotiation: Given that CCM-MAC protocol uses the handshaking method to select a data channel, all of the nodes must listen/tune to the control channel for making channel selection. In addition to regular RTS and CTS messages, CCM-MAC introduces new coordination messages and keeps other problems at bay. These messages are: Decide Channel Send (DCTS), which indicates the

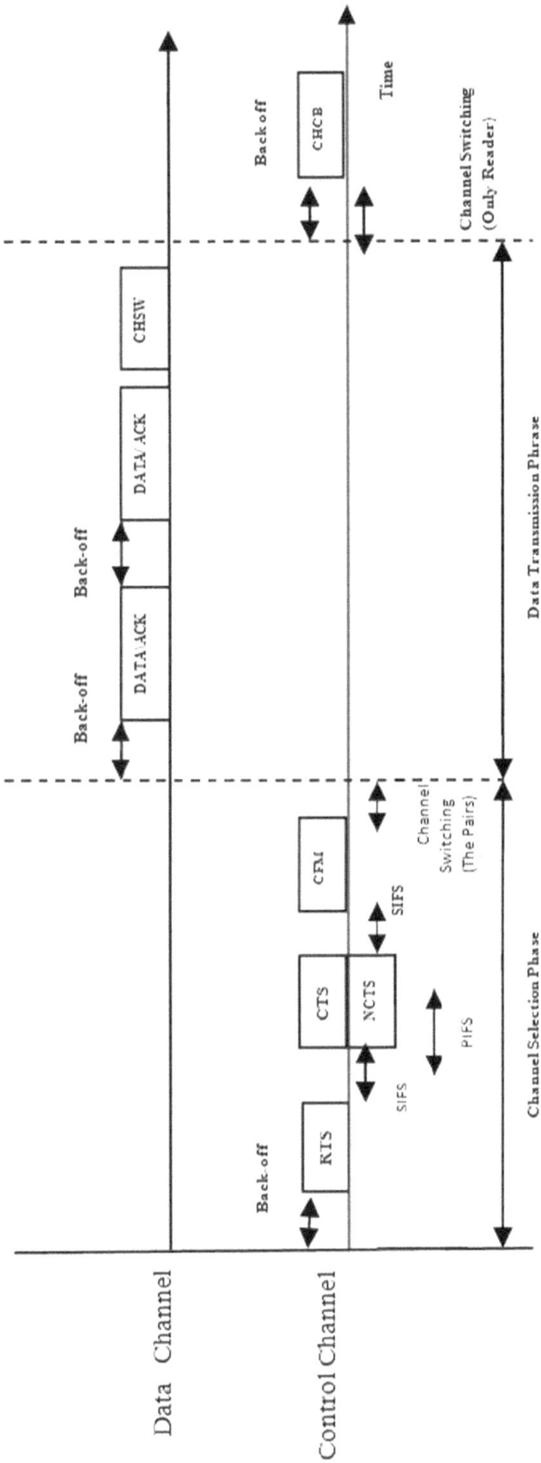

FIGURE 5.10 Working method of RcMAC.

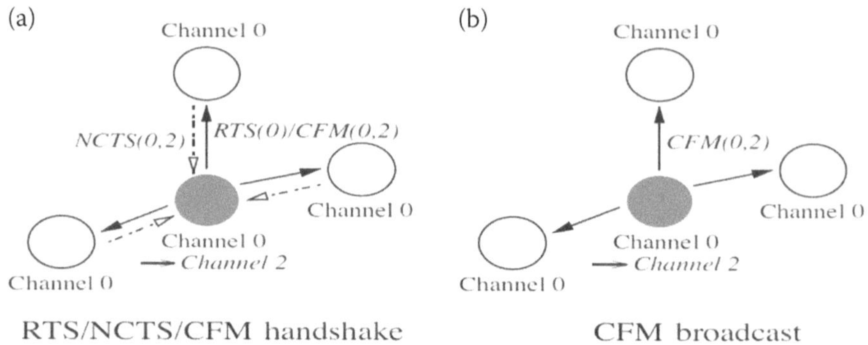

FIGURE 5.11 (a) Illustration of receiver-oriented channel switching (b) Channels number 0 and 2 are the control and data channels.

selected channel, Information to Inform (ITI) message, and co-operating nodes that use it to help decision making and confirm (CFM) messages; CFM messages are used to notify all the nodes about the decision made, i.e., channel selection. These messages are depicted in Figure 5.12, which also explains the channel-selection process. Communication from node B to node C is shown in Figure 5.12 and IEEE 802.11 DCF is followed.

The formula can derive the maximum outcome of CCM-MAC:

$$\textit{Throughput}\ (\text{CCM-MAC}) = \frac{M \times L}{\frac{L}{B} + T_{ACK}},$$

Where

$$M = \begin{cases} H_{max} & \text{if} \quad H_{max} \le K \times N, \\ K \times N & \text{if} \quad H_{max} > K \times N. \end{cases}$$

1. DPC [11] is an essential protocol called Dynamic MC-MAC for Ad-Hoc LAN and supports the use of multiple transceivers. It uses a single dedicated control channel, and every gadget is consistently tuned to this single control channel for channel negotiation & synchronization. Rest channels are data channels and are solely used for the transfer of data. The channel-allocation process is dynamic, and each device can use any available data channel for communication. Figure 5.13 presents the model of this scheme. There exists one dedicated control channel labeled (CCH) & many other data channels denoted by DCHs. The single control channel is shared among all the nodes, and one transceiver is always dedicated to it. Due to this continuous tuning to the control channel, all the nodes in the range can see the status of all the participating devices.

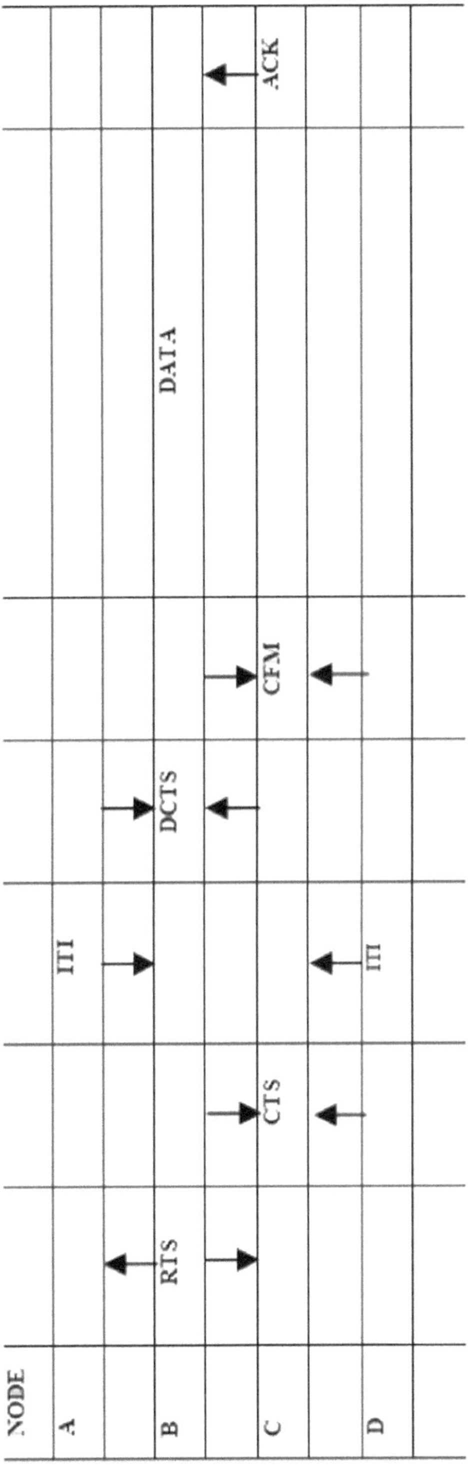

FIGURE 5.12 Example of channel negotiation with no collision among ITI control packets.

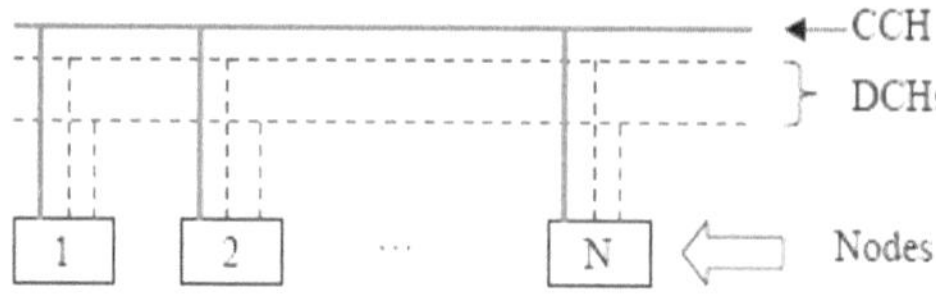

FIGURE 5.13 DPC and its multichannel system.

Data Communication Process: At the initial stage, when host A wants to send some data to B, one of the data channels (if accessible) will be allocated to them for a limited duration. After this time interval is over, the data channel will become free and allocated to other communicating entities. Before the allocation of the data channel, the nodes must make a reservation through the control channel.

The simulations for this approach exhibit that for up to four channels, the channel utilization increases; above that, it starts degrading. This degradation occurs due to the blocking when communicators have already reached the consensus to use separate communication channels with different hosts.

2. Jain et al. [15] have proposed a Novel Multichannel CSMA/CA-based MAC Protocol along with Receiver Centric Channel Selection method for Multiple hop wireless networks (MAC-RBCS). This particular protocol makes use of a single control channel & many other information channels. In this policy, the receiver plays a critical role in channel selection. A receiver selects a channel depending upon enhancing signal-to-interference with the addition of noise ratio, i.e., the most acceptable channel is selected for communication. The simulation results show that this protocol offers higher throughput and lower delay in communication when employed in fixed/static or grid structure-based networks, whereby the likelihood of interference from hidden hosts is very high. This protocol has the drawback of underutilizing multiple channels since multiple channels remain idle for a significant time. This problem can be overcome by allowing multiple transmissions on each channel to exploit its full bandwidth benefits.
3. DCA [13] presents an MC-MAC Protocol with the feature of On Demand Channel Allocations; it was supported for Multi-Hop MANET. The authors proposed a protocol that uses dynamic channel assignment. The operation of this protocol is coordinated by a dedicated control channel. One of the transceivers is always tuned to this channel to provide channel access and synchronization. The second transceiver is used for data communication, and this transceiver can hop within different available multichannel. Figure 5.14 illustrates the operation of the protocol.

This protocol uses two significant data structures to carry out the work: CUL, FCL, which can be abbreviated as channel Usage List, Free Channel list.

CUL: The CUL table contains the current channel usages of all the channels in the network. Each ith list entry is represented by CUL[j], which keeps records and

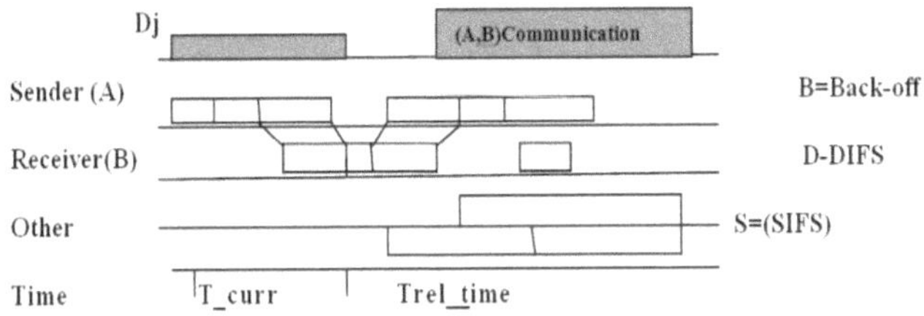

FIGURE 5.14 Timing diagram of the process to find out the status of a channel exchange (RTS and CTS).

information on when a node neighboring or near X uses a channel. In CUL data structure, each entry i.e., CUL[j] has three major fields/attributes: CUL[j]. node is a neighbor node of node X, CUL[j].channel is the data channel used by CUL[j].node, and CUL[j].release-time is the time by when channel CUL[j].channel will be given up by CUL[j].node. This data structure, i.e., CUL, is maintained distributed by each mobile node, i.e., each node maintains its own CUL table and thus may not contain precise or highly accurate information.

FCL List (Free Channel List): This data structure is also maintained by all the hosts and represents the free channels available. The information in FCL is dynamically from the CUL list by subtraction, and FCL is populated with free channel information.

The DCA protocol ensures that the channel usages are limited to two-hop neighbors to remove exposed node and hidden node problems. One disadvantage of this approach is the bottleneck posed by the control channel, i.e., the control channel can become saturated when the number of participating nodes increases rapidly.

4. DCA-PC [14] is a power-oriented protocol and is called: Dynamic Channel Assignment method with PC (power control). This protocol is an improvement over simple DCA in terms of power consumption. As an extension to DCA, the DCA-PC protocol resolves and eases the channel allocation, media access, and control of power efficiently, well-integrated, and systematically. The critical control packets, i.e., RTS/CTS/RES are transferred using the highest energy in this protocol.

The power-control mechanism has several advantages, i.e., increased battery life and reduced co-channel disturbance or interference with neighboring nodes. The power-control mechanism can significantly remove the "near and far" problem in CDMA-based systems and negatively impact the system throughput. Another advantage of this approach is channel reuse in a particular geographically bounded area.

The protocol, i.e., DCA-PC calculates transmission power using simple equations e.g.,

Assume mobile nodes X and Y wish to transfer with each other one data packet. Let node X transmit a packet with power Pti, which is listened to by Y with power Pri. Then the following equation holds for this situation.

$$P_{ri} = P_{ti}\left(\frac{\lambda}{4\pi d}\right)^{n} g_{ti}\, g_{ri} \ldots\ldots\ldots\ldots\ldots\ldots \quad (5.1)$$

Where λ is the CW (carrier wavelength), d is the gap between source and destination, n is the coefficient, i.e., path loss coefficient. The other parameters, i.e., g_{ti} and g_{ri} are the antenna gains at the source and the destination. The λ, g_t and g_r are constants in standard environments/situations.

In the given setup, the value taken for n, i.e., path loss coefficient, is typically two, but it may swing between two and six depending on the physical situations/environment, like obstructions/obstacles, etc. Now assume that node Y wishes to reply a data packet to X such that it **gets** the packet with a chosen energy Pxi.

In this way, then Yi's power of transmission will satisfy the following equation:

$$P_{Yi} = P_{Xi}\left(\frac{\lambda}{4\pi d}\right)^{n} g_{t}\, g_{r} \ldots\ldots\ldots\ldots\ldots\ldots\ldots \quad (5.2)$$

Though the values of the setup/environment-oriented or environment-reliant parameters n and d are unknown, these values can be treated as constant for minimal durations. So, Equation (5.2) is divided by Equation (5.1), which gives:

$$\frac{P_{Xi}}{P_{ri}} = \frac{P_{Yi}}{P_{ti}} \ldots\ldots\ldots\ldots\ldots\ldots\ldots \quad (5.3)$$

In this way, Y_i can find its power of transmission Pyi if the other powers parameters are known.

5. To handle the MC hidden-terminal problem, the MMAC [8] protocol was proposed. MMAC stands for MC-MAC for Ad Hoc Networks. All the participating nodes switch to the control channel to ensure channel allocation and synchronization. The protocol uses ATIM (Ad hoc traffic indication message) on the control channel for granting channel access. The control channel timing is split into ATIM interval and data interval. While ATIM interval channel negotiations are done, and after the ATIM period is over, the control channel is used for data exchange.

In MMAC, all hosts uphold a PCL (i.e., Preferable Channel List), indicating a node's suitable or best channel preference. The PCL has been categorized into three types: high, medium, low preference and indicates the priorities of channels. While requesting a channel, a node (either sender or receiver) will prefer a high-priority channel, and if the high channel is not available, then a medium-priority channel will be considered for allocation. Low-priority channels will be chosen when both of the other priority level channels are not available.

All the hosts are tuned to a common control channel at the beginning of the ATIM slot. While transmission of the ATIM window, the receiver & sender negotiate communication channels using handshaking, i.e., connection establishment by three-way handshake system which denotes their corresponding PCLs.

After the expiry of the ATIM window, all the nodes change over to their concerning data channel and start a regular cycle for communication i.e., RTS/CTS/data/ACK regular cycles for transmission of data. Through interleaving, the CCH is also utilized for sending information after the expiry of the AT1IM slot period. The simulations carried out present considerable improvements over the standard IEEE-802.11. The authors also affirm that the M-MAC protocol performs better or is comparable (i.e., performs equally) to other multichannel protocols, including DCA. Figure 5.15 presents the process of data exchange and channel negotiation in MMAC protocol.

6. Li et al. [11] invented a new and efficient multichannel framework and corresponding protocol. This presented protocol is an improved part of standard 802.11 MAC supporting multiple communications channels. In this environment, all nodes are also equipped with a half-duplex transceiver, and channels are assumed to be the same number of nodes. One of the channels is a particular channel dedicated for control and synchronization purposes and is called control channel. The rest of the n channels is assumed to be data channels. The extensive simulation results exhibit improvements over the single-channel approach and other MC protocols in multi-hop ad-hoc networks. The working of this protocol is presented in Figure 5.16 and Figure 5.17.
7. CHMA [5] is a Channel-Hopping Multiple Access protocol and CHAT [13] is also a similar kind of protocol. CHMA protocol performs its function by hopping among multiple available channels. Every node in the setup/network follows a controlled channel-hopping (also common hopping) sequence. If a host has some data to transfer in its current hop, and if the destination is on the same hop, RTS is sent to it, and then it transfers CTS to the source node. After this control information exchange, the nodes stand over the same channel hop as long the data-transfer session is ongoing. Other nodes switch to other hops and keep following the standard

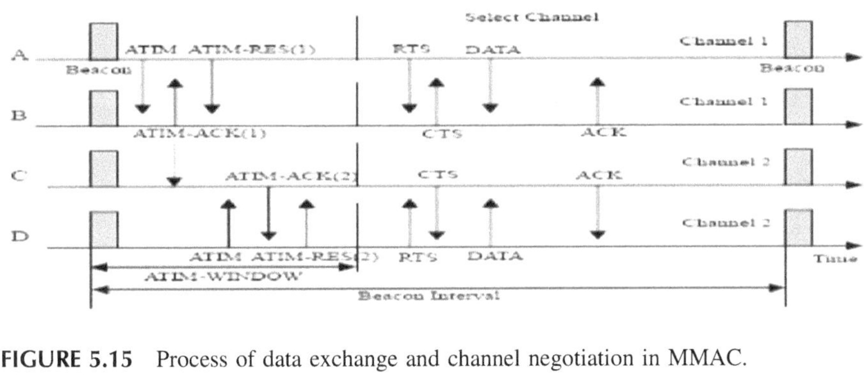

FIGURE 5.15 Process of data exchange and channel negotiation in MMAC.

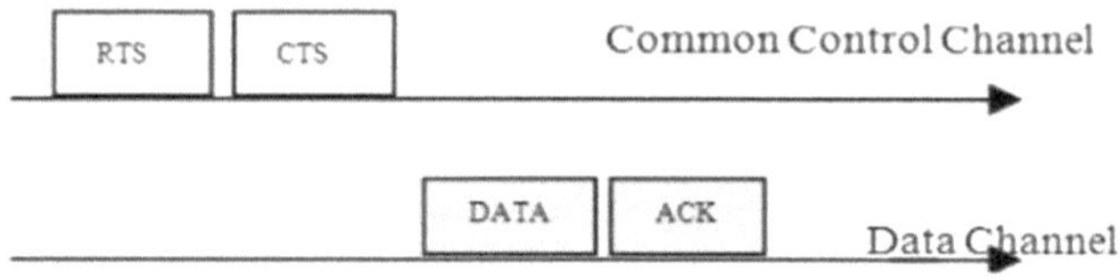

FIGURE 5.16 Working of the modified protocol.

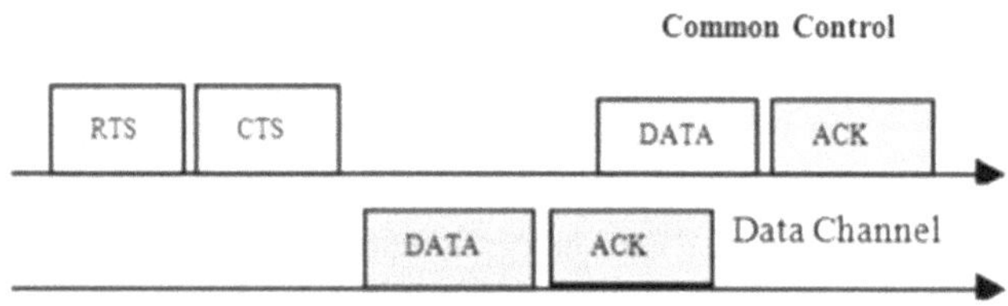

FIGURE 5.17 The procedure of retransmission.

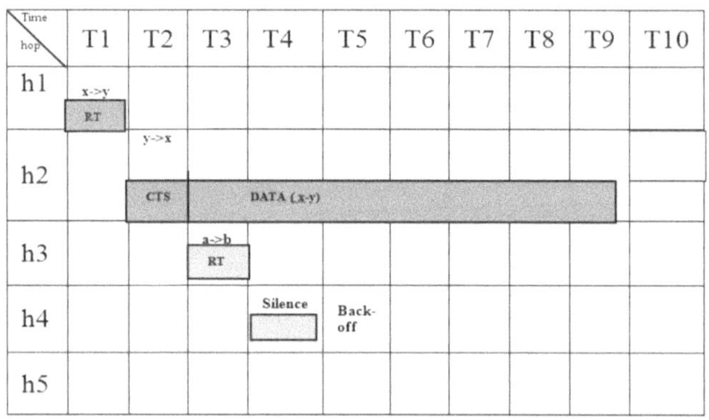

FIGURE 5.18 Illustration of CHMA.

sequence. After communication is over, both the communicators are required to synchronize themselves with the present joint hop. Figure 5.18 shows the usage scenario of the CHMA protocol.

One other protocol called CHAT is one such protocol that is way similar to CHMA but exhibits significant progress in terms of throughput for an ad-hoc network. The protocols have been simulated with uni-cast, broadcast, and hybrid traffic consisting of both unicast and broadcast transmissions.

8. J. G. L. Aceves et al. proposed another RICH-DP protocol [19], which is initiated at the receiver end and employs channel hopping with a dual polling scheme. It is similar to CHAT (Channel-Hopping Multiple Access protocols). A receiver-initiated and receiver-controlled collision avoidance

policy does not sense the carries or assign any uniquely identified codes or another coding system. The strategy is based on a standard frequency hop sequence, and each node is assumed to follow the same.

9. SSCH [9] is an IEE1E 8021.11 standard-based protocol that stands for "Slotted Seeded Channel Hopping," and this protocol has been tested to provide significant ability and improvement in IEEE1 802.11. For the active part of this single protocol, radio is required with each node. The protocol is highly appropriate for a multi-hop environment and does not need changes to the standard IEEE 8012.11. In addition, the critical feature of this approach is that it does not necessitate the use of multiple transceivers.

This method allows control and synchronization traffic for disseminating among all available channels and therefore eliminates the common CCH saturation problem, which is a crucial bottleneck posed by dedicated control channel-based protocol.

The technique also presents a second novel method called the partial-synchronization method. This method permits a given dispatching host to moderately align with the originator and a given recipient node. The synchronization method permits communication work-load for a single multi-hop data flow to be dispersed diagonally many channels and results in increased transmission rate.

The SSCH protocol in a node handles and cares for three essential aspects of channel hopping mechanisms.

1. It implements the host's channel-hopping schedule, which is to be followed by the host.
2. It transmits the channel-hopping information and schedule to neighboring hosts.
3. It changes the host's hopping roster to adjust to varying traffic plans.

The current I1EEE 802.11a MA1C protocol offers many orthogonal disjoint channels. SSCH try to utilize these channels efficiently. This protocol uses time slots to define a channel, e.g., in one slot, a transceiver may be tuned to channel 2 and in the next slot, it may switch to channel 0. Each node maintains its channel switching schedule, i.e., a node will switch over to a given channel. Each node uses a formula:

$$x_i = (x_i + a_i)\ mode\ (no.\ of\ channels) \ldots\ldots\ldots\ldots\ldots\ldots\ldots\ldots \quad (5.4)$$

To calculate its schedule. Here, x_i (in parenthesis) represents the present channel and a_i indicates the seed. The newly calculated value of x_i represents the new channel chosen by a node. For simplicity, the formula can be written as follows:

$$\text{New} - \text{Channel} = (\text{Old} - \text{Channel} + \text{Seed})\text{Mod}(\text{Number} - \text{of} - \text{Channels})\ldots\ldots \quad (5.5)$$

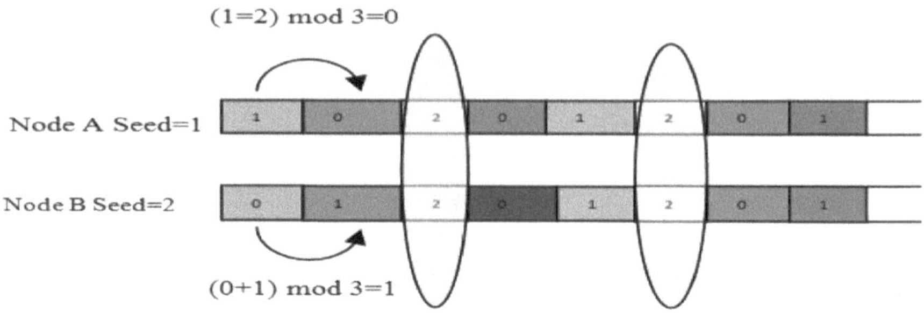

FIGURE 5.19 Channel hopping in SSCH.

This seed is an integer and is always one less than all available channels, i.e., if the channels are 13, the seed will be 12. Figure 5.19 presents the idea of SSCH. The schedule of a node is frequently propagated to neighboring nodes to know the schedule of neighboring nodes. The nodes can utilize this schedule to gain channel access to communicate with a particular node. Nodes do not need a control channel for channel access as they can independently make decisions regarding channel access. This protocol supports multiple communications to existing simultaneously.

10. ODC [6] is an on-demand channel switching protocol and works under the assumption that all hosts are always fitted with only one half-duplex transmitter and receiver. This particular protocol exhibits dynamic nature and operates according to the present traffic patterns. The channel-selection process is dynamic, and care is taken to choose the best channel for the communication. The switching overhead is kept to a minimum by choosing a suitable dwell time controlled by present traffic. A node will settle on a channel till the channel's traffic does not fall below a given threshold.

Simulations carried out present that OCD protocol has fewer packet delays, increased energy savings compared to increased bandwidth utilization, compared to MMAC and IEEE 802.11, especially in one-to-one strategy under the condition of heavy load traffic.

11. J. Walrand et al. presented McMAC [20] a robust multichannel protocol. McMAC stands for "MC Medium Access Control Proposal for Ad Hoc Networks." The McMAC assumes that every host is bundled with one transmitter/receiver, which can tune and hops/jump across all accessible channels purely random or pseudo-random fashion. When a particular node is idle, i.e., it does not have any data to transfer, then there is a default sequence that this node will follow. All the participant nodes put their seed in every packet they want to transfer to other nodes. This seed helps other nodes to identify the hopping sequence. Different measures have been taken to reduce unnecessary switching and similar to ODC, hopping has been made less recurrent than regular hopping. When a

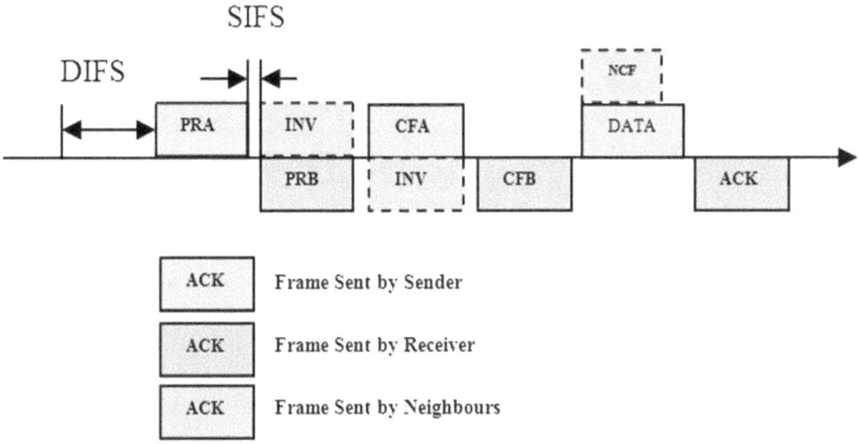

FIGURE 5.20 CAM-MAC and handshaking procedure.

given host X has some data to transfer to host Y, the host X1 flips a coin and transfers with some known probability p during every period. If host Y does not respond with CTS, then node X tries over later by coin flips. On the other hand, if node B responds with CTS, both node A1 and node B stop hopping1 to interchange data. When the 1 data transfer is complete, node A, B reverts to their original/primary regular hopping sequence.

12. 1 CAM-MAC [21] protocol is one of the initial protocols that uses a single transceiver for multichannel communications. This protocol is dynamic, and each host has the capability for switching to another channel to transfer data. This protocol assumes n data channels along with a control channel. Different nodes cooperate with each other to provide channel condition information to participating nodes. Figure 5.20 shows the connection establishment process in CAM-MAC.
13. PSM-MMAC [22] is a proficient protocol developed for wireless local area networks. PSM-MMAC stands for "Power Saving Multi radio MC-MAC." This protocol is capable of working with multiple independent NICs. In this protocol, every host has at least one primary default network interface card to streamline, coordinate, and control traffic indication and other traffic features and perform the channel-negotiation process. In this protocol, i.e., PSM-MMAC, time is divided into similar kinds of beacon intervals.

The beacon period is further separated into three phases. In the first phase, every node remains awake, tunes to the primary channel, and selects a small time slot to transfer a busy signal if it has some information. In the second part, each host indicates traffic on the channel & performs negotiations on the default control channel. In the third phase, nodes transfer data on a data channel and switch to doze mode when idle.

This power-saving PSM-MMAC protocol results in power saving by guessing and calculating the number of live and active links in the underlying network. Secondly, if a node has no data to transmit, the respective transceivers will remain in sleep mode to save power. Figure 5.21 shown below illustrates show the time structure in the PSM-MMAC protocol.

14. PCAM [12] was proposed by A.K. et al. works under the assumptions that every node is fitted with three half duplex transceivers called (a) primary/basic, (b) secondary, and (c) the third transceiver. Among the primary and the secondary (first two transceivers) are used for actual data transfer. The last transceiver is used to support broadcast communication specifically for transferring and receiving broadcast messages. The PCAM protocol does not need time synchronization and can enhance the available channel bandwidth utilization. Figure 5.22 presents the channel allotment procedure in PCAM.
15. FMC-MAC [23–28] is a flexible protocol, and FMC-MAC stands for "Flexible Multichannel Coordination MAC." This protocol works using a single transceiver for multi-hop mobile ad-hoc networks. Though it is an asynchronous multichannel protocol, it also favors asynchronous channel conciliation, i.e., using a back-off process when the channel is occupied. This protocol also uses a control channel for negotiations and synchronization. This protocol efficiently tackles the head-of-the-line blocking problem. When the severe head-of-line blocking problem occurs, the node switches to the control channel to find a new communication channel. The simulation has also shown that this protocol solves the head-of-the-line blocking problem more effectively than other synchronous protocols.
16. In addition to just discussed multichannel protocols, many other multichannel MAC protocols also exist with different variations [19,29,30]. Currently, the researchers are working in the field of Cognit1ive Radio (C1R) networks, and many Cognitive Radio-based M1AC protocols have been proposed and implemented [31–65]. Few protocols have concentrated primarily on enhancing the capability of wireless networks [66–69], by MAC layer in association with other layers has been proposed to optimize the network performance [21,68,70,71].

5.4 CHARACTERISTICS OF MULTICHANNEL PROTOCOLS AND THEIR COMPARISON

Multichannel MAC protocols provide remedies for all the problems that exist in single-channel MAC protocols. All the major performance parameters get a boost by multichannel. However, the design and development of multichannel protocols are not easy since they have to tackle new challenges posed by the underlying medium and newer interference. In this paper, all these methods have been elaborately discussed, focusing on the efficiency and challenges of these methods. All

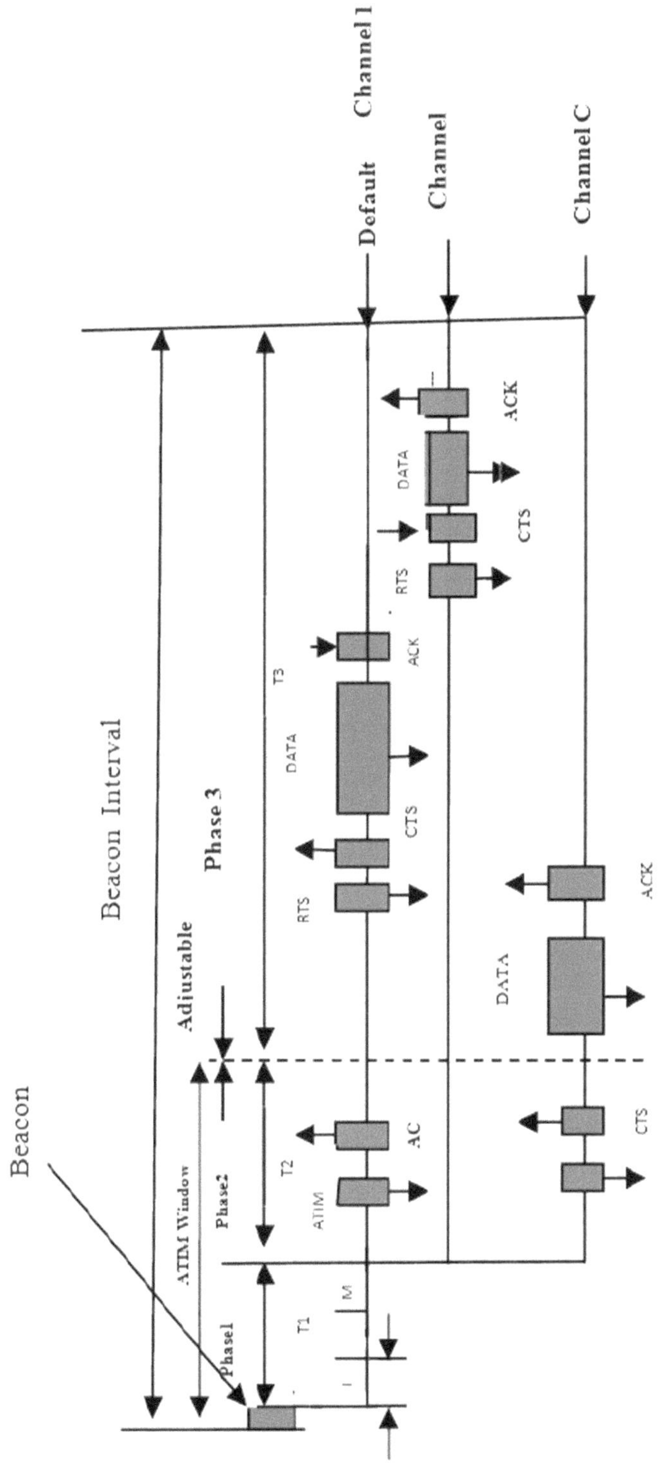

FIGURE 5.21 PSM-MMAC and its time structure.

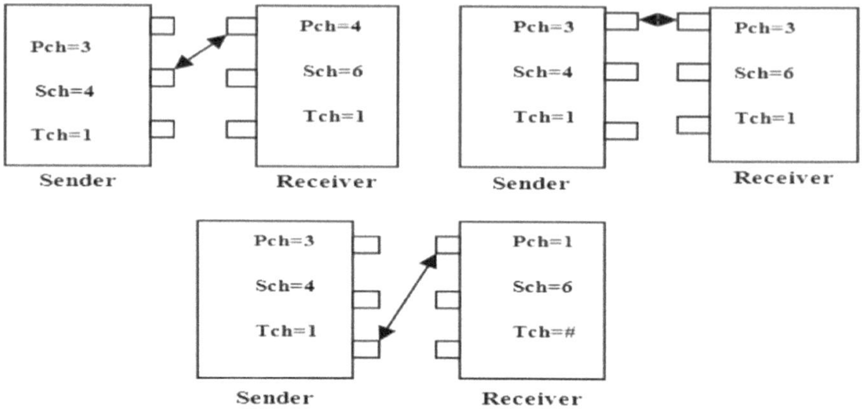

FIGURE 5.22 PCAM and channel assignment.

the performance parameters, ranging from minor to major, have been rigorously discussed and compared in tabular form for all the protocols discussed.

Various governing factors take the lead on the performance of the network; they include: availability of dedicated/ shared channels, transceivers deployed, synchronization method, type of rendezvous, etc. There have been various challenges resolved by the protocols discussed. Table 5.1 presents the problems and challenges that have been tackled by the protocols discussed.

Numerous parameters can be taken for comparison, and performance evaluation is done based on all such parameters. Different multichannel protocols discussed in this paper w.r.t. different parameters are summarized in Table 5.2 & following facts are being concluded.

1. Single transceivers protocols are easier to implement than multi transceivers-based protocols, but they face hidden channel problems, and exposed, near-far and channel-switching delay is also incurred, which causes degradation in the performance.
2. On the other side, multi-transceiver protocols are complex and costly to implement. However, possessing the capability of parallel transmission and also better QoS than that of single-channel based protocols. Also, the congestion problem is easily alleviated by multiple rendezvous protocols in multi-hop ad-hoc networks.

5.5 CONCLUSIONS AND FUTURE SCOPE

Many multi-channel (MC) MAC protocols have been invented and used, but no such protocol has become a de-facto standard. One reason for this can be application-specific development where these protocols have to work. The majority of the developed ad-hoc networks MAC protocols work on a single channel only.

TABLE 5.1
Problems resolved by the protocols discussed

Name of Protocol	Year	Hidden Terminal	Deafness	Synchronization	Partitions	Channel Switching	Broadcast Support	Network Joining	The Method Used to Cope Up with Interference (intra –inter)	Head of line Blocking Existence	Near Far Problem Existence
MMAC-DA	2016	Yes	Yes	Through Control channel	No	Yes	yes	At ATIM window	With use of Directional Antenna	No	No
RcMAC	2013	YEs	Yes	Through Control channel and receiver co-operation	No	Yes	Yes	Anytime	Hop count based	No	No
CCM-MAC	2009	Yes	Not known	Through Control channel	No	No	Yes	during control channel active cycle	Using different codes similar to CDMA	Not Known	Yes
SSCH	2004	Yes	Yes	Controlled Channel Hopping	No	Yes	Yes	At System	IEEE 802.11 coupled with channel hopping	No	

(*Continued*)

TABLE 5.1 (Continued)
Problems resolved by the protocols discussed

Name of Protocol	Year	Hidden Terminal	Deafness	Synchronization	Partitions	Channel Switching	Broadcast Support	Network Joining	The Method Used to Cope Up with Interference (intra –inter)	Head of line Blocking Existence	Near Far Problem Existence
MMAC	2004	YES	YES	Control Channel	No	Yes	YES	Anytime	Not Known	No Very bad at it	YES
Li et al.	2003	No	No	Control Channel	No	Yes	Yes	Anytime	Same as IEEE 802.11	No	No
DPC	2002	Not Known	Not known	Through Control Channel	NO	YES	YES	Anytime	Hop Count Based	Not Known	Not Known
MAC-RBCS by Jain et al.	2001	YES	NO	Through Control Channel	NO	YES	YES	Anytime	Hop Count Based	No	No
DCA	2000	Yes	Yes	Through Control Channel	NO	YES	YES	Any time	Hop Count Base	Yes	Yes
CHMA	2000	Yes	Yes	Using Codes	No	Automatic	Yes	During system setup	Not Required	Yes	Yes

ODC	2000	Yes	Yes	Through Enhanced IEEE 802.11 handshaking	No	Yes	Yes	Anytime	Hop count based	Yes	Yes
McMAC	2005	Yes	Yes	Through Control Channel	No	Yes	Yes	Anytime	Hop Count Based	Yes	Yes
CAM-MAC	2006	Yes	Yes	Controlled Channel hopping	no	yes	Yes	Anytime	Hop Count Based	Yes	yes
RICH-DP	2001	Yes	Yes	Controlled Channel hopping	No	Not Required	Yes	At System Setup	Not Required	Yes	Yes
PSM-MMAC	2006	Yes	yes	Through Control Channel with AITM messaging	no	Yes	Yes	anytime	Hop Count Based	Yes	yes
PCAM	208	Yes	Yes	Using the third transceiver	No	Yes	Yes	Anytime	Hop Count Based	Yes	Yes
FMC-MAC	2007	Yes	Yes	Through Control Channel	Yes	Yes	Yes	Anytime	Hop Count Based	Yes	Yes

TABLE 5.2
Key features of multichannel MAC protocols

Protocols	Transceiver	Mobility Support	Medium Access	Dedicated Control Channel	Type of Rendezvous	Synchronous	Evaluation Method	Objective	Implementation
MMAC-DA	Single Half-duplex	Yes	CSMA/CA With handshaking	Yes	Single	No	MATLAB Simulation	To use multiple channels with interleaving of data cycle on the control channel.	Distributed
RcMAC	Single	Yes	Yes	Yes	Multiple	No	NS-2	Efficient utilization of multichannel with minimizing switching	Distributed
CCM-MAC	Single Half-duplex	Yes	CSMA/CA With handshaking	Yes	Single	No	MATLAB Simulation	Efficient utilization of multi-channels	Distributed
SSCH	Single	Yes	Channel Hopping	No	Multiple	Yes	QualNet	To overcome the control channel saturation problem	Centralized / Inbuilt
MMAC	Single	Yes	CSMA/CA	Yes	Single	Yes	NS-2	Solving Multichannel hidden problem using a single transceiver	Distributed
Li et al.	Single	Yes	CSMA/CA	Yes	Single	No	OPNET	Improved performance over IEEE 802.11	Distributed

DPC	Multiple	Yes	CSMA/CA	Yes	Single	No	Not Known	Power Savings in Multi-channel Wireless networks	Distributed
MAC-RECS by Jain et al	Multiple	Yes	CSMA/CA	Yes	Single	No	Network Simulator	Dynamic and receiver controlled channel allocation	Distributed
DCA	Two	Yes	CSMA/CA	Yes	Single	No	Not Known	To provide dynamic channel allocation	Distributed
CHMA	Single	Yes	Channel Hopping	No	Multiple	Yes	Self Created Testbed	To provide a fully multichannel environment without carrier sensing	Centralized /Inbuilt
ODC	Single	Yes	CSMA/CA	No	Multiple	No	Ns-2	To provide Dynamic channel allocation	Distributed
McMAC	Single	Yes	Channel Hopping	No	Multiple	Yes	Self Created Test Bed	To increase the throughput of the Multi-Channel Network	Distributed
CAM-MAC	Single	Yes	CSMA/CA	Yes	Single	No	NS-2	To increase co-operation among nodes to select a channel	Distributed
RICH-DP	Single	Yes	Channel Hopping	Yes	Multiple	Yes	OPNET	To use Multichannel capability without carrier sensing and any codes	Centralized/ Inbuilt

(Continued)

TABLE 5.2 (Continued)
Key features of multichannel MAC protocols

Protocols	Transceiver	Mobility Support	Medium Access	Dedicated Control Channel	Type of Rendezvous	Synchronous	Evaluation Method	Objective	Implementation
PSM-MMAC	Multiple	Yes	CSMA/CA	Yes	Single	Yes	Discreet event-based testbed	To provide increased power savings in the Multichannel network	Distributed
PCAM	Three	Yes	CSMA/CA	No	Single	No	NS-2	To remove the bottleneck of the control channel	Distributed
FMC-MAC	Single	Yes	CSMA/CA	Yes	Single	Yes	Discreet Event network Simulator	To increase the flexibility of operation in multichannel networks	Distributed

In these protocols, the critical consideration is energy efficiency. However, in MC-MAC, little work has been done toward energy efficiency, and their crucial focus has been on the development of channel access methods.

Many of the MC-MAC protocols studied and discussed are theoretical and have not been implemented practically, e.g., they assume that the interference area and a node's radio range are the same, but practically, intervention range is usually 2–3 times compared to that of radio range. So the development of practical MC-MAC is essential for setting up performance benchmarks and comparative analysis of such protocols. Much MC-MAC protocol tackles only intra-network interference, thus limiting their applications in practical situations; several networks can exist, and methods must be developed to tackle internetwork interference. Similarly, one other problem of network joining must be dealt with practically as it restricts a new node to join at a particular time slot only, and some of the older MC-MAC protocols do not allow network joining after the initial network setup thus, limiting the access and usability of such networks. So the protocols must make provision for smooth joining at any point in time.

From the survey, it can be concluded that accurate and efficient MC-MAC development is an open research area. Provision of security, node mobility, reliability, and energy efficiency are essential and work should be done in this direction. Except for the MAC layer, other layers can be refined (i.e., cross-layer development) to makes MC protocols more usable and efficient. In addition, MC-MAC protocols for newer kinds of ad-hoc networks like underwater mobile ad-hoc networks, VANETs, and similar kinds of other networks need extra attention and precision to make them practically viable.

REFERENCES

1. H. T. Le et al., "A Proficient MC-MAC Protocol for Wireless Ad Hoc Networks," *Ad Hoc Networks Issue*, vol. 44, pp. 46–57, 2016.
2. G. H. Ekbatani Fard and R. Monsefi, "A Detailed Review of Multichannel Medium Access Control Protocols for Wireless Sensor Networks," *International Journal of Wireless Information Networks*, vol. 19, no. 1, pp. 1–21, 2012.
3. C. H. Cheng and C. C. Ho, "Implementation of Multichannel Technology in ZigBee Wireless Sensor Networks," *Computers and Electrical Engineering*, vol. 56, pp. 498–508, 2016.
4. K. Ding, Y. Li a, D. E. Quevedo, S. Deyc, and L. Shi, "A Multichannel Transmission Schedule for Remote State Estimation under DoS Attacks-channel MAC Protocol for Wireless Ad Hoc Networks," *Automatica*, vol. 78, pp. 194–201, 2017.
5. J. G. L. Aceves et al., "Channel is Hopping Multiple Access," Proceedings of IEEE Conference ICC, pp. 1–7, 2000.
6. P. Maria et al., "On-Demand Channel Switching for Multichannel Wireless MAC Protocols," IEEE 802.11 Technical Report, pp.1–65, 2003.
7. S. Das et al., "A Multichannel Protocol with Receiver Based Channel Selection for Multi-Hop Wireless Networks," Proceedings of IEEE Conference, pp. 25–32, 2001.
8. N.V et al., "Multichannel MAC System for Ad-Hoc Networks: Handling Multi-Channel Hidden Terminals Using a Single Transceiver," Proceedings of ACM Conference, pp. 1–8, 2004.

9. J. Dunagan et al., "SSCH Slotted Seeded Channel Hopping for Capacity Improvement in Ad-Hoc Wireless Networks,", Proceedings of IEEE Conference Mobicom, pp. 232–240, 2004.
10. Y. Chen et al., "Performance Evaluation of Modified 802.11 MAC for Multi Channel Multi-Hop Ad Hoc Network," Proceedings of AINA, pp. 423–430, 2003.
11. L. Garcia et al.," A Dynamic Multi-Channel MAC for Ad Hoc LAN," Proceedings of 21st IEEE Conference, pp. 31–35, 2002.
12. A.K. et al., "Primary Channel Assignment-Based MAC a Multichannel MAC Protocol for Multi-Hop Wireless Networks," Proceedings of IEEE Conference WCNC, pp. 168–175, 2004.
13. J. P. Sheu et al., "A Multichannel MAC Protocol with On-Demand Channel Assignment for Multi-Hop Mobile Ad Hoc Networks," Proceedings of IEEE conference I-SPAN, pp. 1–8, 2000.
14. J. P. Sheu et al., "A New Multi-Channel MAC Protocol with Power Control for Multihop Mobile Ad-hoc Networks," *The Computer Journal*, vol. 45, no. 1, pp. 101–110, 2002.
15. S. Das et al., "A Novel Multichannel CSMA MAC Protocol with Receiver based Channel Selection for Multihop Wireless Networks," Proceedings of IEEE Conference ICCCN, pp. 102–108, 2001.
16. J. G. L. Aceves et al., "Channel Hopping Multiple Access with Packet Trains for Ad Hoc Networks," Proceedings of IEEE Conference, pp. 101–109, 2000.
17. J. Hwang et al., "A Receiver-Oriented Multichannel MAC Protocol for Wireless Networks," *Computer Communication*, vol. 2, no. 36, pp. 431–444, 2013.
18. Y. Moon et al., "A cooperative CDMA-Based MC-MAC Protocol for MANET," *Computer Communications*, vol. 4, 32, pp. 1810–1819, 2009.
19. J. G. L. Aceves et al., "Receiver-Initiated Collision-Avoidance Protocol for Multi-Channel Networks," Proceedings of IEEE Conference Infocom, pp. 459–465, 2001.
20. J. Walrand et al., "A Multichannel MAC Proposal for Ad-Hoc Wireless Networks", Technical Report, pp. 1–89, 2005.
21. V. Srinivasan et al., "A Cooperative Asynchronous Multi-Channel MAC Protocol for Ad Hoc Networks," Proceedings of International Conference BROAD-NETS, pp.25–32, 2006.
22. D. Wu et al., "Power Saving Multi-Radio Multichannel MAC Protocol for Wireless Local Area Networks," Proceedings of IEEE Conference Infocom, pp. 110–116, 2006.
23. M. Ghosh et al., "Cognitive PHY and MAC Layers Protocol for Dynamic Spectrum Access and Sharing of TV Bands," Proceedings of International Conference on Signals, Systems and Computers, pp. 606–612, 2004.
24. Gupta et al., "Cross-Layer Architecture to Exploit Multichannel Diversity with a Single Transceiver," Proceedings of IEEE Conference INFOCOM, pp. 27–35, 2007.
25. J. M. et al., "An Integrated Agent Architecture for a Software-defined Radio," PhD. Dissertation, Royal Institute of Technology Sweden, pp. 1–198, 2000.
26. S. H. et al., "Brain-Empowered Wireless Communications," *IEEE JSAC*, vol. 23, no 2, pp. 126–136, 2005.
27. J. Ma et al., "Multi-Channel Coordination MAC for Multi-Hop Ad Hoc Network," Technical Report, pp. 1–1318, 2007.
28. J. Mo et al.," Comparison of MC MAC Protocols," Proceedings of IEEE Conference MSWiM'05, pp. 2109–2219, 2005.
29. T. Nandagopal et al., "Characterizing the Capacity Region in Multi-Radio and Multi-Channel Wireless Mesh Networks," Proceedings of ACM Conference, pp. 112–120, 2005.
30. L. Li et al., "Channel Assignment and Routing for throughput Optimization in Multi-Radio Wireless Mesh Networks," Proceedings of ACM Conference MobiCom, pp. 122–128, 2005.

31. J. Walrand et al., "Comparison of Multichannel MAC Protocols," Proceedings of ACM Conference, pp. 173–179, 2005.
32. A. K. Pathmasuntharam et al., "Primary Channel Assignment-Based MAC a Multichannel MAC Protocol for Multi-Hop Wireless Networks," Proceedings of IEEE Conference WCNC, pp. 410–416, 2004.
33. J. Hwang, T. Kim, J. So, and H. Lim, "A Receiver-centric Multichannel MAC Protocol for Wireless Networks," *Computer Communications*, vol. 36, pp. 431–444, 2013.
34. Y. Wu and M. Cardei, "Multi-Channel and Cognitive Radio Approaches for Wireless Sensor Networks" *Computer Communications*, vol. 94, pp. 30–45, 2016.
35. S. Holzera, T. Locher, Y. A. Pignolet, and R. Wattenhofer "Deterministic Multichannel Information Exchange," *Journal of Computer and System Sciences*, vol. 87, pp. 84–103, 2017.
36. P. J. Wu and C. N. Lee "Connection-Oriented Multichannel MAC Protocol for Ad-hoc Networks" *Computer Communications*, vol. 32, pp. 169–178, 2009.
37. J. B. Lim, B. Jang, and M. L. Sichitiu, "MCAS-MAC: A Multichannel Asynchronous Scheduled MAC Protocol for Wireless Sensor Networks," *Computer Communications*, vol. 56, 98–107, 2015.
38. Z. Askari, A. Avokh, and M. D. Farzanegan "Low-Interference Multicast Routing in Multi-Radio Multi-Channel Wireless Mesh Networks Using Adaptive Directional Antennas," *Computer Communications*, vol. 104, pp. 175–190, 2017.
39. K. H. Phung, B. Lemmens, M. Goossens, A. Nowe, L. Tran, and K. Steenhaut, "Schedule-Based Multichannel Communication in Wireless Sensor Networks: A Complete Design and Performance Evaluation," *Ad Hoc Networks*, Vol. 26, pp. 88–102, 2015.
40. R. Block and B. Van Houdt, "Spatial Fairness in Mmultichannel CSMA Line Networks" *Performance Evaluation*, vol. 103, pp. 69–85, 2016.
41. H. J. Lei, C. Gao, Y. C. Guo, and Z. Z. Zhang, "Survey of Multichannel MAC Protocols for IEEE 802.11-Based Wireless Mesh Networks," *The Journal of China Universities of Posts and Telecommunications*, vol. 18, no. 2, pp. 33–44, 2011.
42. S. S. Wang, C. C. Li, H. W. Lin, and K. P. Shih, "A Passive Self-configuration MAC Protocol for Supporting Network Management in IEEE802.11-Based Multi-Hop Mobile Ad Hoc Networks." *Journal of Network and Computer Applications*, vol. 56, pp. 149–157, 2015.
43. J. Zhang, Y. P. Chen, and I. Marsic "MAC-Layer Proactive Mixing for Network Coding in Multi-Hop Wireless Networks," *Computer Networks*, vol. 54, pp. 196–207, 2010.
44. A. A. K. Jeng, R. H. Jan, C. Y. Li, and C. Chen "Release-Time-Based Multichannel MAC Protocol for Wireless Mesh Networks," *Computer Networks*, vol. 55, pp. 2176–2195, 2011.
45. C. Nam and S. Bahk, "ΔSNR-MAC: Apriority-Based Multi-Round Contention Scheme for MU-MIMO WLANs," *Computer Networks*, vol. 92, pp. 24–40, 2015.
46. L. Le a and I. Rhee, "Implementation and Experimental Evaluation of Multichannel MAC Protocols for 802.11 Networks," *Ad Hoc Networks*, vol. 8, pp. 626–639, 2010.
47. O. D. Incel, L. van Hoesel, P. Jansen, and P. Havinga, "MC-LMAC: A Multi-Channel MAC Protocol for Wireless Sensor Networks," *Ad Hoc Networks*, vol. 9, pp. 73–94, 2011.
48. G. S. Uyanik and S. Oktug, " Cognitive Channel Selection and Scheduling for Multichannel Dynamic Spectrum Access Networks Considering QoS Levels," *Ad Hoc Networks*, vol. 62, pp. 22–34, 2017.
49. M. J. Shih, G. Y. Lin, and H. Y. Wei, "A Distributed Multi-Channel Feed backless MAC Protocol for D2D Broadcast Communications," *IEEE Wireless Communications Letters*, vol. 4, no. 1, pp. 102–106, 2015.

50. S. Zhuo, Z. Wang, Y. Q. Song, Member, Z. Wang, and L. Almeida, "A Traffic Adaptive Multi-Channel MAC Protocol with Dynamic Slot Allocation for WSNs," *IEEE Transactions on Mobile Computing*, vol. 15, no. 7, pp. 1600–1614, 2016.
51. H. Chen and L. Cui, "DS-MMAC: A Delay-sensitive Multichannel MAC Protocol for Ambient Assistant Living Systems," *China Communications*, vol. 5, pp. 38–47, 2016.
52. Y. Zhang, L. Lazos, K. Chen, B. Hu, and S. Shivaramaiah "Multi-Channel Medium Access without Control Channels: A Full Duplex MAC Design," *IEEE Transactions on Mobile Computing*, vol. 16, no. 4, pp. 1032–1047, 2017.
53. P. Murdiyat, K. S. Chung, and K. S. Chan "A Multi-Channel MAC for Multi-Hop Wireless Sensor Networks Minimizing Hidden Node Collision," Proceedings of IEEE Conference (APCC2016), pp. 535–541, 2016.
54. B. Yang, B. Li, Z. Yan, and M. Yang, "Channel Reservation based Multi-Channel MAC Protocol with Serial Cooperation for the Next Generation WLAN," Proceedings of IEEE Conference, ICSPCC-2016, pp. 1–6, 2016.
55. M. F. Caetano and J. L. Bordim, "On the Performance Evaluation of Multichannel MAC Protocols," Proceedings of IEEE 2016 Fourth International Symposium on Computing and Networking, pp. 671–676, 2016.
56. O. Dabeer, "Fast Distributed Multi-Channel MAC for Achieving FDM in Unlicensed Spectrum," Proceedings of IEEE 2016 Information Theory and Applications Workshop (ITA), pp. 1–4, 2016.
57. Y. Li, G. Zhou, N. Zheng, and L. Hong, "An Adaptive Backoff Algorithm for Multichannel CSMA in Wireless Sensor Networks," *Neural Computing & Applications*, vol. 25, pp. 1845–1851, 2014.
58. Z. Wei, T. Zhang, F. Wu, X. Gao, G. Chen, and P. Yi "A Truthful Auction Mechanism for Channel Allocation in Multi-Radio, Multichannel Non-cooperative Wireless Networks," *Personal and Ubiquitous Computing*, vol. 18, pp. 925–937, 2014.
59. A. Rangnekar and K. M. Sivalingam, "QoS Aware Multi-Channel Scheduling for IEEE 802.15.3 Networks," *Mobile Networks and Applications*, vol. 11, pp. 47–62, 2006.
60. H. Yu, P. Mohapatra, and X. Liu, "Channel Assignment and Link Scheduling in Multi-Radio Multi-Channel Wireless Mesh Networks," *Mobile Network Application*, vol. 13, pp.169–185, 2008.
61. T. Liu and W. Liao, and J. F. Lee, Distributed Contention-Aware Call Admission Control for IEEE 802.11 Multi-Radio Multi-Rate Multi-Channel Wireless Mesh Networks," *Mobile Network Applications*, vol. 14, pp. 134–142, 2009.
62. M. Zhao, M. Ma, and Y. Yang, "Applying Opportunistic Medium Access and Multiuser MIMO Techniques in Multi-Channel Multi-Radio WLANs," *Mobile Network and Applications*, vol. 14, pp. 486–507, 2009.
63. Q. Liu, X. Jia, and Y. Zhou, "Topology Control for Multi-Channel Multi-Radio Wireless Mesh Networks Using Directional Antennas," *Wireless Networks*, vol. 41–51, pp. 41–51, 2011.
64. S. H. Lim, Y. B. Ko, C. Kim, and N. H. Vaidya, "Design and Implementation of Multicasting for Multi-Channel Multi-Interface Wireless Mesh Networks," *Wireless Networks*, vol. 17, pp. 955–972. 2011.
65. M. Won, C. Yang, W. Zhou, and R. Stoleru, "Energy-Efficient Multichannel Media Access Control for Dense Wireless Ad Hoc and Sensor Networks," *Wireless Networks*, vol. 19, pp. 1537–1551, 2013.
66. T. Chiueh et al., "Centralized Channel Assignment with Routing Algorithms for MultiChannel Wireless Mesh Networks." *Journal of Mobile Computing and Communications Review*, vol. 8, no. 2, pp. 50–65, 2004.

67. T. Chiueh et al., "Architecture and Algorithms for the 802.11-Based Multichannel Wireless Mesh Network," Proceedings of IEEE Conference Infocom, pp. 625–632, 2005.
68. P. K. et al., "Routing and Interface Assignment in Multi-Channel Multi-Interface Wireless Networks," Proceedings of IEEE Conference WCNC, pp. 1–7, 2005.
69. P. K et al., "Capacity of Multichannel Wireless Networks Impact of Several Channels and Interfaces," Proceedings of IEEE Conference Mobicom, pp. 10–16, 2005.
70. K. Challapali et al., "A Cognitive MAC Protocol for Multi-Channel Wireless Networks," Proceedings of IEEE Conference, pp. 1–7, 2007.
71. C.W et al., "Channel Selection Strategies for Multi-Channel MAC Protocols Wireless Ad Hoc Networks," Proceedings of IEEE Conference WiMOB, pp. 264–270, 2006.

6 Maximum Utilization of WDM-PON Resources to Enhance Reachability of Access Network

Dinesh Kumar Verma
PhD Scholar, Deenbandhu Chhotu Ram University of Science and Technology, Sonipat, India

Amit Kumar Garg
Professor, Deenbandhu Chhotu Ram University of Science and Technology, Sonipat, India

CONTENTS

6.1 INTRODUCTION

The internet is the easiest way of communication for today's generations to transfer emails, files, and messaging and use online gaming, video streaming, cloud computing services, etc. People do not quickly get water in some areas, but it is easy to play with the internet. This is only possible because of new fiber access technologies. The latest research in the optical domain makes it possible to use new research horizons like capacity boost up of the fiber, coherent transmission, receiver and transmitter technology development, and different algorithms to handle the massive data. It offers bandwidth access in Gbps (gigabits per second) using different time division and wavelength multiplexing techniques and the overall capacity reached up to tens of terabits per second. The highest capacity of the core-fiber access networks is only one

DOI: 10.1201/9781003230526-6

side of the significant developments in broadband access technologies. End-user can access the data in high volumes using two significant fields [1]. The first one is wireless technology, but the second one, i.e., optical fiber, is still the most demanding technology to fulfill the present and tremendous futuristic demand of the data traffic [1–4].

6.1.1 ITU Standards and Recommendations

One step further in the optical domain is the induction of passive optical network (PON) technology for the access network. Several generations of the PON technology are there, and each has its own benefits concerning different technologies and time. For future compliances, different standardization has been established by IEEE special taskforce 802.3ca for next-generation PON (NG-PON)]. NGPON 1 & 2 can operate at a 100 Gbps or more data rate to provide high bandwidth to every user. Full-Service Access Network (FSAN) is a leading telecommunication forum responsible for independently working with research labs, manufacturers, and service providers [5,6]. FSAN plays a significant role in promoting the different projects and research and other standard bodies (like ITU-T). In 2016, with a collective discussion of forum members, FSAN published the Standards Roadmap 2.0, depicted in Figure 6.1 [2].

This roadmap works as a blueprint and fulfills all needs of the service providers, manufacturers to standardize a technology like the passive optical network to fulfill the future need for bandwidth. Further, it elaborated the PON evolution and the different generations of the PON technology through a roadmap [7–12]. In 2018, ITU-T finally concluded and focused on the next-generation PON technology beyond 10 Gbps data rate and working out 50 Gbps using NGPON technology [13]. It

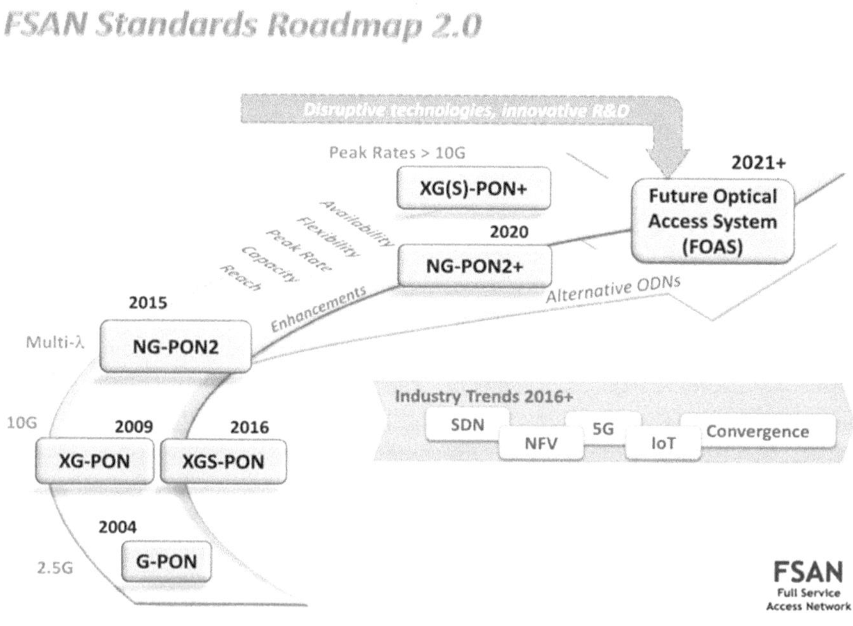

FIGURE 6.1 FSAN standards roadmap 2.0 for the PON technology.

takes care of the development of the physical layer and standards. The main recommendations of ITU areas:

1. **High-Speed NGPON (G.9804.1):** It was focused on the evolution, overall requirements of resources, coexistence of other technologies, services, and different interfaces for NGPON systems.
2. **High-Speed NGPON(G.hsp.50GPMD):** 50G PMD: Physical Media Dependent (PMD) Layer: It was a 50 Gbps standard project, which elaborated the physical addressing of layered architecture and other parameters using TDM (time-division multiplexed) PON.
3. **High-Speed NGPON (G.hsp. TWDM PMD):** It focused on about 50 Gbps TWDM (time-wavelength division multiplexing) PON architecture and elaborated on the interfaces of optical layer parameters.
4. **High-Speed NGPON (G.hsp. COMTC):** This common transmission convergence (COMTC) discussed the TC layer and other significant definitions.

6.1.2 Evolution of PON Technology

Different generations of the PON technologies are exiting, which has been deployed in the past and present. These access network technologies are APON (A-Active/Asynchronous), BPON (B-Broadband), EPON (E-Ethernet), GPON (G-Gigabits), 10G EPON, Next-generation PON1 (NG-PON1), and wavelength division multiplexing PON (WDM-PON). All technologies have their own pros and cons. However, the main concern of the future is to get the technology that should be able to fulfill the demand of the data rate used by different applications, especially in video streaming and cloud computing at the data centers [14–17]. The old generation includes APON, BPON. Presently deployed PON technologies are EPON and GPON. Next-generation is categorized in different phases as 10GEPON, XGPON and NGPON 1. Here, NGPON 2 (i.e., WDM PON) comes under the future generation of the passive optical network. The evolution of the PON technologies has been classified into different generations, as shown in Table 6.1.

All technologies have their own pros and cons. However, the main concern of the future is to get the technology that should be able to fulfill the demand of the data rate as required by different applications, especially in video streaming and cloud computing at the data centers.

TABLE 6.1
Different generations of PON

Old Generations	APON & BPON
Current Generations	EPON & GPON
Next Generations	10G EPON, XG PON, NGPON 1
Future Generation	NGPON 2(WDM PON)

Different technologies of the PON system can be described as:

A. APON
   This is the first generation (1G) of the PON technology standardized by the FSAN group (ITU-T) in the 1990s. Here, asynchronous transfer mode (ATM) technology is used to expand the end-to-end connection. It utilizes the topology called point-to-multipoint, i.e., tree topology. APON technology supports 32 subscribers with an upstream (US) of 155 Mbps and downstream (DS) of 622 or 155 Mbps. IEEE 802.3 standards define that the maximum speed supported by APON is 1.244 Gbps.
B. BPON
   The advancement of APON technology is defined in ITU-T G.983. It provides a cost-effective solution for the access network. It offers many different services like Ethernet and video services, etc. BPON utilizes ATM as the primary signaling protocol, as used in APON technology. It is more secure than APON technology. Advancement in the BPON technology allows video streaming, while APON did not focus on video streaming [20]. It operates at ATM rates of 155 Mbps, 622 Mbps and 1244 Mbps.
C. EPON
   Ethernet PON is an IEEE standard (also called IEEE 802.3 ah). Frames (Ethernet) are used in EPON for data transmission. The integration of internet protocol (IP) and Ethernet technologies plays a significant role. Since high security is used in Ethernet frame format (due to variable-length packets), it is widely used [15,16]. It operates in the US and the DS at a speed of 1.25 Gbps (symmetrical) with a maximum coverage area of 20 km.
D. 10G-EPON
   It was first time used in 2009 with the name IEEE 802.3av. It is capable of increasing the speed of the PON system from 1 Gbps to 10 Gbps. These standards came into existence to increase the data rates & to have compatibility with the previous version of PON in the same network. 10G-EPON also benefits from both types of data rates, i.e., symmetrical and asymmetrical. In symmetrical, upstream and downstream both have the same data rate of 10 Gbps. However, in asymmetric systems, upstream and downstream have different data rates. Here, the US has 1 Gbps at a wavelength of 1310 nm and DS has 10 Gbps at 1270 nm.
E. GPON (Gigabit PON)
   It supports both symmetrical and asymmetrical data rates in the US and DS, respectively. In symmetrical, the data rate for US and DS is 2.488 Gbps for each. It facilitates the transportation of Ethernet and ATM. It has a robust operation, administration, maintenance, and provisioning capabilities (OAM & P). GPON provides more bandwidth than its legacy PON technologies like APON, BPON and EPON etc. It uses the GPON Encapsulation Method (GEM) for enhancing security. Both GPON and EPON is a good competitor of each other. Different countries utilize these standards as needed, either GPON or EPON [4–6].

F. NG-EPON (Next-generation EPON)
It transmits data using various wavelengths provided by the system. Due to this, network capacity will enhance up to 100 Gb/s and improve the network's flexibility since it utilizes the four wavelengths with a capacity of 25 Gbps each, which collectively forms 100 Gbps. However, the dynamic bandwidth allocation (DBA) algorithm provides a better solution for improving the spectral efficiency of NG-EPON [7–9].

G. NG PON (Next Generation-PON)
FSAN and ITU predicted that NG PON would offer high bandwidth and divided it into two parts: NG PON 1 & NG PON 2. NG-PON1 would be interoperable with its legacy networks, while NGPON2 is a near-future solution. NG-PON1 defines both symmetric and asymmetric bandwidths of 10 Gbps each. The following standard defined by the ITU-T is XG-PONs (X means Symmetric), which enhances the GPON network that provides an asymmetrical data rate for both DS and US [9,10].

H. WDM PON
WDM-PON, a future PON system, probably will take some time after XG-PON for its imminent existence in the market. WDM-PON standards are still in the development phase; however, manufacturers and service providers admit that it will be compatible with its legacy architecture PON networks that can feed 32 ONTs (optical network termination). Each ONT will be needed by individual wavelength so that customers will get the higher bandwidth. This is the significant difference between the earlier explained technologies [18]. WDM-PON has many benefits over its legacy technology because it is catching the eyes of everyone in society and service providers. Some of these are no bandwidth scheduling is needed as required in GPON, longer reach is possible (>20 km), utilization of AWG (Arrayed Waveguide) which have low losses as a comparison to the splitter used in GPON [15].

6.1.3 Quality Parameters

Bit error rate (BER) measures the error at the receiver side by comparing the input bits with received bits in a telecommunication system. If BER is 10^{-9}, only 1 bit will have an error out of 10^9. It can mainly measure the latency of the data and errors in transmission. It can be improved by using the following factors:

- High SNR (Signal to Noise) value
- Modulation schemes
- Channel coding schemes
- FEC (Forward Error Correction) etc.

However, noise levels can be expressed in SNR, but it is common practice to utilize the Eb/No (energy bitwise/noise) power spectral density [4–6].

The process of transmitting information from the transmitter (source) to receiver (destination) using high-power carrier signals is called modulation. Different modulation schemes exist, but in the case of optical transmission, two modulation techniques are mainly used: 1: Electro-Absorption modulator (EAM) and 2. Mach-Zehnder modulator (MZM). EAM device is used for modulating intensity by variation in electric voltage of a LASER beam. EAM is mainly used in external modulation in telecommunications systems [1–3]. This can be operated at high speed, and it can achieve the modulation bandwidth in the range of gigahertz (GHz), which is the required quality parameter of the present and future. MZM is also an intensity-modulating device that uses a simple driving circuit for modulating voltage. MZM has two waveguides in comparison to multiple quantum well (MQW) structures. The phase can be changed using both EAM and MZM, which will give the light output as intensity-modulated [4].

Further, in PON technologies, different modulation schemes can be used for both uplink and downlink. Here, NRZ is used. The main concern of using NRZ is that it has a high-sensitivity benefit. Interoperability is the primary attribute between the OLT and ONU. Some other types of coding can get the higher bandwidth; they include Return-to-Zero (RZ), Alternate Mark Inversion (AMI), Manchester, Differential Manchester and Multi-state Coding etc.

6.1.4 Optical Signal Loss

Optical fibers are the leading resource of optical communications, which have the capacity of long-distance communication providing higher data rates than other access networks available. Two types of fibers exist.

a. SMF – Single-Mode Fiber
b. MMF – Multimode Fiber

These can be used in all-optical windows. Generally, for longer distance (>2 km) communication, SMF is used, while for small distances (<2 km), MMF is used. Observe the bit error in optical fiber, and first, calculate the attenuation using the power budget equation [2–5]. The power budget equation states that the difference between transmitted power and receiver sensitivity should be >= to the summation of different power losses available and power margin, as shown by the equation.

$$P^T - S^R = AL^F + L^C + L^A + M \tag{6.1}$$

P^T – for the transmitter, S^R – for the receiver's sensitivity, A – attenuation, L^F – length of the fiber, L^C – losses due to coupling, L^A – other loss and M – corresponding power margin [2].

PON access technology provides enhanced bandwidth compared with traditionally used access networks, and the need for getting higher bandwidth wavelength-division multiplexing (WDM) has to be employed.

6.1.5 Comparison of Different PON Systems

The main hindrance to WDM-PON technology is the cost. This is very critical to bear the high cost since it directly affects the budget of the end-user. The cost of PON technologies can be minimized by utilizing a simple architecture rather than a complex one. This can enhance the bandwidth of the WDM PON system. This paper includes a comprehensive review of WDM-PON technologies first. Different characteristics of the different PON technologies have been compared, as shown in Table 6.2. It includes the different comparative quality parameters, like the IEEE or ITU standards, bandwidth for the upstream (US) and downstream (DS), splitting ratio, and distance covered for the different PON technologies. As per the roadmap shown above in Figure 6.1, NGPON has better results in terms of all quality parameters. 100 Gbps of bandwidth can be achieved, while APON has only 622 Mbps for the US and 155 Mbps for DS. Nowadays, GPON and EPON have been extensively deployed, but future technology is only the NGPON, i.e., WDM PON [15,16]. This comparison must be handy to all the researchers in the field of passive optical networks.

This paper has been organized as follows. Section 6.1 includes different ITU recommendations, PON technology evolution, quality parameters, and losses in optical fiber. Section 6.2 describes basic WDM-PON architecture along with a comparative study. Section 6.3 elaborates results and discussion, whereas Section 6.4 concludes the research paper.

6.2 WDM-PON SYSTEM ARCHITECTURE

The beauty of the WDM PON system is to provide a higher bandwidth using optical fiber with a low maintenance and installation cost compared to other traditional access services through copper cable. Since complexity in the WDM-PON architecture is not so high, the cost is less (for the long term). It utilizes the power splitters or arrayed waveguide (AWG) for splitting the incoming optical signal. Several end-users depend on the passive splitter or AWG [1,17–19]. So, optical network terminal (ONUs) is limited by the splitting ratio and bit rate of the transceiver available at the central office (CO) and in ONUs. The maximum distance supported is 20 km, as per the comparison shown in Table 6.2. It provides scalability, a higher bit rate and is prone to power losses (due to passive equipment).

The basic WDMPON system architecture is shown in Figure 6.2. Optical Network Terminal (ONT) is the primary device at the transmitter side that can communicate between service providers and end-users. The ONUs is located on the customer side or the last mile of the access network. ONT signal is divided by the splitter (depends on splitting ratio) accordingly as 1:32, 1:64, 1:128 and 1: 256 or even more. However, till now, only 1:32 and 1:64 split ratios have been deployed. The splitting of the signal depends on the quality of the splitter employed in the network.

To build a better WDM-PON system is to use a separate wavelength for each communication from OLT to ONU, as shown in Figure 6.2. WDM-PON is based on the point-to-point type of communication, while other TDM-PON technologies use a point-to-multipoint type of communication. The different data rates can be achieved since it uses a separate wavelength, per the requirement of the end-user (ONU). A passive AWG router is deployed in WDM PON, while a passive splitter is used in TDM-PON [1,13,20]. The AWG is a unique passive device with periodicity characteristics through which multiple signals are routed.

TABLE 6.2
Comparative study of various PON technologies

Technology Parameters	APON	BPON	GPON	EPON	10-G EPON	XG PON (10G PON)	NGPON 2
Standard	IEEE (802.3)	ITU-T (G.983)	ITU-T (G.984)	IEEE (802.3ah)	IEEE (802.3av)	ITU-T	ITU-T
Bandwidth (downstream)	622 Mbps	622–1244 Mbps	2.488 Gbps (1490 nm)	1.25 Gbps	10 Gbps	10 Gbps (1570 nm)	80 Gbps (1596 nm)
Bandwidth (upstream)	155 Mbps	155 Mbps	1.244 Gbps (1310 nm)	1.25 Gbps	10 Gbps	2.5 Gbps (1270 nm)	80 Gbps (1532 nm)
Splitting Ratio	1:16	1:32	1:32–64	1:64	1:64	1:128	1:32–256
Distance	–	–	20 km	20 km	20 km	20 km	20 km

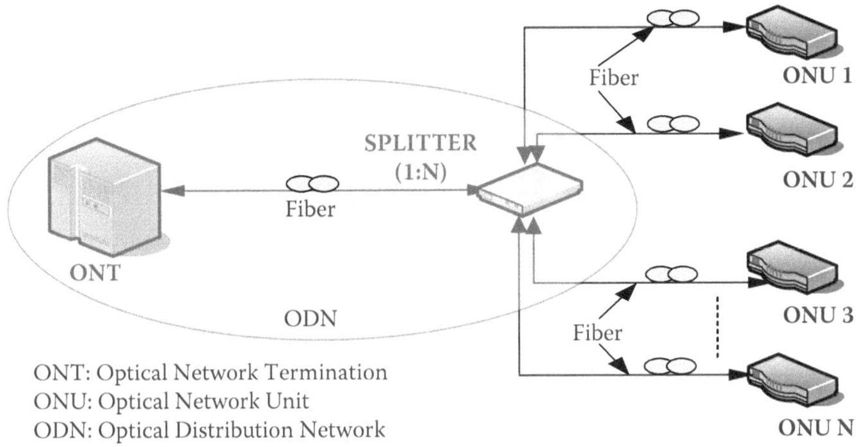

FIGURE 6.2 General architecture of WDM PON.

For the US, OLT uses a WDM de-multiplexer to receive the incoming signal from the CO or ONU. DS and US occur at different optical wavelengths that can be separated by using coarse WDM (CWDM) or dense WDM (WDM).

6.3 RESULTS AND DISCUSSION

A novel WDM-PON architecture has been proposed. The cost of ONU or the whole system can be reduced by designing a simple architecture. Different quality parameters are used to predict the quality of the WDM-PON system [3]. BER is the main parameter that measures the crosstalk and the error present in the received signal. The simulation of the WDM-PON has predicated results at OPTISYSTEM 18 software. Different coverage areas have been considered by using the variation in the length of the optical fiber, as shown in Table 6.3. This performance comparison has been carried out using 40 km, 50 km, 60 km and 70 km, respectively, as shown in Table 6.3. On the receiver side i.e., ONU has the avalanche photodetector (APD) receiver type. It has been simulated that results using APD receiver by the distance variation, shown in Figures 6.3 and 6.4. Here, it has been observed that eye

TABLE 6.3
Performance comparison of WDM PON [40 km, 50 km, 60 km and 70 km]

Receiver Type	Distance (km)	Max. Quality Factor	Min. BER	Eye Height	Threshold
APD	70	11.35	5.11e-347	1.55e-005	7.68e-006
APD	60	20.96	7.61e-098	2.92e-00	1.44e-005
APD	50	39.35	0	5.46e-005	2.52e-005
APD	40	52.27	0	9.58e-005	3.11e-005

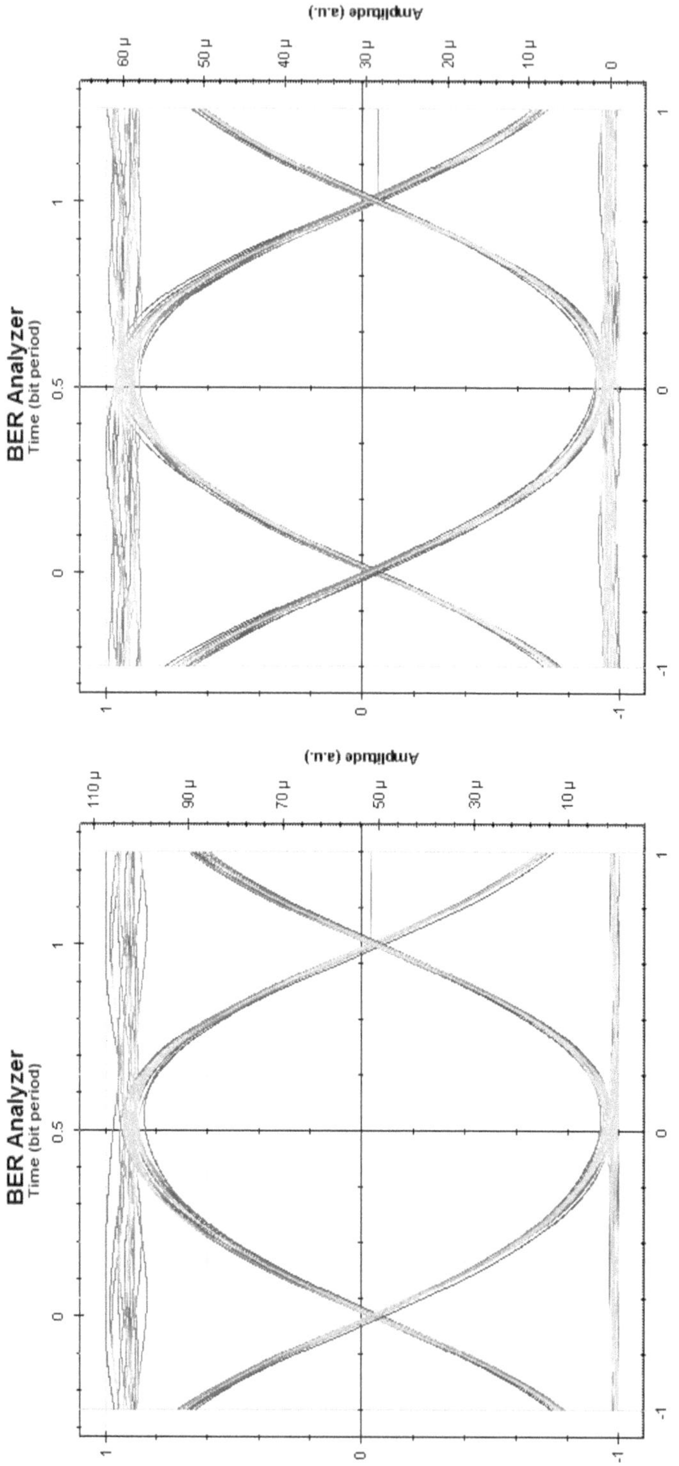

FIGURE 6.3 Simulation result of APD at a distance of (a) 40 km (b) 50 km.

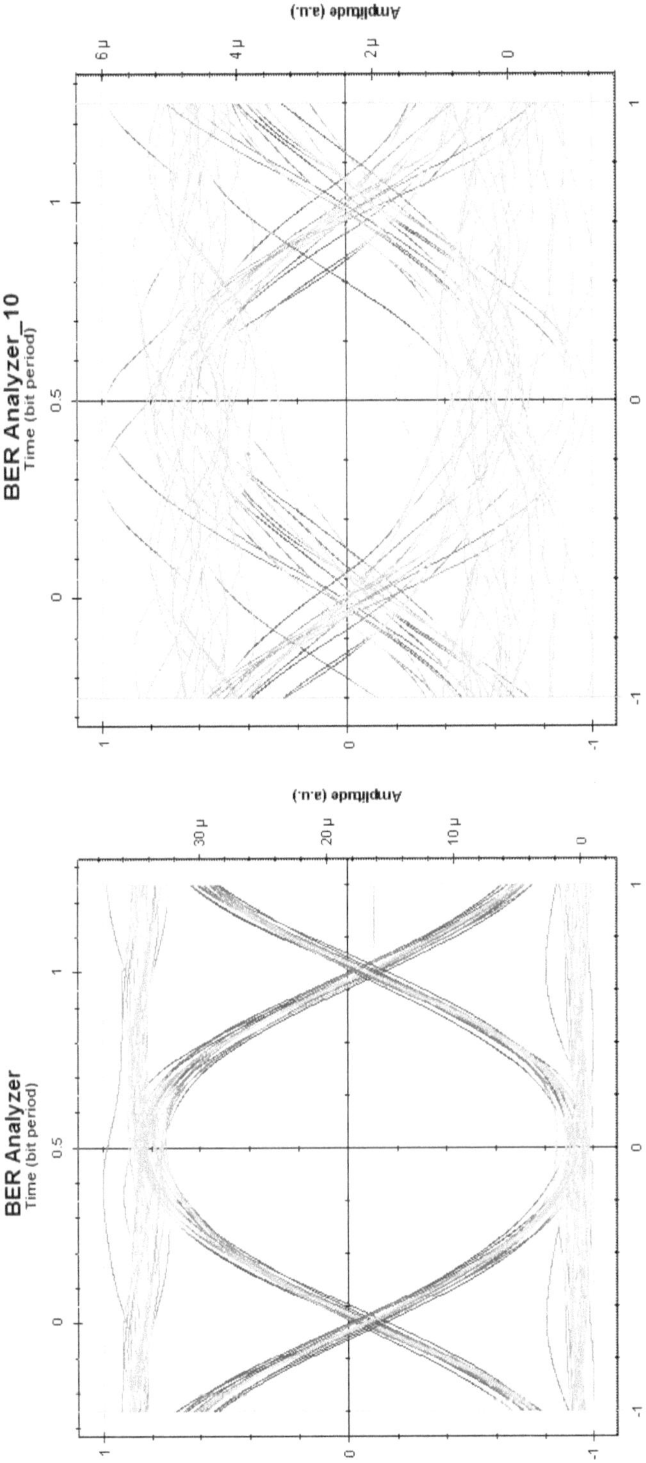

FIGURE 6.4 Simulation result of APD at a distance of (a) 60 km (b) 70 km.

height is good at a distance of 40 km in Figure 6.3(a), 50 km in Figure 6.3(b) and 60 km in Figure 6.4(a), respectively. However, we got the worst result when we were beyond this limit, as shown in Figure 6.4(b) for 70 km.

It has the worst value of BER (5e-34) along with Q-factor (11.35), while at a distance <60 km, it has the better results, as shown in Table 6.3.

The threshold value is almost constant for all the cases. So, we have elaborated the result based on the other three parameters like Q-Factor, BER and eye height. In comparing both (PIN and APD), APD got the best result in terms of Q-Factor, eye-opening, and BER value. In this, parameters, especially BER, Q-factor and eye height, have the worst design-quality parameters. So, it has been concluded that we have a good result in APD (for 40 km and 60 km) (BER $< e^{-10}$). All results have been validated using the simulation in optisystem 18.0, as shown in Figures 6.3 and 6.4.

6.4 CONCLUSION

This paper has elaborated on the different PON technologies along with a comparative study. Different quality parameters are there to explain the quality of a PON system. WDM PON is the most promising technology for the near future. A comprehensive discussion of various aspects of different recourses needed for designing a WDM-PON network has been presented, like different types of modulation schemes, coding, and quality parameters. The excellent quality of the WDM-PON system has been depicted using the quality parameters as BER, quality factor, and eye height, which are justified with simulated results. It has been concluded that the APD receiver has better results with maximum feasible distance covered of 60 km with acceptable BER and Q-factor. With the increase in the distance of more than 60 km, the Q-factor and eye height got the worst value. So feasible distance is 60 km where WDM-PON can be used effectively. In the future, other hybrid PON systems that may include combining two or more PON technology or OFDM can be deployed.

REFERENCES

1. T. Horvath, P. Munster, V. Oujezsky, and N.H. Bao, "Passive Optical Networks Progress: A Tutorial," *Electronics (MDPI)*, vol. 9, pp. 1–3, 2020.
2. S. M. Jahangir Alam, M. Rabiul Alam, G. Hu, and M. d. Z. Mehrab, "Bit Error Rate Optimization in Fiber Optic Communications," *International Journal of Machine Learning and Computing*, vol. 1, no. 5, pp. 435–440, December 2011.
3. D. Zhang, D. Liu, X. Wu, and D. Nesset, "Progress of ITU-T Higher Speed Passive Optical Network (50G-PON) Standardization," *Journal of Optical Communications and Networking*, vol. 12, no. 10, pp. D99–D108, October 2020.
4. A. Banerjee et al., "Wavelength-Division-Multiplexed Passive Optical Network (WDM-PON) Technologies for Broadband Access: A Review [Invited]," *Journal of Optical Networking, Optical Society of America*, vol. 4, no. 11, pp. 738–758, November 2005.
5. H. Zhang, X. Cheng, Z. Xu, and Y.K. Yeo, "A Novel Combined WDM-PON with a Single Shared DI Using Downlink DPSK and Uplink Remodulated OOK Signals," *Optics Communications*, vol. 2, no. 6, pp. 992–996, 2012.

6. F. Hou, and M. Yang, "The Analysis of System Performance of WDM/OCDMA-PON Based on DQPSK/OOK," *Optik – Int. J. Light Electron Opt*, 2013, pp. 1–3. 10.1016/j.ijleo.2012.12.007
7. Z. Zhang, J. Wang, L. Wang, and X. Chen, "Bidirectional 40 Gb/s/k, 100 km-Reach, Channel-reuse WDM-PON Employing Tunable Optical Transceiver with Optical Intensity Detection-Based Wavelength Management," *Optical Fiber Technology*, vol. 25, pp. 51–57, 2015.
8. A. E. A. Eltraify, S. H. Mohamed, and J. M. H. Elmirghani, "VM Placement over WDM-TDM AWGR PON Based Data Centre Architecture," in IEEE, ICTON, 2020, pp. 1–5.
9. W. Wang, W. Guo, C. Li, W. Hu, and M. Xia, "ONU Aggregation Schemes for TWDM PONs With Multiple Tuning Ranges," *Optical Society of America*, pp. 319–326, 2017.
10. H. S. Gill, S. S. Gill, and K. S. Bhatia, "A Novel Chaos-Based Encryption Approach for Future-Generation Passive Optical Networks Using SHA-2," *J. Opt. Commun. Netw*, vol. 9, no. 12, pp. 1184–1190, 2017.
11. S.M. Faizan Shah, M. A. Khan, and M. K. Shahid, "Reach Extendibility of Passive Optical Network Technologies Optical Switching and Networking,", vol. 2, no. 18, pp. 211–221, 2015.
12. V. Kachhatiya, and S. Prince, "Downstream Performance Analysis and Optimization of the Next Generation Passive Optical Network Stage 2 (NG-PON2)," *Optics and Laser Technology*, 104 2018, pp. 90–102.
13. www.wikipedia.org/wiki/Passive_optical_network#Enabling_technologoes
14. E. Wong et al., "Broadband Passive Optical Access Networks," *IEEE/OSA J. Lightw. Technol*, vol. 30, no. 4, pp. 597–608, 2018.
15. Y. Zheng et al., "Low Latency Passive Optical Node for Optical Access Network" in International Conference on Optical Communications and Networks (ICOCN), IEEE, 2017.
16. N. Ghani, A. Shami, C. Assi, and M. Y. A. Raja, "Intra-ONU Bandwidth Scheduling in Ethernet Passive Optical Networks," *IEEE Communications Letters*, vol. 8, no. 11, pp. 683–685, 2004.
17. C. H. Yeh, C. W. Chowb, and S. Chi, "Using 10 Gb/s Remodulation DPSK Signal in Self-Restored Colorless WDM-PON System," *Optical Fiber Technology*, vol. 15, pp. 274–278, 2009.
18. [https://www.itu.int/en/ITU-T/studygroups/2017-2020/15/Documents/OFC2018 Q2_v5.pdf
19. D. Nesset, "PON Roadmap [Invited]," *J. Opt. Commun. Netw*, vol. 9, no. 1, pp. A71–A76, January 2017.

7 Observations on Learning and Tracking Challenges of Online Microwave Engineering Teaching

V. A. Sankar Ponnapalli
Department of Electronics and Communication Engineering, Sreyas Institute of Engineering and Technology, Hyderabad-68, India

CONTENTS

7.1 INTRODUCTION

Due to the current pandemic announced by the World Health Organization in 2020 [1], offline or physical education has been closed by academic institutions and universities, and online teaching has gained more popularity. Various faculties have been engaged in how easy it is to present online material, engage students, and evaluate performance. This is the greatest challenge in the teaching and learning process. Engaging the student community through the online mode is one of the most significant challenges to the faculty members for offline or blended teaching because various methods have been involved to grab the students' attention; these ways are depicted in Figure 7.1. Blended learning involves classroom teaching, traditional labs, informal learning, mobile learning, discussion boards, email, virtual labs, blogging, web conferencing, audio/video conferencing. Handling core engineering courses, classic engineering courses, and pure mathematical courses is a challenging thing in the regular or offline mode, and it requires some more attentiveness in the online class mode to grab the student's attention. For this, faculty and students should have a good idea about the

DOI: 10.1201/9781003230526-7

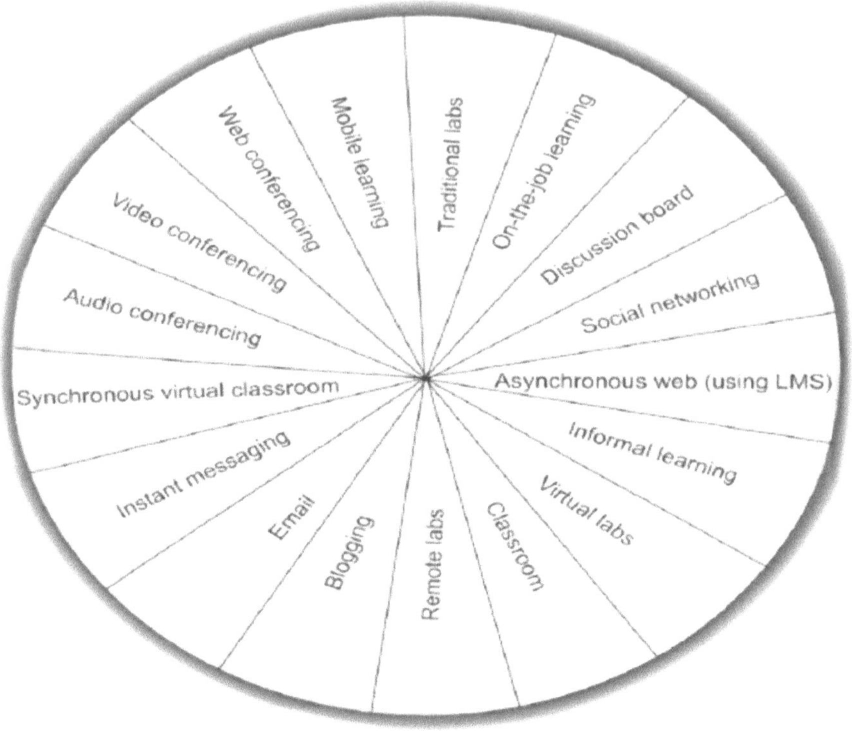

FIGURE 7.1 Blended teaching and learning [7].

online learning environment [2–6]. The evolution of distance or online learning is depicted in Figure 7.2 [7]. It is clear from Figure 7.2 that the initial distance learning happened through the printed materials, followed by trained films or recorded videos and radio broadcasting or T.V. broadcasting, followed by computer-based training, and at present, most of the distance learning happens through the e-learning platforms.

Online learning can happen through synchronous and asynchronous methods of teaching [8]. This book chapter concentrated on the synchronous and asynchronous teaching methods of microwave engineering and related courses and made some observations on learning and tracking challenges of microwave engineering and related courses. In this book chapter, the author has considered his affiliating university (J N T University Hyderabad) curriculum to explain the observations on this course [9]. The basic outline of online learning has presented in the introduction section. Program and course details of microwave engineering, electromagnetic, antennas, and radar engineering have been discussed in the course and outcomes section. In the asynchronous teaching method section, the author has considered the national technology-enhanced learning (NPTEL) details and statistics to showcase how this type of learning happened at the institutional and country levels [10]. In the synchronous teaching section, the author has presented his own experiences with microwave engineering in the current pandemic situation through the online mode and also presented the student feedback on the same. Finally, the conclusion

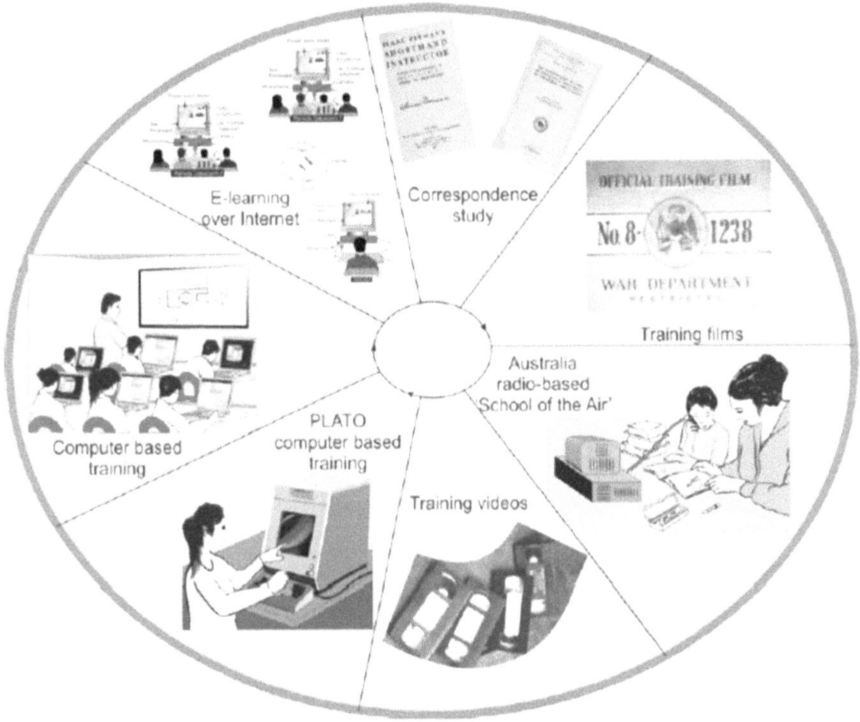

FIGURE 7.2 Evolution of distance or online learning [7].

is addressed in the final section. This type of study paper on microwave engineering and related courses is the first of its kind, and the literature available on this particular type of paper is also less or not there.

7.2 COURSE AND OUTCOMES

EC701PC: Microwave engineering is generally offered for the undergraduate final-year electronics and communication-engineering students. This course covers the introduction to microwaves, the basics of passive microwave devices like rectangular waveguides and a particular type of waveguides and their parameters, active microwave devices like microwave tubes, and solid-state microwave devices for the generation and amplification microwaves. Students are also learning microwave measurements theoretically in the course as well-practiced in the respective laboratory. The learning outcomes and mapping with program outcomes are listed in Table 7.1. The ECE601PC: Antenna and wave-propagation course is offered for undergraduate third-year electronics and communication-engineering students. This course focuses on and introduces basic antenna terminology, analysis of antenna arrays with isotropic elements, wire antennas, microstrip antennas, and aperture antennas. Students are also learning wave-propagation concepts, but there is no practical lab available for this course. The course, EC402PC: Electromagnetic fields

TABLE 7.1
Microwave engineering course outcomes and mapping with program outcomes

Sl. No.	Course Outcomes
CO:1	Apply Electromagnetic wave theory to analyze microwave parameters.
CO:2	Analyze passive microwave components such as rectangular waveguides, directional couplers, power dividers/combiner etc.
CO:3	Study the performance of specialized microwave tubes such as Klystron and reflex klystron.
CO:4	Study the performance of specialized microwave tubes such as traveling wave tubes.
CO:5	Study the performance of specialized microwave tubes such as M-type tubes and understand microwave solid-state devices' operation.
CO:6	Measure various Microwave parameters (power, reflection coefficient, VSWR etc.)

(a)

COs	Program Outcomes (POs)											
	1	2	3	4	5	6	7	8	9	10	11	12
CO:1	3	3	1	–	–	–	1	–	–	–	–	1
CO:2	3	3	1	–	–	–	–	–	–	–	–	1
CO:3	3	3	1	–	–	–	–	–	–	–	–	1
CO:4	3	3	1	–	–	–	–	–	–	–	–	1
CO:5	3	3	1	–	–	–	–	–	–	–	–	1
CO:6	3	3	1	–	–	–	–	–	–	–	–	1

(b)

and waves, is offered for second-year electronics and communication students. This course is one of the classic courses in electrical engineering, and it includes electrostatics, magnetostatics, Maxwell's equations for time-varying fields, and wave propagation concepts. The course outcomes and mapping with antennas and electromagnetic program outcomes are listed in Tables 7.2 and 7.3. In the author, affiliating universities, and most Indian universities or institutions are offered the same or equivalent patterns. The online courses of the NPTEL program are also offered, mainly in the same pattern with some advanced topics.

7.3 ASYNCHRONOUS METHOD OF TEACHING

This method of teaching is student-based, which means learners can learn at their convenience. The major drawback of this learning is missing social interaction. Otherwise, the student can explore from top universities and highly cited faculty members. Here, NPTEL related data of microwaves and related subjects is presented. NPTEL is an online-learning environment based in India for engineering education and higher education in science and mathematics subjects. This platform is developed by the IITs and IISC Bangalore and the MHRD, Government of India fund this program. The courses considered are microwave engineering, microwave

TABLE 7.2
Antenna and wave propagation course outcomes and mapping with program outcomes

Sl. No.	Course Outcomes
CO:1	Define the parameters in the design of antenna and field evaluation under various conditions and formulate the electric as well as magnetic field equations.
CO:2	Understand the design issues and the operation of fundamental antennas like Yagi – Uda, Horn antenna and Helical structure.
CO:3	Understand the designs of RF and Microwave antennas.
CO:4	Analyze the structure and working of the parabolic-reflector antenna for a given specification.
CO:5	Define the array system for different antennas and field analysis.
CO:6	Understand the behavior of nature on EM wave propagation.

(a)

COs	Program Outcomes (POs)											
	1	2	3	4	5	6	7	8	9	10	11	12
CO:1	3	1	1	1	–	–	–	–	–	–	–	1
CO:2	2	2	2	1	–	–	1	–	–	–	–	1
CO:3	2	2	3	1	–	–	1	–	–	–	–	1
CO:4	2	2	3	1	–	–	1	–	–	–	–	1
CO:5	3	2	3	1	–	–	–	–	–	–	–	1
CO:6	1	2	2	1	–	–	2	–	–	–	–	1

(b)

theory and techniques, antennas, and applied electromagnetics for engineers. Only a few sets of samples have been considered from 2019 to 2021.

Microwave engineering [11]: The IIT Guwahati, India, offered this course from June to September 2019 as a part of NPTEL asynchronous online learning and certification. In this course, 3115 learners have enrolled, 144 have registered for the exam, and finally, only 48 are eligible for the certification. Among these 48 learners, only four learners have scored between 75–89 marks band, 17 learners have scored between 60–74 marks band and remaining learners have scored between 40–59 marks band. Microwave Theory and techniques [12]: The IIT Bombay, India, offered this course from September to December 2020 as part of NPTEL asynchronous online learning and certification. In this course, 2556 learners have enrolled in this exam, 147 learners have registered, and 115 learners are eligible for the certification. Among these 115 learners, only six learners obtained above 90 marks, 34 learners obtained 75–89 marks, 40 learners received 60–74 marks, and 35 learners received 40–59 marks. Antennas [13]: The IIT Bombay, India, offers this course from January–April 2020 as part of NPTEL online learning. In this course, 6161 learners have enrolled, 141 registered for the final assessment exam, and only 102 qualified for the certification. Among 102 learners, ten learners secured above 90 marks, 38 learners secured 75–89 marks, and the remaining learners got

TABLE 7.3
Electromagnetics course outcomes and mapping with program outcomes

Sl. No.	Course Outcomes
CO:1	Get the Knowledge of Basic Laws, Concepts and proofs related to Electrostatic Fields.
CO:2	Get the Knowledge of Basic Laws, Concepts and proofs related to Magnetostatic Fields.
CO:3	Distinguish between static and time-varying fields, establish the corresponding sets of Maxwell's Equations.
CO:4	Distinguish Boundary Conditions and provide solutions to communication engineering problems.
CO:5	Analyze the wave equation for good conductors, suitable dielectrics and evaluate the UPW characteristics for several practical media of interest.
CO:6	Analyze the rectangular waveguides, their mode characteristics, and design waveguides for solving practical problems thoroughly.

(a)

COs	Program Outcomes (POs)											
	1	2	3	4	5	6	7	8	9	10	11	12
CO:1	3	3	1	–	–	–	1	–	–	–	–	1
CO:2	3	3	1	–	–	–	–	–	–	–	–	1
CO:3	3	3	1	–	–	–	–	–	–	–	–	1
CO:4	3	3	1	–	–	–	–	–	–	–	–	1
CO:5	3	3	1	–	–	–	–	–	–	–	–	1
CO:6	3	3	1	–	–	–	–	–	–	–	–	1

(b)

40–74 marks. Applied electromagnetics for engineers [14]: The IIT Madras, India, offers this course from September–December 2020. In this course, 4425 learners have enrolled. One hundred forty-one learners have registered for the final assessment exam, and only 73 have secured the certificate. At the institute level, learner interests in microwave engineering and related courses have also been observed and depicted in Figure 7.3. Most of the students and faculty members have shown interest in basic electronics, advanced electronics, and computer-related subjects.

Observations and learning challenges on microwave engineering asynchronous teaching methods:

- For in-country and institutional levels, the number of learners enrolled and registered for the exam is significantly less when compared to other electronic and computer courses.
- Most learners have cleared the exam with minimum marks in these courses, and very few students have cleared with maximum or top band marks.
- Slow learners cannot understand some of the derivations due to the PowerPoint presentations or no live derivation of equations.
- Interaction between the teacher and learner, social interaction and peer discussions are missing in this method.

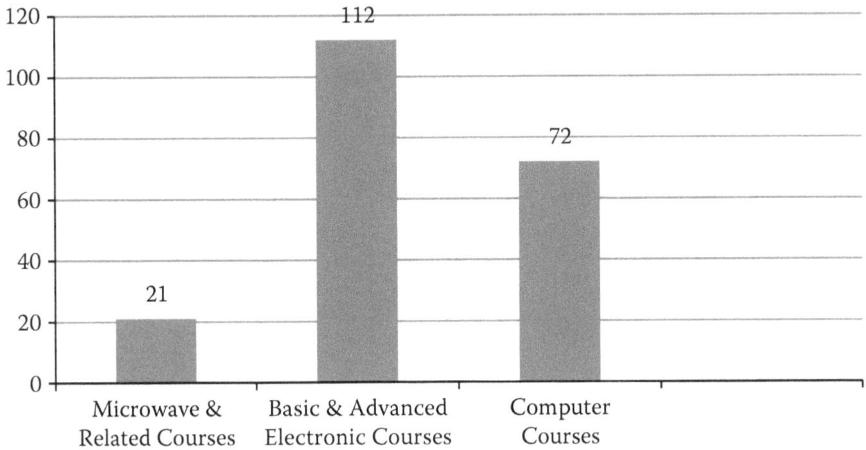

FIGURE 7.3 NPTEL certificates and participation of learners from Sreyas Institute of Engineering and Technology.

7.4 SYNCHRONOUS METHOD OF TEACHING

In this teaching method, both students and faculty members or learners and instructors are simultaneously placed. However, in the current pandemic situation, students and teachers are meeting simultaneously but through the online mode using learning-management systems. Microsoft Team is one of the popular learning-management systems since it is readily available for educational institutions. The author has used 'MS teams' for the online microwave teaching and 'one note' digital board for the derivations. Each session lasts 60 minutes, and the instructor has to engage 60 students. In each session, one survey or a quick test on the present class must be conducted to understand the students' levels. A sample copy of the digital whiteboard teaching class is depicted in Figure 7.4.

Observations and learning challenges of synchronous online microwave teaching:

- Difficult for instructors to assess the student's actual learning understandings.
- In microwave online teaching, slow learners cannot understand the concepts and mathematical derivation of rectangular waveguide concepts.
- Rural area students have faced internet connectivity issues, and economically backward students have no proper computers or personal electronic gadgets to attend the regular classwork.
- The problem-solving capability of the students is not good in some cases.

Some recommendations to improve online microwave and related online courses:

- The incorporation of open-source software in the online teaching of these courses may improve the learner interaction with the course [15–18].
- The digital whiteboard is recommended for microwave, antennas, and electromagnetics teaching because so many derivations are involved.

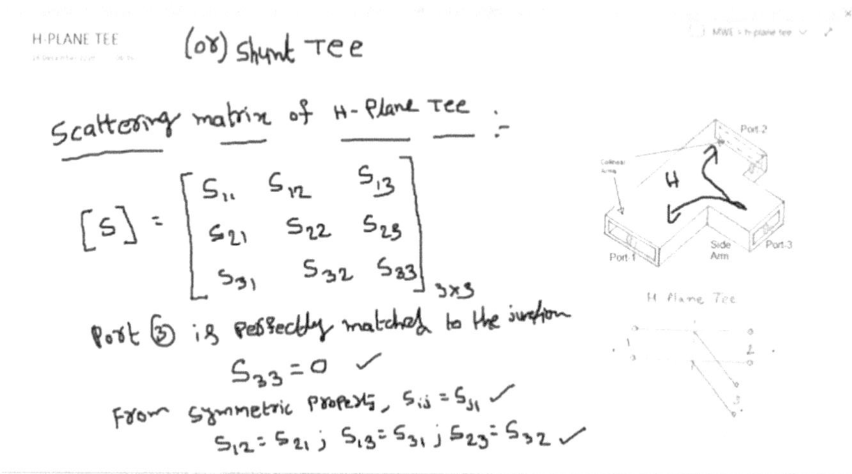

FIGURE 7.4 A sample of digital whiteboard class of microwave engineering online class.

- Surveys and quiz exams may help to attain learner interaction.
- Usage of virtual labs may also help effective student engagement.

7.5 CONCLUSIONS

Observations on the learning and tracking challenges of online microwave-engineering teaching and related courses have been presented in this chapter. Asynchronous and synchronous microwave teaching methods have been discussed, and specific observations on these teaching methods are also discussed here. Some recommendations to improve these courses are also made in this book chapter. Open-source software may be effectively helpful for online engineering education and engage the students effectively.

REFERENCES

1. WHO: Coronavirus disease (COVID-2019) situation reports. Available from: https://www.who.int/emergencies/diseases/novel-coronavirus-2019/situation-reports. [Last assessed on August 14, 2020].
2. F. Colace, M. De Santo, and A. Pietrosanto, "Evaluation Models for e-Learning Platform: An AHP Approach," Frontiers in Education Conference, 36th Annual. San Diego, CA, Institute of Electrical and Electronics Engineers, pp. 1–6, 2006.
3. D. Lawton, J. Bransford, N. Vye, M. C. Richey, V. T. Dang, and D. E. French, "Learning Science Principles for Effective Online Learning in the Workplace," Proceedings of the 40th ASEE/IEEE Frontiers in Education Conference. Retrieved March 5, 2012, from http://fie-conference.org/fie2010/, 2010.
4. P. Britt, "Elearning on the rise: Companies Move Classroom Content Online," *EContent*, vol. 27, no. 11, pp. 36–40, 2004.
5. J. Clarey, "E-learning 101: An Introduction to E-learning, Learning Tools and Technologies," Retrieved August 15, 2009, from brandon-hall.com, 2006.

6. J. Bersin, "Measuring E-Learning's Effectiveness: A Five-Step Program for Success. E-Learning," 2002, March.
7. S. Mackay and D. Fisher, "Practical Online Learning and Laboratories, for Engineering, Science and Technology, IDC Technologies," 2013.
8. M. Poe and M. L. A. Stassen, "Teaching and Learning Online, University of Massachusetts Amherst," 2006.
9. R16 B.Tech (ECE) syllabus. Accessed from: https://jntuh.ac.in/syllabus
10. Microwave Engineering and Related Course Information and Statistics. Accessed from: https://nptel.ac.in
11. Microwave Engineering – NPTEL Online Course. Accessed from: https://nptel.ac.in/noc/courses/noc19/SEM2/noc19-ee68/
12. Microwave Theory and Techniques-NPTEL Online Course. Accessed from: https://nptel.ac.in/noc/courses/noc20/SEM2/noc20-ee63/
13. Antennas-NPTEL Online Course. Accessed from: https://nptel.ac.in/noc/courses/noc20/SEM1/noc20-ee20/
14. Applied Electromagnetics for An Engineers-NPTEL Online Course. Accessed from: https://nptel.ac.in/noc/courses/noc20/SEM2/noc20-ee93/
15. V. S. Ponnapalli, and A. Praveena (2021). "Various Antenna Array Designs Using Scilab Software: An Exploratory Study," *Next-Generation Antennas: Advances and Challenges*, vol. 2, no. 3, 49–59.
16. K. Manohar, K. Sravani, and V. S. Ponnapalli, "An Investigation on Scilab Software for the Design of Transform Techniques and Digital Filters," In 2021 International Conference on Computer Communication and Informatics (ICCCI), pp. 1–5, IEEE, 2021, January.
17. V. S. Ponnapalli and V. Babu, "A Study on Scilab Free and Open-Source Programming for Antenna Array Design," In 2020 IEEE International IoT, Electronics and Mechatronics Conference (IEMTRONICS), pp. 1–3. IEEE, 2020, September.
18. D. Sarkar, "Antenna Learning and Research for Students and Young Professionals in the Post-COVID-19 Era," *IEEE Antennas and Propagation Magazine*, vol. 63, no. 3, pp. 146–148, 2021.

8 Performance Analysis of OOK and *L*-PPM Modulated VLC System with Imperfect CSI over LOS and Non-LOS Channel

Vipul Dixit and Atul Kumar
Department of Electronics and Communication Engineering, PDPM-Indian Institute of Information Technology, Design and Manufacturing, Jabalpur, India

CONTENTS

8.1 INTRODUCTION

Over the last decade, wireless communication devices have proliferated. The current radio frequency (RF) spectrum cannot satisfy this massive demand for wireless

DOI: 10.1201/9781003230526-8

data traffic. Visible light communication (VLC) has attracted much interest from researchers, academia, and industry as it provides dual use of light-emitting diode (LED), illumination, and communication. In particular, visible light, which occupies the broad spectrum of 430–790 THz, is used by VLC systems for communication and thus, overcomes the limitation of low bandwidth in RF communication [1]. In addition to unregulated license-free huge spectrum, VLC systems offer easy accessibility, high durability, compact size, power efficiency, and no-radiation [2]. Recently, VLC has been adopted in different fields, like Li-Fi, vehicle to vehicle and underwater communication, healthcare, identification systems and indoor network, etc. [2,3].

The transmitter and receiver can communicate via a line-of-sight (LOS) or non-line-of-sight (Non-LOS) channel. The transmitter and receiver are connected directly in the LOS channel. Hence, the received signal has good strength as any obstruction does not reflect it. Whereas in the Non-LOS channel, there is no direct path between transmitter and receiver, and the received signal is an accumulation of Non-LOS components, which are the multiple reflections through walls, tables, and other objects of the room. Thus, compared to the LOS signal, the strength of the received signal over the Non-LOS channel is lower. The LOS channel for the VLC system has been explored to a large extent and is used extensively [1,4,5]. At the same time, the Non-LOS channel becomes crucial under the condition of non-availability of the LOS channel, either due to blocking the link or due to the received signal being out of the receiver's viewing angle [6–9].

The modulation format is a fundamental aspect for performance improvement of a system in terms of probability of error, spectral and energy efficiency, etc. Various modulation schemes are implemented with VLC systems, e.g., on-off keying (OOK) [10], pulse position modulation (PPM) [10], orthogonal division multiple access (OFDM) [11] and color shift keying (CSK) [12], etc. OK is the most popular modulation format used in VLC systems mainly due to its simple implementation [4]. In optical wireless systems, digital PPM was first introduced in [13], and it is also adopted for infrared physical layer standard, i.e., IEEE 802.11. The PPM-modulation format provides various merits, such as power efficiency, dimming control, and immunity against flicker, but at the cost of high modulation bandwidth, which creates problems in implementation. In literature, various variants of PPM-modulation format are proposed, e.g., differential PPM (DPPM) [14], variable PPM (VPPM) [12], dual-header pulse-interval modulation (DHPIM) [15], etc., to improve the efficiency of PPM. However, all these variants suffer from the nonlinearity of the channel when compared to simple PPM. Thus, OOK/L-PPM are the most popular modulation formats for the VLC system.

8.2 LOS CHANNEL

A general VLC system is considered to be a LOS dominant technology. In the LOS channel, light-emitting diode (LED) and photodetector (PD) have a direct link between them, and hence, the received signal has good strength since any obstruction does not reflect it.

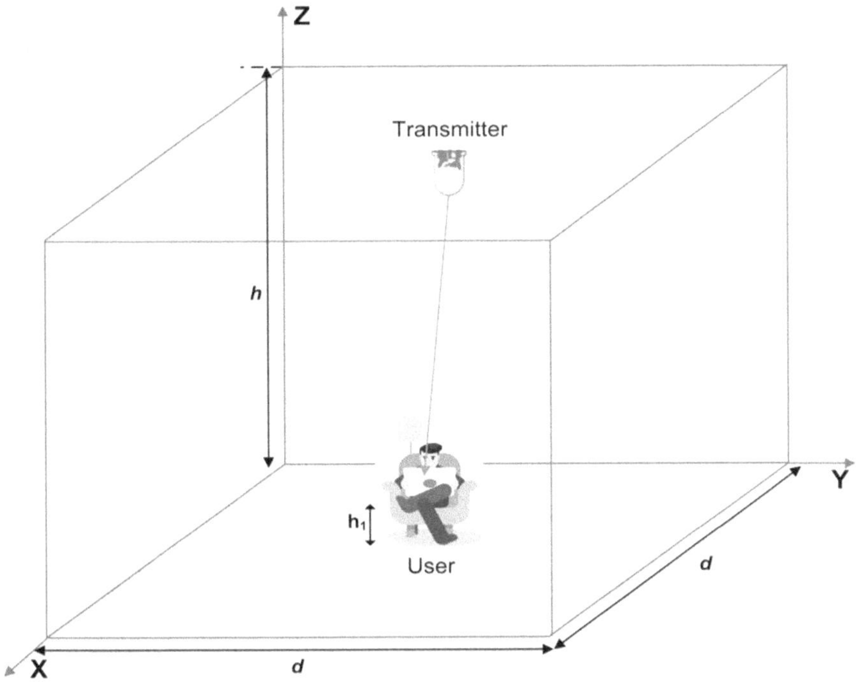

FIGURE 8.1 A LOS channel-based VLC system.

8.2.1 System Model

An AVLC system with intensity modulation and direct detection (IM/DD) is considered, which consists of a single LED as transmitter and a single PD as a receiver.

A room of dimension (d, d, h)m is considered in Figure 8.1 with a LED $\left(\frac{d}{2}, \frac{d}{2}, h\right)$m and a PD located at $\left(\frac{d}{2}, \frac{d}{2}, h_1\right)$m a height h_1 above the floor. It is assumed that the communication is carried out over the LOS channel.

The gain of the LOS channel between the LED and PD is given by [4]

$$h_{LOS} = \begin{cases} \frac{(m+1)A_r}{2\pi d^2} \cos^m \phi \; T(\psi) g(\psi) \cos(\psi), & \text{for } \psi \le \psi_c \\ 0, & \text{otherwise} \end{cases} \tag{8.1}$$

where A_r is PD's active area, d is the distance from LED to PD, ψ is incidence angle at PD, $g(\psi) = n^2/\sin^2 \psi_c$ is gain of optical concentrator with refractive index n, ψ_c is a field of view (FOV), $T(\psi)$ is gain of an optical filter, ϕ is the angle of emission and $m = -\ln 2/\ln\left(\cos \phi_{1/2}\right)$ is Lambertian order related to semi-angle at half power $\phi_{1/2}$.

The PD's output at time instant t is given as:

$$y(t) = h_{LOS} R_p x(t) + n(t), \tag{8.2}$$

where $x(t)$ is transmit signal using OOK/L-PPM modulation, R_p is PD's responsivity, $n(t)$ is additive white Gaussian noise (AWGN) at PD with mean zero and variance σ_n^2. Since, $n(t)$ consists of shot noise and thermal noise, σ_n^2 can be written as [1,4]

$$\sigma_n^2 = \sigma_{shot}^2 + \sigma_{thermal}^2, \tag{8.3}$$

where σ_{shot}^2 is the variance of shot noise and $\sigma_{thermal}^2$ is the variance of thermal noise.

The parameter σ_{shot}^2 is given as [4,5]

$$\sigma_{shot}^2 = 2qI_{bg}I_2B + 2qRP_tBh_{LOS}, \tag{8.4}$$

where q is the electron charge, B is the bandwidth of noise equal to modulation bandwidth, $I_{bg} = 5100\,\mu A$ background current when exposed to sunlight, P_t optical power transmitted, h_{LOS} channel gain, and $I_2 = 0.562$ bandwidth factor of noise.

Similarly, $\sigma_{thermal}^2$ it is given as [4,5]

$$\sigma_{thermal}^2 = \frac{8\pi kT_k A_r I_2 B^2 \eta}{G} + \frac{16\pi^2 kT_k A_r^2 I_3 B^3 \eta^2 \Gamma}{g_m}, \tag{8.5}$$

where Γg_m is channel noise factor, the transconductance of FET, kT_k Boltzmann's constant andabsolute temperature, G open-loop voltage gain, η PD's capacitance per unit area, and $I_3 = 0.0868$ model the thermal noise, following values are used: $G = 10 \eta = 112pF/cm^2 T_k = 295K \Gamma = 1.5$ and $g_m = 30\ mS$.

8.2.2 OOK Modulation

In the literature, OOK is the most popular modulation format used in VLC systems mainly due to its simplicity in implementation. Let us consider the transmitted symbol by the LED as 'S'. An optical pulse represents $S = 1$ and the absence of optical pulse represents $S = 0$ as shown in Figure 8.2. A non-return to zero (NRZ) OOK-modulated signal is written as:

$$x(t) = \begin{cases} 2P_{avg} & for\ \ t \in (0, T_b) \\ 0 & otherwise \end{cases} \tag{8.6}$$

where T_b is bit interval, P_{avg} is average optical power and factor '2' is used to keep the average power P_{avg}.

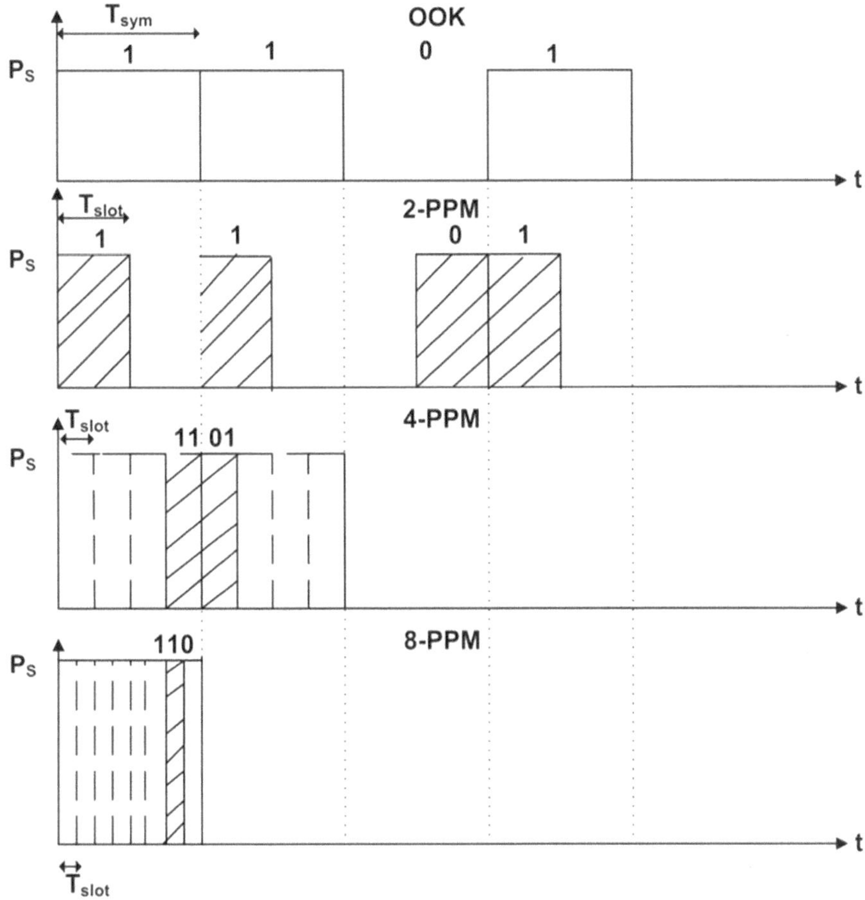

FIGURE 8.2 Waveforms of OOK and L-PPM.

8.2.2.1 Performance Analysis with Imperfect CSI

From Equation (8.2), assuming $\tau \ll T_b$, the PD's output can be written as

$$y = h_{LOS} R_p 2P_{avg} S + n, \tag{8.7}$$

where $x = 2P_{avg}S$ is the transmit signal using OOK modulation?

When $S = 1$ the (8.7) can be represented as

$$y_1 = h_{LOS} R_p 2P_{avg} + n, \tag{8.8}$$

where $h_{LOS} R_p 2P_{avg} = I_P$ is generated photocurrent at PD. The output y_1 at PD is a Gaussian random variable with mean $\mu_1 = 2P_{avg} R_p h_{LOS}$ and variance $\sigma_1^2 = \sigma_n^2$.

Similarly, when $S = 0$, the (8.7) can be represented as $y_0 = n$ with mean $\mu_0 = 0$ and variance $\sigma_0^2 = \sigma_n^2$.

In a practical scenario, the VLC system suffers from various impairments like outdated CSI and noisy CSI [16–18]. Under noisy CSI conditions, the estimated channel coefficient is given as

$$\widehat{h}_{LOS} = h_{LOS} + \varepsilon, \tag{8.9}$$

where h_{LOS} is the actual gain of the LOS channel and ε is the error added during the channel estimation process modeled as $N(0, \sigma_\varepsilon^2)$.

Thus, $\widehat{h}_{LOS}$ it is a Gaussian distributed random variable modeled as $N(h_{LOS}, \sigma_\varepsilon^2)$. The probability density function (PDF) $\widehat{h}_{LOS}$ can be written as $f_{\widehat{h}_{LOS}} = \frac{1}{\sqrt{2\pi\sigma_\varepsilon^2}} e^{-(\widehat{h}_{LOS} - h_{LOS})^2/2\sigma_\varepsilon^2}$. BER of VLC system using OOK modulation with imperfect CSI can be written as

$$P_e = P(S = 0)P(y_0 = 1|S = 0) + P(S = 1)P(y_1 = 0|S = 1). \tag{8.10}$$

As y_1 is Gaussian distributed, it $P(y_1 = 0|S = 1)$ can be written as

$$P(y_1 = 0|S = 1) = \int_{-\infty}^{\widehat{I}_{th}} f_{y_1}(i)di, \tag{8.11}$$

where $f_{y_1}(i) = \frac{1}{\sqrt{2\pi\sigma_1^2}} e^{-(i-\mu_1)^2/2\sigma_1^2}$ is PDF of y_1 and $\widehat{I}_{th}$ is the estimated symbol detection threshold at the PD.

By putting the value of $f_{y_1}(i)$ in (8.11), we get

$$\begin{aligned} P(y_1 = 0|S = 1) &= \frac{1}{\sqrt{2\pi\sigma_1^2}} \int_{-\infty}^{\widehat{I}_{th}} e^{-(i-\mu_1)^2/2\sigma_1^2} di, \\ &= Q\left(\frac{\mu_1 - \widehat{I}_{th}}{\sigma_1}\right) \end{aligned} \tag{8.12}$$

Similarly, it $P(y_0 = 1|S = 0)$ can be written as

$$P(y_0 = 1|S = 0) = \int_{\widehat{I}_{th}}^{\infty} f_{y_0}(i)di, \tag{8.13}$$

where $f_{y_0}(i) = \frac{1}{\sqrt{2\pi\sigma_0^2}} e^{-(i-\mu_0)^2/2\sigma_0^2}$ is the PDF of y_0? By putting the value of $f_{y_0}(i)$ in (8.13), we get

$$\begin{aligned} P(y_0 = 1|S = 0) &= \frac{1}{\sqrt{2\pi\sigma_0^2}} \int_{\widehat{I}_{th}}^{\infty} e^{-(i-\mu_0)^2/2\sigma_0^2} di, \\ &= Q\left(\frac{\widehat{I}_{th} - \mu_0}{\sigma_0}\right) \end{aligned} \tag{8.14}$$

For equiprobable symbols, by putting the values from (8.12) and (8.14) in (8.10), we get

$$P_e = \frac{1}{2}Q\left(\frac{\hat{I}_{th} - \mu_0}{\sigma_0}\right) + \frac{1}{2}Q\left(\frac{\mu_1 - \hat{I}_{th}}{\sigma_1}\right). \tag{8.15}$$

$\hat{I}_{th} = \hat{I}_P/2$ Where is the estimated symbol detection threshold at the PD and $\hat{I}_P = 2P_{avg}R_p\hat{h}_{LOS}$ is the estimated photocurrent at PD.

The average BER can be calculated by averaging P_e over the PDF of $\hat{h}_{LOS}$ as

$$\begin{aligned} P_{e,avg} &= \frac{1}{2}\int_{-\infty}^{\infty} Q\left(\frac{P_{avg}R_p\hat{h}_{LOS}}{\sigma_n}\right) f_{\hat{h}_{LOS}} d\hat{h}_{LOS} \\ &+ \frac{1}{2}\int_{-\infty}^{\infty} Q\left(\frac{2P_{avg}R_p h_{LOS} - P_{avg}R_p\hat{h}_{LOS}}{\sigma_n}\right) f_{\hat{h}_{LOS}} d\hat{h}_{LOS}. \end{aligned} \tag{8.16}$$

Using [19, (1.3)], Equation (8.16) can be future solved as

$$\begin{aligned} P_{e,\,avg} &= \frac{1}{2}Q\left(\frac{P_{avg}R_p h_{LOS}}{\sqrt{\sigma_n^2 + (P_{avg}R_p)^2\sigma_\varepsilon^2}}\right) + \frac{1}{2}Q\left(\frac{2P_{avg}R_p h_{LOS} - P_{avg}R_p h_{LOS}}{\sqrt{\sigma_n^2 + (P_{avg}R_p)^2\sigma_\varepsilon^2}}\right). \\ &= Q\left(\frac{P_{avg}R_p h_{LOS}}{\sqrt{\sigma_n^2 + (P_{avg}R_p)^2\sigma_\varepsilon^2}}\right) \end{aligned} \tag{8.17}$$

Thus, BER of VLC system using OOK modulation with perfect CSI i.e., $\sigma_\varepsilon^2 = 0$ can be written as

$$P_e = Q\left(\frac{P_{avg}R_p h_{LOS}}{\sigma_n}\right). \tag{8.18}$$

8.2.3 L-PPM Modulation

In L-PPM, a symbol period, i.e., T_{sym} is divided into L time slots. Figure 8.2 shows that a symbol is represented by a pulse transmitted in a particular time slot and from symbol to symbol, which changes its position along the time axis. In L-PPM, N bits of the generated message are encoded in a particular symbol $L = 2^N$.

8.2.4 Performance Analysis with Imperfect CSI

From Equation (8.2), assuming, $\tau \ll T_{slot}$ the PD's outputcan be written as

$$y = h_{LOS}R_p LP_S S_\ell + n, \tag{8.19}$$

where P_S is transmitted symbol power, L are several slots per symbol period and S_ℓ is slot data and $x = LP_S S_\ell$ is transmit signal using L-PPM modulation.

In L-PPM, only one slot out of L slots will be represented by an optical pulse with $S_\ell = 1$ and other $(L - 1)$ slots are represented by the absence of a pulse $S_\ell = 0$. It is important to note that in L-PPM, the slot noise power is L times the noise power of OOK modulation format, i.e., $\sigma_{n'}^2 = L\sigma_n^2$ which is due to the increase in slot bandwidth by a factor of L.

When $S_\ell = 1$ the (8.20) can be represented as

$$y_1 = LP_S R_p h_{LOS} + n, \tag{8.20}$$

where $LP_S R_p h_{LOS} = I_P$ is photocurrent generated at PD. The output y_1 at PD is a Gaussian random variable with $\mu_1 = LP_S R_p h_{LOS}$ and $\sigma_1^2 = \sigma_{n'}^2$.

Similarly, when $S_\ell = 0$, the (8.20) can be represented as $y_0 = n$ with mean $\mu_0 = 0$ and variance $\sigma_0^2 = \sigma_{n'}^2$.

Under noisy CSI conditions, the estimated channel coefficient is given as

$$\widehat{h}_{LOS} = h_{LOS} + \varepsilon, \tag{8.21}$$

where h_{LOS} is the actual gain of the LOS channel and ε is error added during the channel-estimation process modeled as $\mathcal{N}(0, \sigma_\varepsilon^2)$.

Thus, $\widehat{h}_{LOS}$ it is a Gaussian random variable modeled as $\mathcal{N}(h_{LOS}, \sigma_\varepsilon^2)$. The PDF of $\widehat{h}_{LOS}$ can be written as $f_{\widehat{h}_{LOS}} = \frac{1}{\sqrt{2\pi\sigma_\varepsilon^2}} e^{-(\widehat{h}_{LOS} - h_{LOS})^2/2\sigma_\varepsilon^2}$.

Thus, the slot error probability (SLER) of the considered system is written as

$$P_e^{slot} = P(S_\ell = 0)P(y_0 = 1|S_\ell = 0) + P(S_\ell = 1)P(y_1 = 0|S_\ell = 1). \tag{8.22}$$

As y_1 is Gaussian distributed, it $P(y_1 = 0|S_\ell = 1)$ can be written as

$$P(y_1 = 0|S_\ell = 1) = \int_{-\infty}^{\widehat{I}_{th}} f_{y_1}(i)di, \tag{8.23}$$

where $f_{y_1}(i) = \frac{1}{\sqrt{2\pi\sigma_1^2}} e^{-(i-\mu_1)^2/2\sigma_1^2}$ is PDF of y_1 and $\widehat{I}_{th}$ is the estimated-symbol detection threshold.

By putting the value of $f_{y_1}(i)$ in (8.23), we get

$$\begin{aligned} P(y_1 = 0|S = 1) &= \frac{1}{\sqrt{2\pi\sigma_1^2}} \int_{-\infty}^{\widehat{I}_{th}} e^{-(i-\mu_1)^2/2\sigma_1^2} di, \\ &= Q\left(\frac{\mu_1 - \widehat{I}_{th}}{\sigma_1}\right) \end{aligned} \tag{8.24}$$

Similarly, it $P(y_0 = 1|S_\ell = 0)$ can be written as

$$P(y_0 = 1|S = 0) = \int_{\hat{I}_{th}}^{\infty} f_{y_0}(i)di, \tag{8.25}$$

where $f_{y_0}(i) = \frac{1}{\sqrt{2\pi\sigma_0^2}} e^{-(i-\mu_0)^2/2\sigma_0^2}$ is the PDF of y_0? By putting the value of $f_{y_0}(i)$ in (8.25), we get

$$\begin{aligned} P(y_0 = 1|S = 0) &= \frac{1}{\sqrt{2\pi\sigma_0^2}} \int_{I_{th}}^{\infty} e^{-(i-\mu_0)^2/2\sigma_0^2} di, \\ &= Q\left(\frac{\hat{I}_{th} - \mu_0}{\sigma_0}\right) \end{aligned} \tag{8.26}$$

The occurrence probability of a slot S_ℓ is given as

$$P(S_\ell = 1) = \frac{1}{L} \tag{8.27}$$

$$P(S_\ell = 0) = \frac{L-1}{L} \tag{8.28}$$

Now by putting the values from (8.24), (8.26), (8.27) and (8.28) in (8.22), we get

$$P_e^{slot} = \frac{L-1}{L} Q\left(\frac{\hat{I}_{th} - \mu_0}{\sigma_0}\right) + \frac{1}{L} Q\left(\frac{\mu_1 - \hat{I}_{th}}{\sigma_1}\right). \tag{8.29}$$

where $\hat{I}_{th} = \hat{I}_P/2\hat{I}_P = LP_S R_p \hat{h}_{LOS}$ and is estimated photocurrent at PD.

Now, the average SLER can be calculated as

$$\begin{aligned} P_{e,avg}^{slot} &= \frac{L-1}{L} \int_{-\infty}^{\infty} Q\left(\frac{0.5LP_S R_p \hat{h}_{LOS}}{\sigma_{n'}}\right) f_{\hat{h}_{LOS}} d\hat{h}_{LOS} \\ &+ \frac{1}{L} \int_{-\infty}^{\infty} Q\left(\frac{LP_S R_p h_{LOS} - 0.5LP_S R_p \hat{h}_{LOS}}{\sigma_{n'}}\right) f_{\hat{h}_{LOS}} d\hat{h}_{LOS}. \end{aligned} \tag{8.30}$$

Using [19, (1.3)], Equation (8.35) can be future solved as

$$\begin{aligned} P_{e,avg}^{slot} &= \frac{L-1}{L} Q\left(\frac{0.5LP_S R_p h_{LOS}}{\sqrt{\sigma_{n'}^2 + (0.5LP_S R_p)^2 \sigma_\varepsilon^2}}\right) + \frac{1}{L} Q\left(\frac{LP_S R_p h_{LOS} - 0.5LP_S R_p h_{LOS}}{\sqrt{\sigma_{n'}^2 + (0.5LP_S R_p)^2 \sigma_\varepsilon^2}}\right). \\ &= Q\left(\frac{0.5LP_S R_p h_{LOS}}{\sqrt{\sigma_{n'}^2 + (0.5LP_S R_p)^2 \sigma_\varepsilon^2}}\right) \end{aligned} \tag{8.31}$$

Thus, the symbol error rate (SER) of L-PPM is given as

$$P_e^{symbol} = 1 - (1 - P_e^{slot})^L. \tag{8.32}$$

The BER of L-PPM is given as

$$P_e = \frac{L}{L-1} P_e^{symbol}. \tag{8.33}$$

Thus, SLER of VLC system using L-PPM modulation with perfect CSI i.e., $\sigma_\varepsilon^2 = 0$ can be written as

$$P_e^{slot} = Q\left(\frac{0.5LP_S R_p h_{LOS}}{\sigma_{n'}}\right). \tag{8.34}$$

8.3 PERFORMANCE ANALYSIS OVER NON-LOS CHANNEL

In the Non-LOS channel, there is no direct path between LED and PD, and the received signal is the summation of Non-LOS components, which are multiple reflections through walls, tables, and other objects of the room. The Non-LOS channel becomes crucial under the condition of nonavailability of the LOS channel either due to blocking the link and due to the received signal being out of the viewing angle of the receiver.

8.3.1 System Model

A VLC system with IM/DD is considered, which consists of a single LED and PD. A room of dimension (d, d, h)m is considered in Figure 8.3 with a LED $\left(\frac{d}{2}, \frac{d}{2}, h\right)$m and a PD located a $\left(\frac{d}{2}, \frac{d}{2}, h_1\right)$m h_1 above the floor. Assuming that there is no LOS link, then communication will occur through the Non-LOS channel. Specifically, this analysis considers only single-order reflection components for the Non-LOS channel.

In this analysis, the jth Non-LOS component is assumed to be the sum of single-order reflected components, i.e., $h_j = \sum_{k=1}^{K} h_{jk} h_{jk}$ the single-order reflected component from k^{th} the wall of jth the Non-LOS component. The h_{jk} can be written as

$$h_{jk} = \begin{cases} \frac{(m+1)A_r}{2\pi^2 d_{k1}^2 d_{k2}^2} \rho_k dA_k \cos^m \phi \cos\alpha_k \cos\beta_k T(\psi) g(\psi) \cos(\psi), & for\ \psi \le \psi_c \\ 0, & otherwise \end{cases} \tag{8.35}$$

where ρ_k is reflection co-efficient of kth the wall, dA_k is a differential area of kth the wall, d_{k1} is the distance between LED and point of contact on k^{th} the wall, d_{k2} is

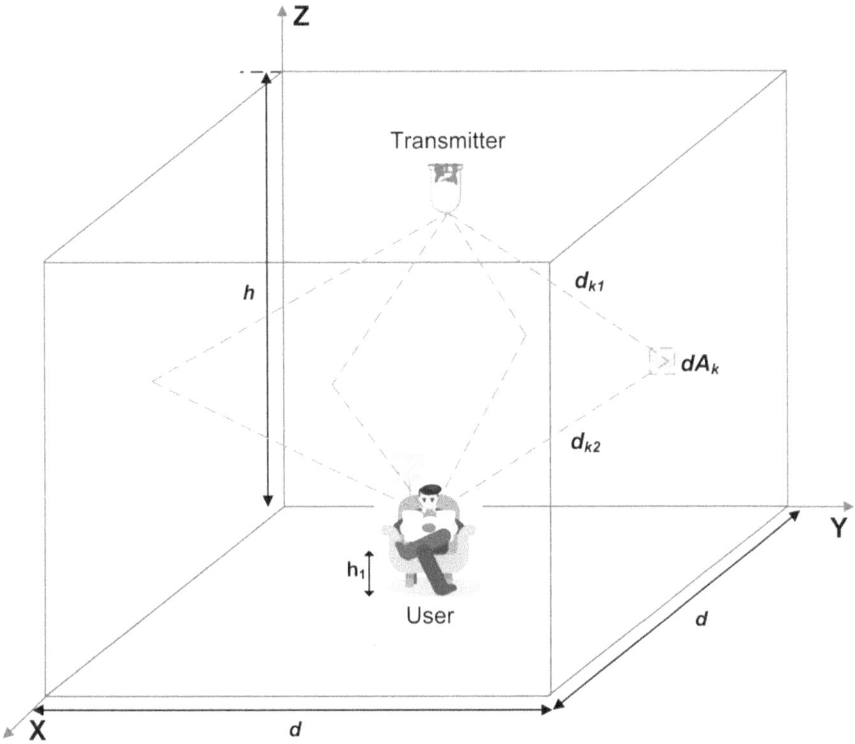

FIGURE 8.3 Non-LOS channel-based VLC system.

the distance between the point of contact on k^{th} wall and PD α_k is the irradiance angle to the point of contact on k^{th} the wall, β_k is irradiance angle from the point of contact on k^{th} the wall.

For uniform luminance all over the room, an ideal Lambertian LED is considered, i.e., $m = 1$ The points of contact on k^{th} the wall are uniformly distributed across the area of kth the wall, but FOV i.e., $\cos\psi \geq \cos\psi_c$ limit the points distribution along with the height of kth the wall as

$$\frac{\tilde{z} - h_1}{\sqrt{\left[\left(\frac{d}{2} - \tilde{x}\right)^2 + \left(\frac{d}{2} - \tilde{y}\right)^2 + (h_1 - \tilde{z})^2\right]}} \geq \cos\psi_c, \tag{8.36}$$

where $(\tilde{x}, \tilde{y}, \tilde{z})$ denotes the coordinates of the point of contact on k^{th} the wall.

$$\tilde{z} \geq \frac{\sqrt{\left[\left(\frac{d}{2} - \tilde{x}\right)^2 + \left(\frac{d}{2} - \tilde{y}\right)^2\right]}}{\tan\psi_c} + h_1. \tag{8.37}$$

TABLE 8.1
Contact points distribution over the walls

Wall Location	$\tilde{x}$	Distribution of $\tilde{y}$	$\tilde{z}$
Front-Wall	d	$\mathscr{U}[0, d]$	$\mathscr{U}[\gamma, h]$
Back-Wall	0	$\mathscr{U}[0, d]$	$\mathscr{U}[\gamma, h]$
Left-Wall	$\mathscr{U}[0, d]$	0	$\mathscr{U}[\gamma, h]$
Right-Wall	$\mathscr{U}[0, d]$	d	$\mathscr{U}[\gamma, h]$

The coordinates of the point of contact on the back wall $(0, \tilde{y}, \tilde{z})$ are where $\tilde{y}$ is distributed uniformly, i.e., $\tilde{y} \sim \mathscr{U}[0, d]$, and using (8.39), $\tilde{z}$ is given as

$$\tilde{z} \geq \frac{\sqrt{\left[\left(\frac{d}{2}\right)^2 + \left(\frac{d}{2} - \tilde{y}\right)^2\right]}}{\tan \psi_c} + h_1. \tag{8.38}$$

$\tilde{z}\tilde{z} \sim \mathscr{U}[\gamma, h]\gamma = \frac{\sqrt{\left[\left(\frac{d}{2}\right)^2 + \left(\frac{d}{2} - \tilde{y}\right)^2\right]}}{\tan \psi_c} + h_1$ Similarly, the distribution of point of contact on other walls can be calculated as given in Table 8.1. The light reflected $\tilde{z} < \gamma$ is not received at PD as it has an angle of incidence at PD greater than FOV, i.e., $\psi > \psi_c$ Thus, following the above analysis h_{jk} can be calculated using (8.35). The jth Non-LOS component $h_j = \Sigma_{k=1}^{4} h_{jk}$ is a random variable with PDF shown in Figure 8.4. Thus h_j is having a mean and variance of 4.8638×10^{-8} and 6.3923×10^{-16} when considered FOV of PD is $\psi_c = 60°$.

The total number of Non-LOS components J accumulated at PD relies on several room conditions and the blocking percentage of PD.

The output of PD at time instant t can be given as

$$y(t) = \sum_{j=1}^{J} h_j R_p x(t) + n(t), \tag{8.39}$$

where $x(t)$ is transmit signalusing OOK and L-PPM modulation, h_j is the jth Non-LOS component, J is the total number of Non-LOS components, R_p and $n(t)$ is defined in previous section 8.2.

8.3.2 OOK Modulation

The OOK modulation format is already defined in the previous section, 8.2.2.

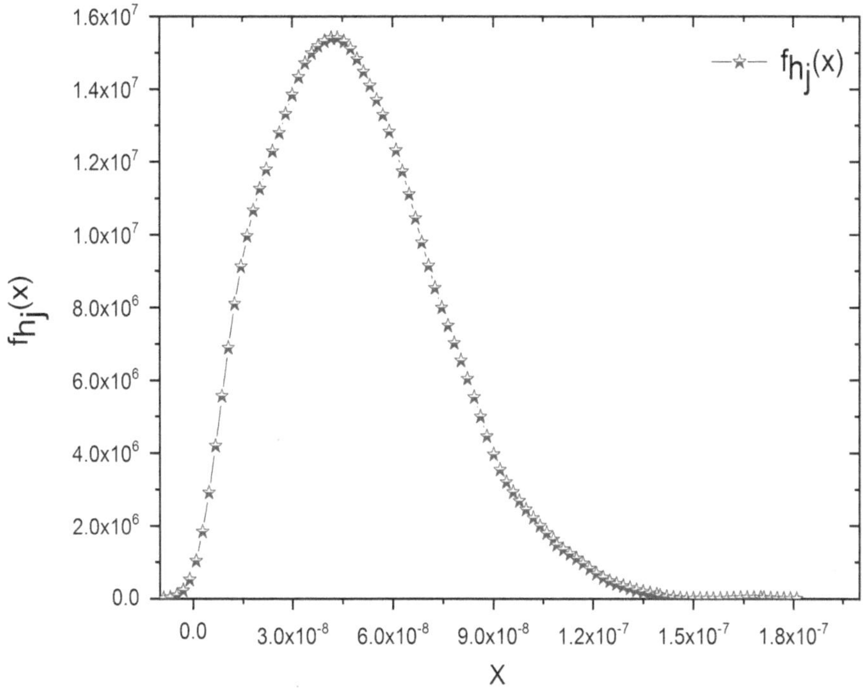

FIGURE 8.4 PDF of jth Non-LOS channel.

8.3.2.1 Performance Analysis with Imperfect CSI

From Equation (8.39), assuming $\tau \ll T_b$, the PD's output can be written as

$$y = \sum_{j=1}^{J} h_j R_p 2P_{avg} S + n, \tag{8.40}$$

where $x = 2P_{avg}S$ is transmit signalusing OOK modulation.

When $S = 1$ the (8.40) can be represented as

$$y_1 = \sum_{j=1}^{J} h_j R_p 2P_{avg} + n, \tag{8.41}$$

where $\sum_{j=1}^{J} h_j R_p 2P_{avg} = I_P$ is photocurrent generated at PD. When J_{it} is large, the output y_1 at PD can be approximated as Gaussian distributed with mean $\mu_1 = 2P_{avg}R_pJh_j^{avg}$ and variance $\sigma_1^2 = (2P_{avg}R_p)^2Jh_j^{\text{var}} + \sigma_n^2$.

Similarly, when $S = 0$, the (8.40) can be represented as $y_0 = n$ with mean $\mu_0 = 0$ and variance $\sigma_0^2 = \sigma_n^2$.

Under noisy CSI conditions, the estimated channel coefficient is given as

$$\widehat{h}_{NLOS} = h_{NLOS} + \varepsilon, \tag{8.42}$$

where $h_{NLOS} = Jh_j^{avg}$ is the actual gain of Non-LOS channel and ε is error added during the channel-estimation process modeled as $\mathcal{N}(0, \sigma_\varepsilon^2)$.

Thus, $\widehat{h}_{NLOS}$ it is a Gaussian random variable modeled as $\mathcal{N}(h_{NLOS}, \sigma_\varepsilon^2)$. The PDF of $\widehat{h}_{NLOS}$ can be written as $f_{\widehat{h}_{NLOS}} = \frac{1}{\sqrt{2\pi\sigma_\varepsilon^2}} e^{-(\widehat{h}_{NLOS} - h_{NLOS})^2/2\sigma_\varepsilon^2}$. The BER of the VLC system using OOK modulation with imperfect CSI is written as

$$P_e = P(S = 0)P(y_0 = 1|S = 0) + P(S = 1)P(y_1 = 0|S = 1). \tag{8.43}$$

As y_1 is Gaussian distributed, it $P(y_1 = 0|S = 1)$ is written as

$$P(y_1 = 0|S = 1) = \int_{-\infty}^{\widehat{I}_{th}} f_{y_1}(i)di, \tag{8.44}$$

where $f_{y_1}(i) = \frac{1}{\sqrt{2\pi\sigma_1^2}} e^{-(i-\mu_1)^2/2\sigma_1^2}$ is PDF of y_1 and $\widehat{I}_{th}$ is symbol detection threshold. By putting the value of $f_{y_1}(i)$ in (8.44), we get

$$\begin{aligned} P(y_1 = 0|S = 1) &= \frac{1}{\sqrt{2\pi\sigma_1^2}} \int_{-\infty}^{\widehat{I}_{th}} e^{-(i-\mu_1)^2/2\sigma_1^2} di, \\ &= Q\left(\frac{\mu_1 - \widehat{I}_{th}}{\sigma_1}\right) \end{aligned} \tag{8.45}$$

Similarly, it $P(y_0 = 1|S = 0)$ can be written as

$$P(y_0 = 1|S = 0) = \int_{\widehat{I}_{th}}^{\infty} f_{y_0}(i)di, \tag{8.46}$$

where $f_{y_0}(i) = \frac{1}{\sqrt{2\pi\sigma_0^2}} e^{-(i-\mu_0)^2/2\sigma_0^2}$ is PDF of y_0. By putting the value of $f_{y_0}(i)$ in (8.46), we get

$$\begin{aligned} P(y_0 = 1|S = 0) &= \frac{1}{\sqrt{2\pi\sigma_0^2}} \int_{\widehat{I}_{th}}^{\infty} e^{-(i-\mu_0)^2/2\sigma_0^2} di, \\ &= Q\left(\frac{\widehat{I}_{th} - \mu_0}{\sigma_0}\right) \end{aligned} \tag{8.47}$$

By putting the values from (8.45) and (8.47) in (8.43), we get

$$P_e = \frac{1}{2} Q\left(\frac{\widehat{I}_{th} - \mu_0}{\sigma_0}\right) + \frac{1}{2} Q\left(\frac{\mu_1 - \widehat{I}_{th}}{\sigma_1}\right). \tag{8.48}$$

For the above analysis, the symbol detection threshold is [20,21]

$$\frac{\widehat{I}_{th}^2}{2\sigma_0^2} = \frac{\left(\widehat{I}_p - \widehat{I}_{th}\right)^2}{2\sigma_1^2} + \ln\left(\frac{\sigma_1}{\sigma_0}\right). \tag{8.49}$$

The last term $\ln\left(\frac{\sigma_1}{\sigma_0}\right)$ in (8.49) is negligible in practical cases [21], and hence, the approximated value of the threshold is

$$\widehat{I}_{th} \approx \frac{\sigma_0 \widehat{I}_p}{\sigma_0 + \sigma_1}. \tag{8.50}$$

By putting the values from (8.50) in (8.48), we get

$$P_e = \frac{1}{2} Q\left(\frac{\widehat{I}_p}{\sigma_0 + \sigma_1}\right) + \frac{1}{2} Q\left(\frac{\mu_1(\sigma_0 + \sigma_1) - \sigma_0 \widehat{I}_p}{\sigma_1(\sigma_0 + \sigma_1)}\right). \tag{8.51}$$

where $\widehat{I}_P = 2P_{avg} R_p \widehat{h}_{NLOS}$ is the estimated photocurrent at the PD.

The average BERcan be calculated by averaging P_e over the PDF of $\widehat{h}_{NLOS}$ as

$$\begin{aligned} P_{e,avg} &= \frac{1}{2} \int_{-\infty}^{\infty} Q\left(\frac{2P_{avg} R_p \widehat{h}_{NLOS}}{\sigma_0 + \sigma_1}\right) f_{\widehat{h}_{NLOS}} d\widehat{h}_{NLOS} \\ &+ \frac{1}{2} \int_{-\infty}^{\infty} Q\left(\frac{2P_{avg} R_p h_{NLOS}(\sigma_0 + \sigma_1) - \sigma_0 2P_{avg} R_p \widehat{h}_{NLOS}}{\sigma_1(\sigma_0 + \sigma_1)}\right) f_{\widehat{h}_{NLOS}} d\widehat{h}_{NLOS}. \end{aligned} \tag{8.52}$$

Using [19, (1.3)], Equation (8.52) can be future solved as

$$P_{e,\,avg} = \frac{1}{2} Q\left(\frac{2P_{avg} R_p h_{NLOS}}{\sqrt{(\sigma_0 + \sigma_1)^2 + \left(2P_{avg} R_p\right)^2 \sigma_\varepsilon^2}}\right) + \frac{1}{2} Q\left(\frac{\sigma_1 2P_{avg} R_p h_{NLOS}}{\sqrt{\sigma_1^2(\sigma_0 + \sigma_1)^2 + \left(2P_{avg} R_p\right)^2 \sigma_0^2 \sigma_\varepsilon^2}}\right). \tag{8.53}$$

Thus, BER of VLC system using OOK modulation with perfect CSI i.e., $\sigma_\varepsilon^2 = 0$ can be written as

$$P_e = Q\left(\frac{2P_{avg} R_p h_{NLOS}}{\sigma_0 + \sigma_1}\right). \tag{8.54}$$

8.3.3 L-PPM Modulation

The L-PPM modulation format is already defined in the previous section 8.2.2.

8.3.3.1 Performance Analysis with Imperfect CSI

From Equation (8.39), assuming, $\tau \ll T_{slot}$ the PD's output can be written as

$$y = \sum_{j=1}^{J} h_j R_p L P_S S_\ell + n, \tag{8.55}$$

where $x = LP_S S_\ell$ is transmit signal using L-PPM modulation.

When $S_\ell = 1$ the (8.55) can be represented as

$$y_1 = \sum_{j=1}^{J} h_j R_p L P_S + n, \tag{8.56}$$

where $\sum_{j=1}^{J} h_j R_p L P_S = I_P$ is photocurrent generated at PD. When J_{it} is large, the output y_1 at PD can be approximated as Gaussian distributed with mean $\mu_1 = LP_S R_p J h_j^{avg}$ and variance $\sigma_1^2 = (LP_S R_p)^2 J h_j^{\text{var}} + \sigma_n^2$.

Similarly, when $S = 0$, the (8.55) can be represented as $y_0 = n$ with mean $\mu_0 = 0$ and variance $\sigma_0^2 = \sigma_n^2$.

Under noisy CSI conditions, the estimated channel coefficient is given as

$$\widehat{h}_{NLOS} = h_{NLOS} + \varepsilon, \tag{8.57}$$

where $h_{NLOS} = J h_j^{avg}$ is the actual gain of Non-LOS channel and ε is error added during the channel-estimation process modeled as $\mathcal{N}(0, \sigma_\varepsilon^2)$.

Thus, $\widehat{h}_{NLOS}$ it is a Gaussian random variable modeled as $\mathcal{N}(h_{NLOS}, \sigma_\varepsilon^2)$. The PDF of $\widehat{h}_{NLOS}$ can be written as $f_{\widehat{h}_{NLOS}} = \frac{1}{\sqrt{2\pi\sigma_\varepsilon^2}} e^{-(\widehat{h}_{NLOS} - h_{NLOS})^2/2\sigma_\varepsilon^2}$. The SLER of the VLC system using L-PPM modulation with imperfect CSI can be written as

$$P_e^{slot} = P(S = 0)P(y_0 = 1|S = 0) + P(S = 1)P(y_1 = 0|S = 1). \tag{8.58}$$

As y_1 is Gaussian distributed, $P(y_1 = 0|S = 1)_{\text{it}}$ can be written as

$$P(y_1 = 0|S = 1) = \int_{-\infty}^{\widehat{I}_{th}} f_{y_1}(i)\,di, \tag{8.59}$$

where $f_{y_1}(i) = \frac{1}{\sqrt{2\pi\sigma_1^2}} e^{-(i-\mu_1)^2/2\sigma_1^2}$ is PDF of y_1 and I_{th} is symbol detection threshold. By putting the value of $f_{y_1}(i)$ in (8.59), we get

$$P(y_1 = 0|S = 1) = \frac{1}{\sqrt{2\pi\sigma_1^2}} \int_{-\infty}^{\hat{I}_{th}} e^{-(i-\mu_1)^2/2\sigma_1^2} di, \\ = Q\left(\frac{\mu_1 - \hat{I}_{th}}{\sigma_1}\right) \tag{8.60}$$

Similarly, it $P(y_0 = 1|S = 0)$ can be written as

$$P(y_0 = 1|S = 0) = \int_{\hat{I}_{th}}^{\infty} f_{y_0}(i) di, \tag{8.61}$$

where $f_{y_0}(i) = \frac{1}{\sqrt{2\pi\sigma_0^2}} e^{-(i-\mu_0)^2/2\sigma_0^2}$ is PDF of y_0. By putting the value of $f_{y_0}(i)$ in (8.61), we get

$$P(y_0 = 1|S = 0) = \frac{1}{\sqrt{2\pi\sigma_0^2}} \int_{\hat{I}_{th}}^{\infty} e^{-(i-\mu_0)^2/2\sigma_0^2} di, \\ = Q\left(\frac{\hat{I}_{th} - \mu_0}{\sigma_0}\right) \tag{8.62}$$

The occurrence probability of a slot S_ℓ is given as

$$P(S_\ell = 1) = \frac{1}{L} \tag{8.63}$$

$$P(S_\ell = 0) = \frac{L-1}{L} \tag{8.64}$$

Now by putting the values from (8.60), (8.62), (8.63) and (8.64) in (8.58), we get

$$P_e^{slot} = \frac{L-1}{L} Q\left(\frac{\hat{I}_{th} - \mu_0}{\sigma_0}\right) + \frac{1}{L} Q\left(\frac{\mu_1 - \hat{I}_{th}}{\sigma_1}\right). \tag{8.65}$$

For the above analysis, the symbol detection threshold is [21]

$$\frac{\hat{I}_{th}^2}{2\sigma_0^2} = \frac{\left(\hat{I}_p - \hat{I}_{th}\right)^2}{2\sigma_1^2} + \ln\left(\frac{\sigma_1}{\sigma_0}\right). \tag{8.66}$$

The last term $\ln\left(\frac{\sigma_1}{\sigma_0}\right)$ in (8.66) is negligible in practical cases [21], and hence, the approximated value of the threshold is

$$\hat{I}_{th} \approx \frac{\sigma_0 \hat{I}_p}{\sigma_0 + \sigma_1}. \tag{8.67}$$

By putting the values from (8.67) in (8.65), we get

$$P_e^{slot} = \frac{L-1}{L} Q\left(\frac{\widehat{I}_p}{\sigma_0 + \sigma_1}\right) + \frac{1}{L} Q\left(\frac{\mu_1(\sigma_0 + \sigma_1) - \sigma_0 \widehat{I}_p}{\sigma_1(\sigma_0 + \sigma_1)}\right). \tag{8.68}$$

where $\widehat{I}_P = LP_S R_p \widehat{h}_{NLOS}$ is estimated photocurrent at PD.

Now, the average SLER can be calculated as

$$P_{e,avg}^{slot} = \frac{L-1}{L} \int_{-\infty}^{\infty} Q\left(\frac{LP_S R_p \widehat{h}_{NLOS}}{\sigma_0 + \sigma_1}\right) f_{\widehat{h}_{NLOS}} d\widehat{h}_{NLOS} + \frac{1}{L} \int_{-\infty}^{\infty} Q\left(\frac{LP_S R_p h_{NLOS}(\sigma_0 + \sigma_1) - \sigma_0 LP_S R_p \widehat{h}_{NLOS}}{\sigma_1(\sigma_0 + \sigma_1)}\right) f_{\widehat{h}_{NLOS}} d\widehat{h}_{NLOS}. \tag{8.69}$$

Using [19, (1.3)], Equation (8.69) can be future solved as

$$P_{e,\,avg}^{slot} = \frac{L-1}{L} Q\left(\frac{LP_S R_p h_{NLOS}}{\sqrt{(\sigma_0 + \sigma_1)^2 + \left(LP_S R_p\right)^2 \sigma_\varepsilon^2}}\right) + \frac{1}{L} Q\left(\frac{\sigma_1 LP_S R_p h_{NLOS}}{\sqrt{\sigma_1^2(\sigma_0 + \sigma_1)^2 + \left(LP_S R_p\right)^2 \sigma_0^2 \sigma_\varepsilon^2}}\right). \tag{8.70}$$

Thus, BER of VLC system using L-PPM modulation for imperfect CSI can be obtained from (8.32), (8.33) and (8.70).

The SLER of VLC system using L-PPM modulation with perfect CSI i.e., $\sigma_\varepsilon^2 = 0$ can be written as

$$P_e^{slot} = Q\left(\frac{LP_S R_p h_{NLOS}}{\sigma_0 + \sigma_1}\right). \tag{8.71}$$

8.4 RESULT AND DISCUSSION

BER of Non-LOS-VLC system using OOK/L-PPM modulation is numericallyevaluated and compared for perfect/imperfect CSI cases. Table 8.2 shows the different parameters of simulation used during the analysis. The total number of Non-LOS components, i.e., J and channel estimation error variance, i.e., σ_ε^2 are considered 450 1×10^{-10} unless otherwise stated.

In Figure 8.5, the BER of the VLC system with OOK/L-PPM modulation is presented over the LOS channel for perfect/imperfect CSI. BER of VLC system with OOK modulation is better than that of VLC system with 2-PPM modulation, but

TABLE 8.2
Parameters of simulation

Specification	Details	Values
Set-up	Room Dimension	$(5 \times 5 \times 3)\ m^3$
	Coefficient of Reflection (ρ_{west}, ρ_{east}, ρ_{south} and ρ_{north})	0.8
	Number of Non-LOS components (J)	250, 350 and 450
Source	LED Co-ordinates	$(2.5 \times 2.5 \times 3)\ m^3$
	The angleof emission at half power ($\Phi_{1/2}$)	60°
Receiver	Receiver Co-ordinates	$P_1(2.5 \times 2.5 \times 0.8)\ m^3$
		$P_2(2.5 \times 1.25 \times 0.8)\ m^3$
		$P_3(1.25 \times 1.25 \times 0.8)\ m^3$
	PD's active area (A_r)	$1\ cm^2$
	Responsivity (R_p)	1
	FOV (ψ_c)	60°
	Optical filter Gain (T_s)	1
	Refractive Index (n)	1.5
	The bandwidth of Noise (B)	50 Hz
	Rate of Transmission (R_b)	50 Mbps

worse than that of VLC system using 4-PPM and 8-PPM modulation for perfect CSI case as shown in the figure. For the perfect CSI case, the BER of the VLC system using L-PPM modulation improves as L (modulation order) increases. On the contrary, for imperfect CSI, the BER of the VLC system with OOK modulation outperforms all the variants of VLC systems with L-PPM modulation $\sigma_\varepsilon^2 = 1 \times 10^{-10}$. The figure also depicts that BER degradation increases as the modulation order increases for imperfect CSI, typically observed at the higher SNR values.

BER vs. σ_ε^2 curves of VLC system with OOK/L-PPM modulation are compared over LOS channel at the SNR of 120 dB in Figure 8.6. BER of VLC system with OOK/L-PPM modulation increases as σ_ε^2 increases. Further, the VLC system using L-PPM modulation is more affected by the channel estimation error if the modulation order is high. BER of VLC system with OOK, 4-PPM and 8-PPM modulation becomes equal at $\sigma_\varepsilon^2 = 1 \times 10^{-10}$.

In Figure 8.7, the BER of the VLC system with OOK modulation over the LOS/Non-LOS channel is plotted against the SNR for perfect/imperfect CSI. The Non-LOS VLC system with OOK modulation outperforms the LOS VLC system with OOK modulation if the total number of Non-LOS components accumulated at

The PD is sufficiently large, i.e., $J = 450$ BER of the VLC system with OOK modulation over the LOS/Non-LOS channel degrades significantly due to channel estimation error. However, the degradation in the BER of the VLC system with OOK modulation over LOS and Non-LOS channels is similar.

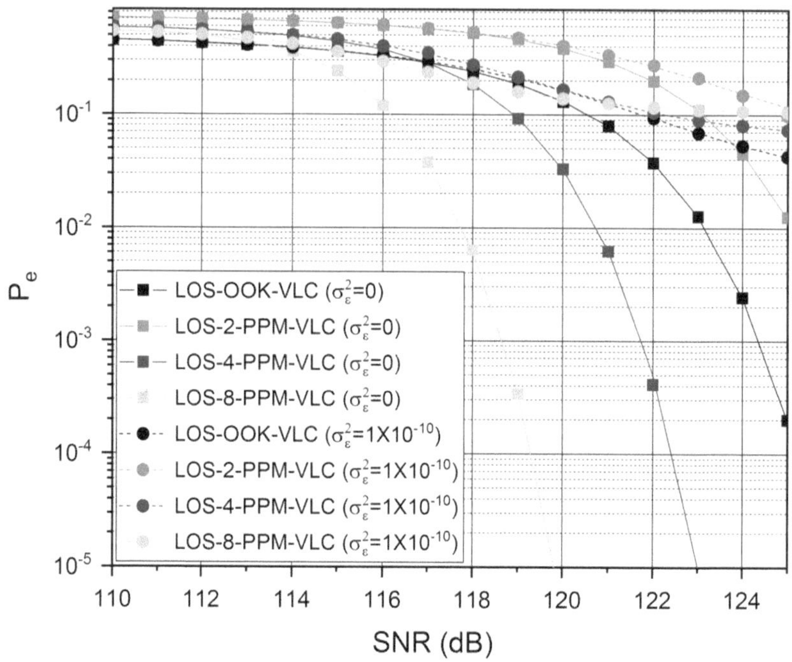

FIGURE 8.5 BER of VLC system with OOK/L-PPM modulation over LOS channel.

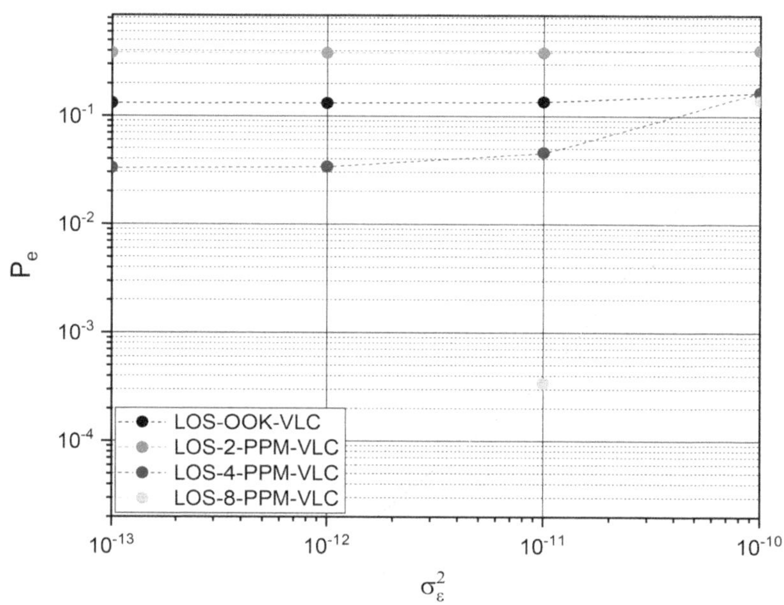

FIGURE 8.6 BER vs. σ_ε^2 curves of VLC system with OOK/L-PPM modulation.

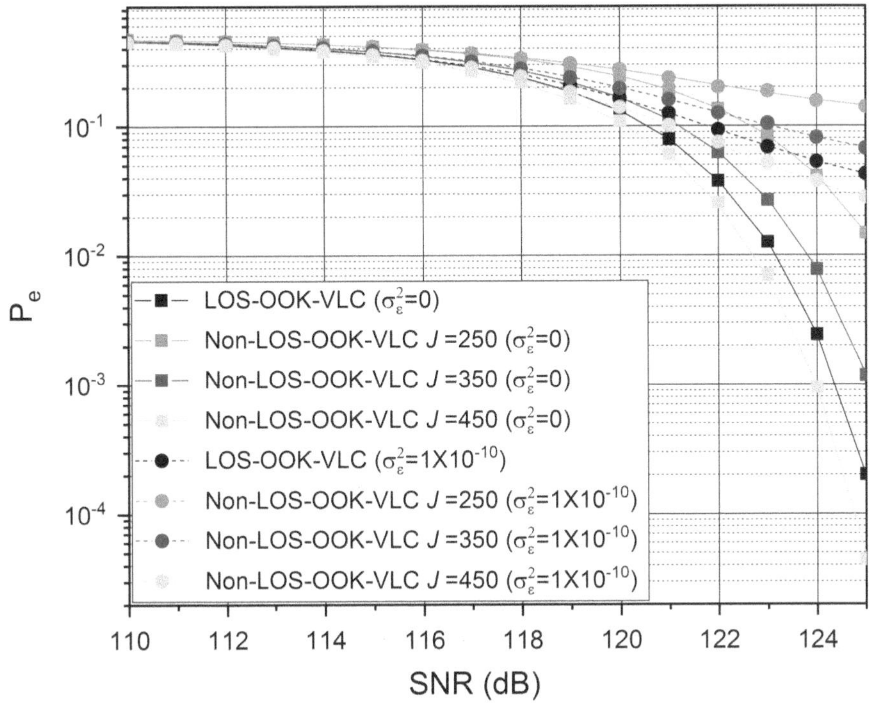

FIGURE 8.7 BER of VLC systemwith OOK modulation over LOS/Non-LOS channel.

BER of VLC system with 4-PPM modulation over LOS/Non-LOS channel is plotted against the SNR for perfect/imperfect CSI in Figure 8.8. The VLC system with 4-PPM modulation over the Non-LOS channel can outperform the system over the LOS channel if the total number of Non-LOS components accumulated at the PD is sufficiently large, i.e., $J = 450$ BER of VLC system with 4-PPM modulation over LOS/Non-LOS channel degrades significantly because of error in channel estimation. However, the degradation in the BER of the VLC system with 4-PPM modulation over LOS and Non-LOS channels is similar.

BER vs. J curves of VLC system with OOK/L-PPM modulation are compared over Non-LOS channel at the SNR of 122 dB in Figure 8.9. BER of VLC system with OOK/L-PPM modulation decreases as the total number of Non-LOS components accumulated at the PD increases. Further, the VLC.

A system using L-PPM modulation with a higher order of modulation, i.e., $(L > 2)$ outperforms the VLC system using OOK modulation in the error probability. The error performance of the VLC system using OOK and L-PPM modulation over Non-LOS channel degrades with the introduction of channel estimation error, i.e., $\sigma_\varepsilon^2 = 1 \times 10^{-10}$.

In Figure 8.10, BER vs.σ_ε^2 curves of VLC system with OOK/L-PPM modulation over Non-LOS channel are compared at the SNR of 122 dB. For the analysis, the total number of Non-LOS components accumulated at PD, i.e., J is considered to be 450. BER of VLC system with OOK/L-PPM modulation degrades as σ_ε^2 it

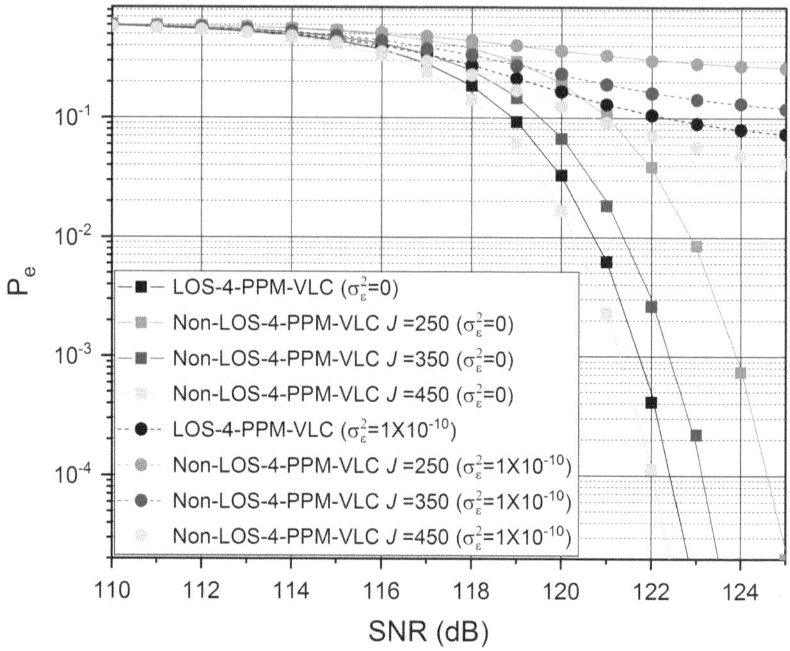

FIGURE 8.8 BER of VLC system with 4-PPM modulation over LOS/Non-LOS channel.

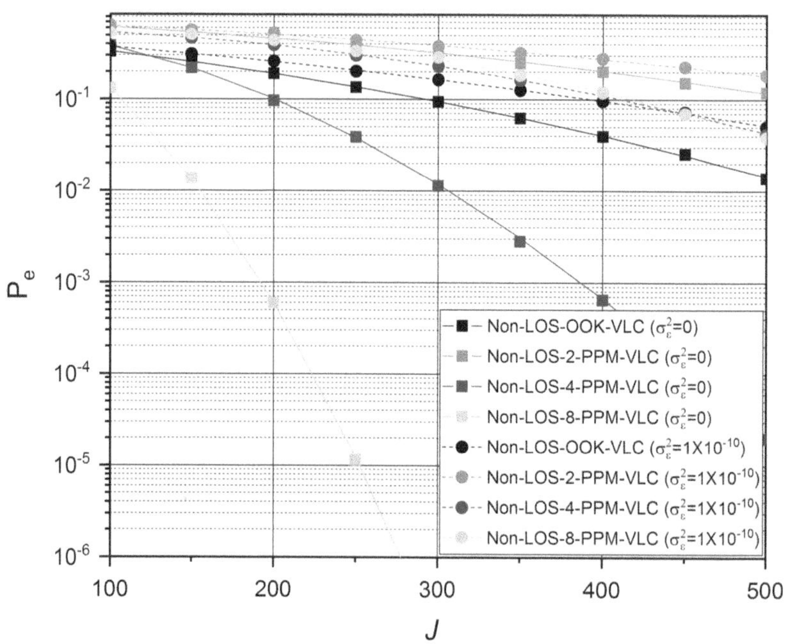

FIGURE 8.9 BER vs. J curves of VLC system using OOK/L-PPM modulation over Non-LOS channel.

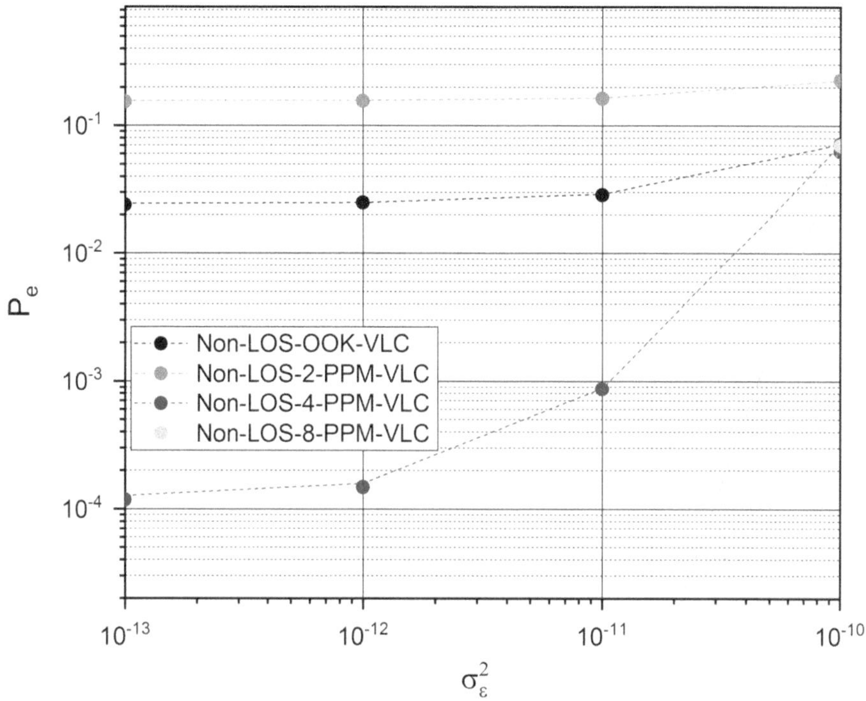

FIGURE 8.10 BER vs.σ_ε^2 curves of VLC system with OOK/L-PPM modulation.

increases. Further, the VLC system using L-PPM modulation over the Non-LOS channel is more affected by the error in channel-estimation error if the modulation order is high. BER of VLC system with OOK, 4-PPM and 8-PPM modulation becomes equal $\sigma_\varepsilon^2 = 1 \times 10^{-10}$.

For a complete system-level analysis, the BER of the VLC system with OOK modulation over the LOS/Non-LOS channel is plotted in Figure 8.11 at different locations of the user. For analysis, three different user locations, i.e., P_1, P_2 and P_3 are considered as mentioned in Table 8.2. BER of VLC system with OOK modulation over LOS channel for perfect CSI degrades when the receiver moves as $P_1 \rightarrow P_2 \rightarrow P_3$ i.e., central to noncentral to a corner location. On the contrary, the BER of the VLC system with OOK modulation over the Non-LOS channel for perfect CSI improves when the receiver shifts $P_1 \rightarrow P_2 \rightarrow P_3$ because the Non-LOS components are dominant at corner locations. Further, the figure shows an increase in BER of the VLC system using OOK modulation over the LOS/Non-LOS channel with imperfect CSI at the receiver (Figure 8.12).

In Figure 8.11, the BER of the VLC system with 4-PPM modulation over LOS/Non-LOS channel is plotted for three different user locations. BER of VLC system with 4-PPM modulation over LOS channel for perfect CSI degrades when the receiver moves as $P_1 \rightarrow P_2 \rightarrow P_3$ i.e., central to noncentral to a corner location. On the contrary, the BER of the VLC system with 4-PPM modulation over the Non-LOS channel for perfect CSI improves as the receiver shifts from $P_1 \rightarrow P_2 \rightarrow P_3$ because the Non-LOS components

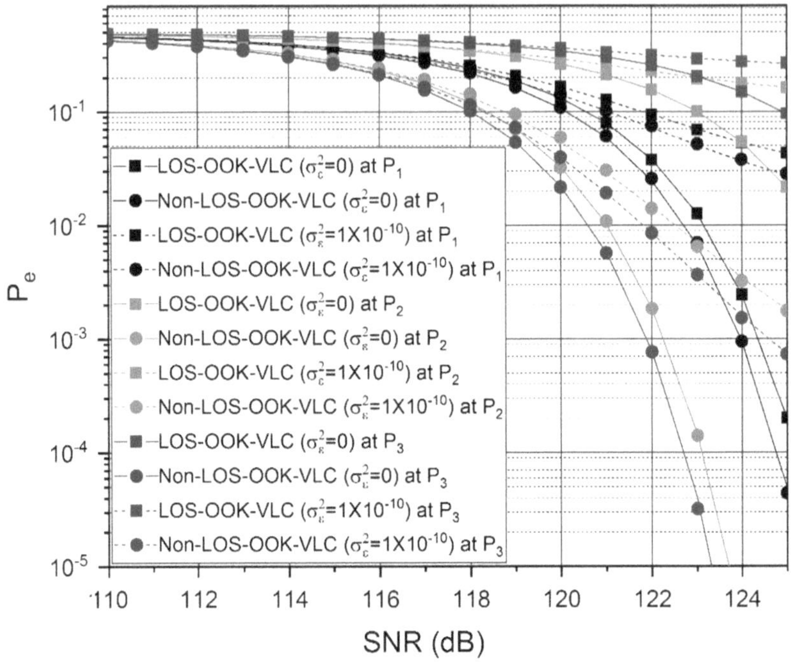

FIGURE 8.11 BER of VLC system with OOK modulation over LOS/Non-LOS channel for different user locations.

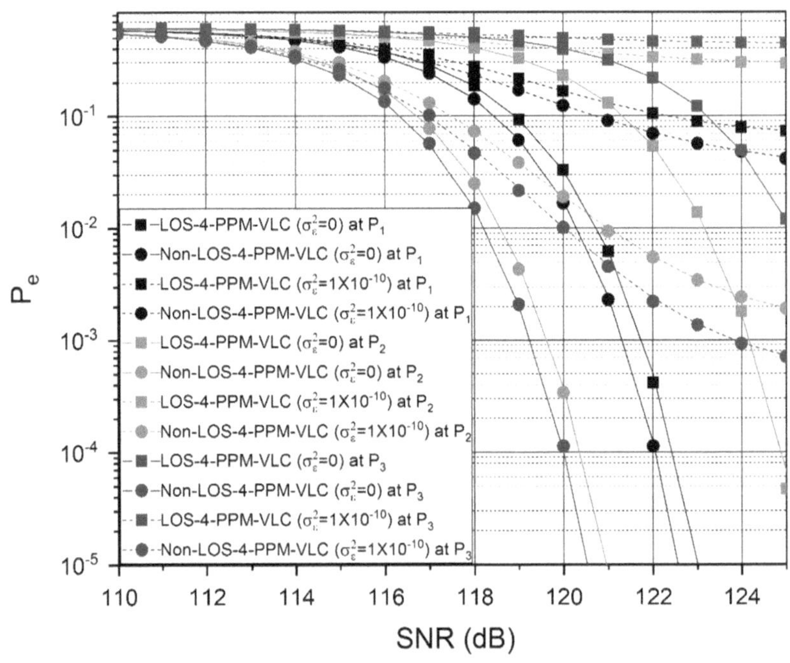

FIGURE 8.12 BER of VLC system with 4-PPM modulation over LOS/Non-LOS channel for different user locations.

are dominant at corner locations. The degradation in the BER of the VLC system with 4-PPM modulation over LOS/Non-LOS channel with imperfect CSI is also observed.

8.5 CONCLUSION

This chapter analyzes the BER of the OOK/L-PPM modulated VLC system over the LOS/Non-LOS channel with imperfect CSI at the receiver. The VLC system using L-PPM modulation with a higher order of modulation, i.e., ($L > 2$) outperforms the VLC system using OOK modulation in error probability over both LOS/Non-LOS channels. Considering imperfect CSI, the BERof the VLC system using L-PPM modulation over the LOS channel increases as the modulation order increases, which is typically observed at the higher SNR values. The error probability of the VLC system using OOK/L-PPM modulation over LOS/Non-LOS channel increases as channel-estimation error variance increases. Further, the VLC system using OOK/L-PPM modulation over the Non-LOS channel can outperform the system over the LOS channel. If the total number of Non-LOS components accumulated at the PD is sufficiently large, i.e., $J = 450$ BER of VLC system using OOK /L-PPM modulation over LOS channel with perfect CSI degrades as the receiver's location is shifted away from the center. On the contrary, the BER of the VLC system using OOK/L-PPM modulation over the Non-LOS channel with perfect CSI improves as the location of the receiver is shifted away from the center.

REFERENCES

1. Z. Ghassemlooy, W. Popoola, and S. Rajbhandari, *Optical Wireless Communications: System and Channel Modeling with MATLAB*, 1st ed., CRC Press, 2013.
2. P. H. Pathak, X. Feng, P. Hu, and P. Mohapatra, "Visible light communication, networking, and sensing: a survey, potential and challenges," *IEEE Commun. Surv. Tutorials*, vol. 17, pp. 172047–172077 Fourth quarter, 2015.
3. H. Le Minh, Z. Ghassemlooy, D. O'Brien, and G. Faulkner, "Indoor gigabit optical wireless communications: Challenges and possibilities," in 12th International Conference on Transparent Optical Networks (ICTON), Munich, pp. 1–6, 2010.
4. T. Komine and M. Nakagawa, "Fundamental analysis for a visible light communication system using LED lights," *IEEE Trans. Consum. Electron.*, vol. 50, no. 1, pp. 100–107, February 2004.
5. V. Dixit and A. Kumar, "Performance analysis of indoor visible light communication system with angle diversity transmitter," in 2020 IEEE 4th Conference on Information & Communication Technology (CICT), Chennai, India, pp. 1–5, 2020.
6. F. Miramirkhani and M. Uysal, "Channel modeling and characterization for visible light communications," *IEEE Photon. J.*, vol. 7, pp. 1–16, 2015.
7. C. Chen, D. A. Basnayaka, X. Wu, and H. Haas, "Efficient analytical calculation of non-line-of-sight channel impulse response in visible light communications," *IEEE J. Lightw. Technol.*, vol. 36, pp. 1666–1682, 2018.
8. V. Dixit and A. Kumar, "Performance analysis of non-line of sight visible light communication systems," *Optics Communication*, vol. 459, 125008, 2020.
9. V. Dixit and A. Kumar, "Performance analysis of L-PPM modulated NLOS-VLC system with perfect and imperfect CSI," *Journal of Optic*, vol. 23, no. 1, 015702, 2020.
10. G. A. Mahdiraji and E. Zahedi, "Comparison of selected digital modulation schemes (OOK, PPM and DPIM) for wireless optical communications," in 4th Student Conf. on Research and Development (SCOReD 2006) (Shah Alam, Selangor, Malaysia), 2006.

11. M. Afghani, H. Haas, H. Elgala, and D. Knipp, "Visible light communication using OFDM," in Proc. 2nd Int. Conf. TRIDENTCOM, pp. 129–134, 2006.
12. S. Rajagopal, R. D. Roberts, and S. K. Lim, "IEEE 802.15.7 visible light communication: modulation schemes and dimming support," *IEEE Commun. Mag.*, vol. 50, pp. 72–82, 2012.
13. M. D. Audeh and J. M. Kahn, "Performance evaluation of L-pulse position modulation on non-directed indoor infrared channels," *Proc. ICC-94 (LA, USA)*, vol. 2, pp. 660–664, 1994.
14. D. Shiu and J. M. Kahn, "Differential pulse position modulation for power-efficient optical communication," *IEEE Trans. Commun.*, vol. 47, pp. 1201–1210, 1999.
15. N. M. Aldibbiat, Z. Ghassemlooy, and R. McLaughlin, "Error performance of dual header pulse interval modulation (DH-PIM) in optical wireless communications," *IEE Proc. Opto. Electron*, vol. 148, pp. 91–96, 2001.
16. V. Dixit and A. Kumar, "Performance analysis of angular diversity receiver-based MIMO–VLC system for imperfect CSI," *Journal of Optic*, vol. 23, no. 8, pp. 085701, 2021.
17. S. Han, S. Ahn, E. Oh, and D. Hong, "Effect of a channel-estimation error on BER performance in cooperative transmission," *IEEE Trans. Veh. Technol.*, vol. 58, pp. 2083–2088, 2009.
18. V. Dixit and A. Kumar, "An exact BER analysis of NOMA-VLC system with imperfect SIC and CSI," *AEU – International Journal of Electronics and Communications*, vol. 138, pp. 153864, 2021. 10.1016/j.aeue.2021.153864
19. H. A. Fayed and A. F. Atiya, "An evaluation of the integral of the product of the error function and the average probability density with application to the bivariate normal integral,"*Math. Comput.*, vol. 83, pp. 235–250, 2014.
20. J. R. Barry, E. A. Lee, and D. G. Messerschmitt, *Digital Communication*, vol. 1, 3rd ed., Springer science + Busines Media, LLC, 2012.
21. G. P. Agrawal, *Fiber-Optic Communications Systems*, Third ed., John Wiley & Sons, Inc, 2002.

9 Narrowband Internet of Things (NB-IoT) – Analysis of Deployment Mode and Radio Coverage

Rasveen and Shilpy Agrawal
Department of Electronics and Communication Engineering,
G. D. Goenka University Gurugram, Haryana, India

Khvati Chopra
Department of Electronics and Communication Engineering,
G. D. Goenka University Gurugram, Haryana, India

CONTENTS

9.1 INTRODUCTION

Narrowband Internet of Things (NB-IoT) has emerged as a new low-power wide-area (LPWA) technology to cover the IoT market. It is a novel option for the cellular Internet of Things introduced by 3GPP in Release 13 and continually upgraded in Release 14 and Release 15. The mobile industry is now inclined to NB-IoT. Members of mobile IoT will prefer NB-IoT networks to share the data and connectivity and choose deployment mode to suit their needs. To take this technology to the next level 5G system, it is essential to amend the design structure and upgrade to fulfill the new IoT applications requirements.

The steps toward mMTC, 3GPP introduced three technologies: extended-coverage GSM IoT (EC-GSM-IoT), long-term evolution for machine type (LTE-M), and the most popular are NB IoT [1]. To handle a massive number of IoT devices over comprehensive coverage, NB-IoT among the three is the most

DOI: 10.1201/9781003230526-9

promising choice [2]. In recent years, NB-IoT has attracted the most attention of all the operators and researchers [3,4]. As the name suggests, NB-IoT uses a very narrow bandwidth and only 180 kHz of frequency to transmitting the data. NB-IoT proposed three deployment options (stand-alone, in-band, guard-band) to operate in this narrow frequency. The deployment option is problematic because it impacts the quality of service (QoS), the total cost of ownership (TCO), and network dimensioning. Further, one of the deployment modes is selected, and the operator comes up with NB-IoT services.

Stand alone is the simplest option and does not need a deep analysis of coverage optimization, but it increases the operational cost. On the other hand, when the operator selects the guard-band option, it leads to an excellent trade-off between QoS and costs. The third option is the in-band option, which coexists with the existing LTE structure and uses the specific primary-resource block (PRB). Moreover, using the full deployment option is not the best optimal resource block usage. On the other hand, if the operator chooses the primary-resource block (PRB), it is best for the infrequent NB-IoT traffic. Moreover, in areas where the coverage is meager (deep basement or rural), the operator must install costly eNBs and verify that the coverage enhancement (CE) technique can be reused.

The essential requirement of NB-IoT devices is to get uninterruptable connectivity in hard-to-reach areas. First, to achieve this, narrow the bandwidth (180 kHz) compared to the existing LTE; this minimizes the transmitted power, reducing the transmitted data rate. Secondly, a repetition technique is utilized to maximize the prospect of data reception. Here for downlink (DL), we repeat the data up to 2048 times, and for uplink (UL), we can repeat it up to 128 times. We use the repetition technique to enhance the coverage up to 20 dB compared to other legacy technology. The repetition number depends on the coverage extension required, channel condition, and operator configuration.

The coverage condition of the device is estimated at the time of the random-access (RA) procedure. The comparison between the reference signal received power (RSRP) and threshold (defined by the operator) helps in estimating the coverage level [5]. The coverage-enhancement level is divided into CE level 0 (like LTE), CE level 1, CE level 2 (applied to remote and harsh areas where more repetition is required for coverage extension).

NB-IoT deployment is now becoming the first choice of researchers and mobile companies to work on it. Many field trials and measurement campaigns are set up to analyze the impact of selecting different deployments for different IoT services. Data-driven analysis is complex for researchers because of deployment selection, challenges in the path of deployment, and new research guidelines. Moreover, the measurement campaigns are costly, scarcely available, and very time-consuming to collect data. That is why simulation-based techniques are used for the analysis of deployment selection.

Considering the above motivations, the article is arranged in the following way: Section 9.2 gives the technological description of the deployment model. Section 9.2 compares the three-deployment mode of NB-IoT. Section 9.3 shows the impact of deployment mode on radio coverage.

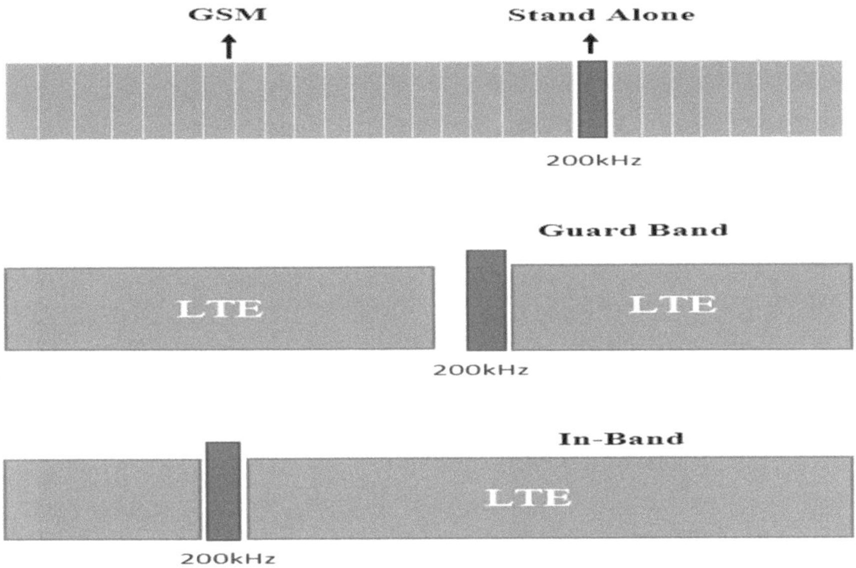

FIGURE 9.1 NB-IoT deployment options.

9.2 DEPLOYMENT MODEL AND PERFORMANCE ANALYSIS

NB-IoT is the upcoming LPWA cellular technology that partially uses the specifications and design structure of existing LTE. NB-IoT has three operational modes, as shown in Figure 9.1:

1. **Standalone:** NB-IoT is separately deployed with 200 kHz bandwidth (similarly used for the GSM carrier). This deployment provides the best performance [6].
2. **Guard band:** The unused space named as guard band present in the system does not carry any information. This space (LTE guard band) is utilized to deploy the NB-IoT carrier, and the center frequency is at 7.5 kHz offset (from 100 kHz raster). It also stops the roll-off of output power (from base stations into adjacent LTE channels).
3. **In-band:** In this mode, the NB-IoT is deployed in the one primary resource block (PRB) space of LTE. However, do not utilize the reserved region (time and frequency resources) of LTE.3GPP; the bandwidth for the NB-IoT is defined, and it is mentioned in Table 9.1 below:

The operator can select the deployment as per the requirement and application usage. After choosing one of the deployment modes, the software upgrade is done and provides NB-IoT services that improve the capabilities of eNodeB and the cell. Table 9.2 below gives the performance results of three of the deployment modes.

Many specifications of NB-IoT are like existing LTE, and others are different to support new IoT applications. NB-IoT supports the following numerologies based on performance evaluation.

TABLE 9.1
Bandwidth defined for specific deployment mode

Required Bandwidth	In-Band	Guard Band	Standalone
Total Channel BW	Per LTE Channel	Per LTE Channel	200
Tx BW (resource block of LTE)	1	1	1
Tx BW (15 kHz)	12	12	12
Tx BW (N-tone 3.75 kHz)	48	48	48

TABLE 9.2
Performance results of deployment modes

	Stand-Alone	In-Band	Guard-Band
Cost of spectrum	High	High	None
Repetition required	Moderate	Excessive	Excessive
Frequency planning	Lots of efforts required to reframe GSM/ UMTS	Upgrade every eNB without planning	Upgrade every eNB, No need for planning
Baseband unit	Upgrade	Upgrade	Upgrade
Regulatory approval	Yes	No	Yes (mostly)
Symmetrical MCL (w/o) Repetitions	151.2–161	138.2–148	138.2–148
Antenna	Upgrade	Reuse	Reuse

- Downlink transmission: The orthogonal frequency-division multiplexing access (OFDMA) modulation technique is utilized with a subcarrier spacing of 15 kHz. The data rate speed is 200 kHz.
- Uplink transmission: In the uplink, both the single-tone and multitone-tone transmission are possible with a data rate of 20 kbps—single-tone support both the frequencies 3.75 kHz and 15 kHz like LTE. Moreover, multi-tone support 15 kHz tone spacing and use an SC-FDMA scheme. When uplink single-tone transmission uses a 3.75 kHz or 15 kHz tone spacing for in-band deployment, interference arises between two networks (NB-IoT and LTE). To avoid this interference, the user can be scheduled with similar SNR and orthogonality between LTE and NB-IoT, respectively.
- UE cost can reduce by using hybrid automatic repeat request (HARQ) because it reduces the peak data rate, 62 kbps, and 26 kbps in uplink and downlink transmission, respectively.
- NB-IoT channels are like LTE. Few changes are required so that they will adjust the 180 kHz frequency. Table 9.3 shows the different channels in uplink and downlink transmission.

TABLE 9.3
Shows the channels utilized in downlink and uplink transmission

Transmission	Channel/Signal	Usage
Uplink	Narrowband Physical Uplink Shared Channel (NPUSCH)	Data transmission, acknowledgment
	Narrowband Physical Random-Access Channel (NPRACH)	Random access
	A demodulation reference signal (DMRS)	Used for the channel estimation
Downlink	Synchronization signal • Narrowband primary synchronization signal (NPSS) • Narrowband secondary synchronization signal (NSSS)	Synchronization of time and frequency Cell ID detection (NSSS)
	Narrowband Physical Broadcast Channel (NPBCH)	Provide Master information for system access
	Narrowband Physical Downlink Shared Channel (NPDSCH)	Common data and downlinks dedicated
	Narrowband Physical Downlink Control Channel (NPDCCH)	Provide Uplink and Downlinkscheduling Information

- The downlink transmission shown in Figure 9.2 comprises channels (NPSS, NSSS, NPBCH, NPDSCH, NPDCCH). NB-IoT comprises two synchronization signals. One is NPSS which is utilized by the user equipment to achieve synchronization in both time and frequency [7–9]. Moreover, NPSS designed so that it can be detectable even when there is a significant frequency offset. On the other hand, the second synchronization signal that is NSSS, is used [7]. NPBCH is used for the delivery of master information block (MIS). For the transmission of downlink, data NPDSCH is utilized.

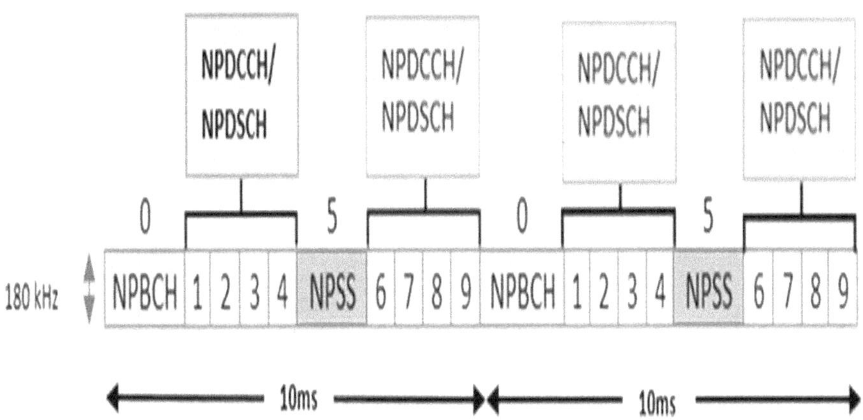

FIGURE 9.2 Downlink transmission.

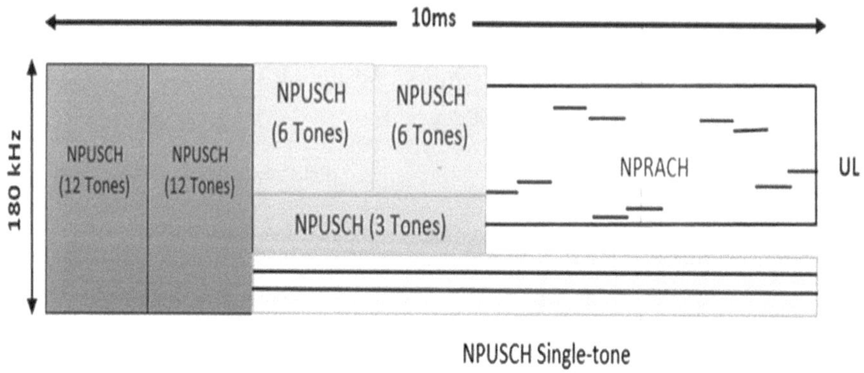

FIGURE 9.3 Uplink transmission.

- The uplink transmission is shown in Figure 9.3. There are two channels in UL; one is NPRACH, which is used to transmit the NB-IoT preamble [10], and it is the foremost step of the random-access procedure. For this, single-tone transmission with frequency hopping is utilized.

The second one is NPUSCH, which can work with both single-tone transmission (can use 3.75 kHz or 15 kHz) and multi-tone (can only use 15 kHz (3, 6, 12 tones). It is used to transmit the control signal and user data and carry HARQ Ack for NPDSCH. When moving to the higher layer, the LTE network is supported by simplified functions, including excellent mode mobility, handle latency, access control, power-saving mode, and optimization of small data transmission.

9.2.1 Deployment of NB-IoT in LTE Guard Band

This deployment option utilized the existing LTE guard band spectrum for the transmission of data. NB-IoT uses 200 kHz of bandwidth, and in many places, it is written as 180 kHz rest of the 20 kHz is used as the guard band of NB-IoT. The guard band of NB-IoT within the LTE bandwidth is less than 5 MHz, and it is incredibly challenging without considering the additional offset. Hence, it would require a long-filter impulse response. However, long-filter impulse response increases the inter-symbol response because of the destruction of cyclic prefix bits.

In different releases, the 3GPP introduced a new requirement for the NB-IoT guard band. In Rel-13 NB-IoT guard band works with a 5 MHz LTE carrier and needs a power-boosting level. Moreover, how much power boosting is required is upgraded in Rel-14 [11]. NB-IoT uses the same 15 kHz of frequency as the LTE system to avoid interference. Here, f_d kHz is considered the NB-IoT center frequency far away from the center frequency of the LTE system. Table 9.4 shows the different sets of f_d system bandwidth.

This frequency is set of LTE and NB-IoT shows that we can deploy NB-IoT quickly in the existing LTE.

TABLE 9.4
Location of NB-IoT carrier within existing LTE carrier [12]

LTE Bandwidth	NB-IoT Carrier Location in Existing LTE Carrier
5 MHz	±2392.5
10 MHz	±4597.5, ±4702.5, ±4807.5, ±4897.5
15 MHz	±6892.5, ±6997.5, ±7102.5, ±7207.5, ±7297.5, ±7402.5
20 MHz	±9097.5, ±9202.5, ±9307.5, ±9397.5, ±9502.5, ±9607.5, ±9697.5, ±9802.5, ±9907.5

9.2.2 Deployment of NB-IoT in LTE PRB (In-Band)

For in-band deployment, the NB-IoT utilizes 180 kHz of bandwidth or one PRB of LTE carrier. The main attractive feature of using in-band deployment is that it does not require additional hardware changes in the radio network because it uses the LTE resources, making it cost-effective for the operators. There is a technical glitch when two of the networks (LTE and NB-IoT) coexist (use the one resource block (PRB) LTE carrier), and this causes inter- PRB interference. To avoid this PRB interference and differentiate between two PRBs (LTE and NB-IoT), 3GPP introduced a feature known as power boosting, shown in Figure 9.4.

In this feature, the LTE PRB should be kept +6 dB higher than in-band. This feature is similarly used for the guard band to avoid inter-PRB interference. 3GPP

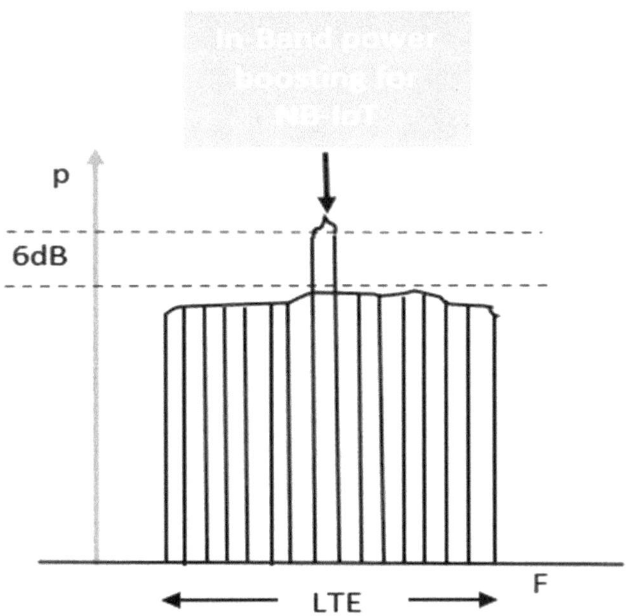

FIGURE 9.4 In-band power-boosting.

FIGURE 9.5 Deployment of NB-IoT in GSM.

also identifies other NB-IoT signal features like error-vector magnitude (EVM) values for control channels and synchronization (measure error performance and modulation quality) and supports the modulation type.

9.2.3 Deployment of NB-IoT as a Stand-Alone

Standalone is one of the promising deployment options. It operates on 200 kHz of frequency on which GSM operates. That is why it is efficiently accelerating in the market. For stand-alone, there are two options for deployment.

1. **Idle spectrum resource:** NB-IoT can be deployed on the idle spectrum of GSM, UMTS, and LTE, as shown in Figures 9.5 and 9.6. A guard-band is needed to avoid interference with the existing LTE networks [13]. In this mode, operators sometimes use their own spectrum, one that does not fulfill some radio-access technique (RAT).
2. **Reframing:** In reframing, the enable part of the RAT is utilized for the NB-IoT deployment. We use the GSM spectrum for the NB-IoT deployment, and to avoid interference between the two networks, we use the guard band. For this purpose, the GSM network is redesigned.

It is required to reframe the entire GSM frequency and to minimize the interference with NB-IoT, inserted a guard band of 100 kHz, as shown in Figure 9.7.

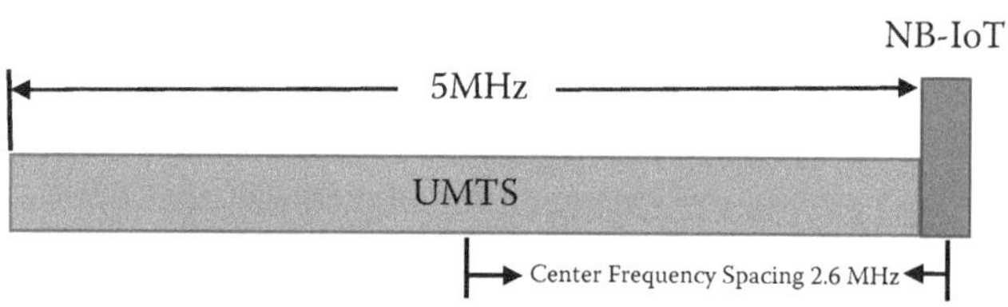

FIGURE 9.6 Deployment of NB- IoT in UMTS.

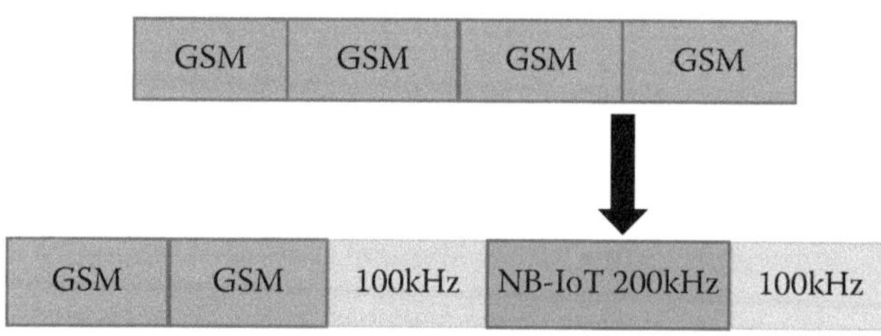

FIGURE 9.7 Reframing deployment for NB-IoT.

9.3 IMPACT OF DEPLOYMENT MODES ON COVERAGE

3GPP introduced a new idea for coverage enhancement, and the mechanism is based on coverage classes (CCs). The receiving sensitivity is improved by using sub-GHz deployment and narrow-band modulation [14,15] to extend the radio coverage. To evaluate the coverage, performance is still based on fundamental analysis method that is maximum coupling loss (MCL) [16,17]. To give a new thought to the coverage enhancement, the researcher focus on adaptive modulation and encoding method [18,19]. This method upgrades the CCs by using dynamic statical multiplexing, which is divided into three steps:

1. Determining the optimum discrimination threshold of CCs for this used RSSI and SINR.
2. CCs adjustment on the bases of HARQ and ACK/NAK.
3. CE set according to retransmission times, transmission power, current state and long-term statistical law, and performance analysis.

In NB-IoT, eNodeB handles and controls scheduling in both uplink and downlink transmission, which is similar to LTE, and for this, there is a requirement of a balanced link budget. Figure 9.8(a) and (b) show the data transfer in uplink and downlink transmission.

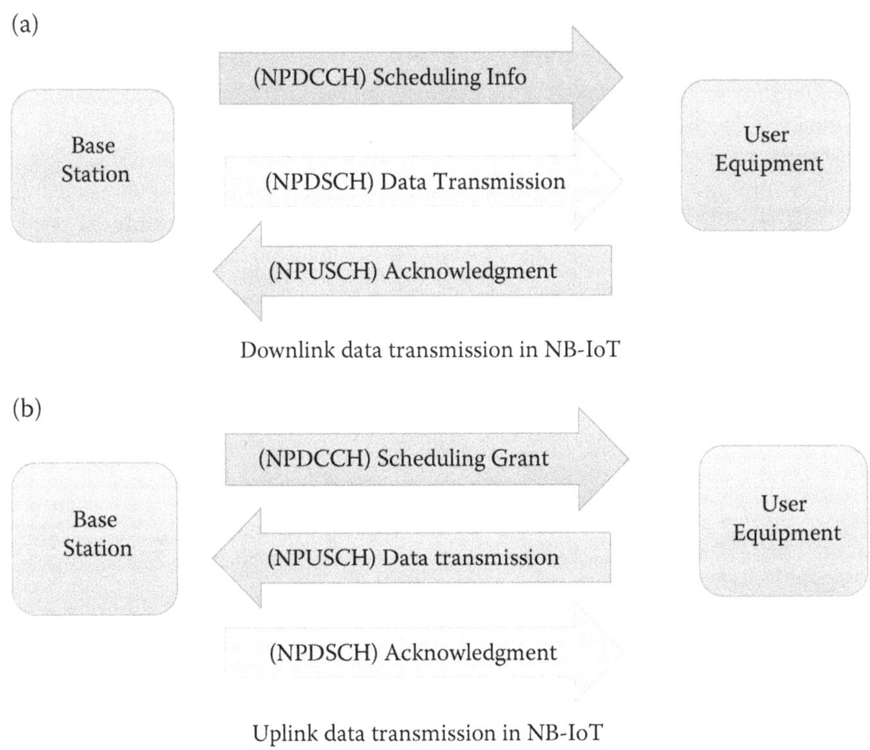

FIGURE 9.8 (a) Downlink data transfer in NB-IoT (b) Uplink data transmission in NB-IoT.

NB-IoT offers a 20 dB coverage extension, but if we talk about the MCL for in-band and guard-band deployment, the link budget is not symmetrical; it is below 164 dB. The asymmetrical link budget in the in-band and guard-band is due to RF module power-sharing between LTE and NB-IoT networks. These RF modules (700/800 MHz) have 2 × 30 W O/P power. On the other hand, it can't be feasible to save 20 W power for NB-IoT. This will have an extreme impact on the LTE coverage. Table 9.5 shows the uplink and downlink transmission link budget in different deployment modes.

To improve the link budget, NB-IoT has to adopt a repetition technique in which the data signal is repeated up to 2048 times in downlink transmission and a maximum of up to 128 times in uplink transmission. NB-IoT supports multiple-coverage enhancement levels (CE level-0, CE level-1, CE level-2). The number of repetitions in each level is different because it all depends on how far the user equipment is placed. Due to the low S/N ratio, more repetitions are required if the device is placed far away from the base station due to the low S/N ratio. Moreover, if the device is placed close to BS, then repetition is also reduced. Each CE level has a particular range of received signal received power (RSRP) associated with it. The device placed in CE level-0 has the higher RSRP, and the device placed in CE level-1 has a lower RSRP. Also, NPRACH is allocated to different CE levels, and NPRACH of lower CE levels has a short duration and vice versa. PRACH for different CE level have different sub-carrier (12, 24, 36, 48) [20]. Figure 9.9 shows the different coverage levels for NB-IoT.

At the initial CE level, the RA transmits a preamble in an NPRACH. To serve three NPRACH, bandwidth is 180 kHz, divided into 48 sub-carrier, and tone-spacing is 3.75 kHz. For all the devices, the preamble sequence remains the same. It is essential to choose the correct repetition number for the particular CE level, and hence, at the receiver, a good quality signal is received with less block error rate. The first repetition is randomly chosen. If the two devices are transmitted on the same NPRACH, the collision takes place [21]. To avoid this, a new sub-carrier is

TABLE 9.5
Receiver sensitivity and MCL in different deployment modes

	Uplink	Downlink Stand-Alone		Downlink In-Band/Guard-Band	
		Negative	Positive	Negative	Positive
Tx Power	23	~43	~43	~30	~30
Receiver Sensitivity (*Device Rx sensitivity may vary, 3GPP reference is –108.2 dBm) (** sensitivity of receiver eNodeB is dependent on the vendor)	–138**	–108.2*	–118*	–108*	–118*
Maximum Coupling Loss (MCL)	161	151.2	161	138.2	148

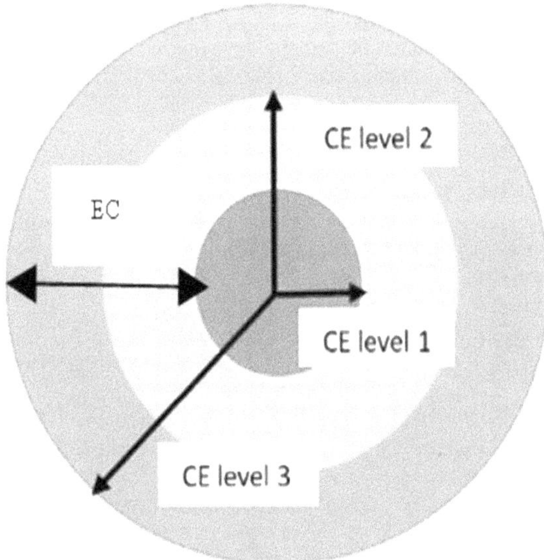

FIGURE 9.9 NB-IoT coverage level.

selected for the next NPRACH after a backoff. If the device fails to achieve the maximum number of attempts, the whole RA procedure is repeated.

9.4 CONCLUSION

This paper discussed the deployment options of the newly introduced NB-IoT LPWA technology. We investigated the problems researchers face while selecting the deployment modes. We differentiated the three deployment options and their technical specifications, and discovered the problems and potential solutions attached with the in-band and guard-band deployment while working with the existing LTE. Finally, it examined the impact of deployment on coverage enhancement.

REFERENCES

1. O. Liberg, et al., *"Cellular Internet of Things: From Massive Deploymentsto Critical 5G Applications, 2nd Edition,"* Academic Press, 2020.
2. U. Raza, P. Kulkarni, and M. Sooriyabandara, "Low power wide area networks: An overview," *IEEE Commun. Surveys Tuts.*, vol. 19, no. 2, pp. 855–873, 2017.
3. W. Yang, et al., "Narrowband wireless access for low-power massive internet of things: A bandwidth perspective," *IEEE Wireless Commun.*, vol. 24, no. 3, pp. 138–145, 2017.
4. M. Chen, et al., "Narrowband internet of things," *IEEE Access*, vol. 5, pp. 20557–20577, 2017.
5. 3GPP, "LTE; Evolved Universal Terrestrial Radio Access (EUTRA); Requirements for support of radio resource management," https://www.etsi.org/deliver/etsi_ts/136100_136199/136133/13.03.00_60/ts_136133v130300p.pdf, Accessed: February 2020.

6. Mozny, R., Masek, P., Stusek, M., Zeman, K., Ometov, A., and Hosek, J. (2019, July). "On the performance of narrow-band internet of things (NB-IoT) for delay-tolerant services," In International Conference on Telecommunications and Signal Processing (TSP) (pp. 637–642). IEEE.
7. Y. Li, S. Chen, W. Ye, and F. Lin, "A joint low-power cell search and frequency tracking scheme in NB-IoT systems for green Internet of Things," *Sensors (Switzerland)*, vol. 18, no. 10, pp. 1–22, 2018, 10.3390/s18103274.
8. H. Kroll, M. Korb, B. Weber, S. Willi, and Q. Huang, "Maximum-likelihood detection for energy-efficient timing acquisition in NB-IoT," In Proceedings of the 2017 IEEE Wireless Communications and Networking Conference Workshops (WCNCW), San Francisco, CA, USA, 19–22 March 2017, pp. 1–5.
9. Intel Corporation. Synchronization and cell search in NB-IoT: Performance evaluations, 3GPP TSG RAN WG1, R1-161898, Intel Corporation: Santa Clara, CA, USA, 2016.
10. M. Kanj, V. Savaux, and M. Le Guen, "A Tutorial on NB-IoT Physical Layer Design," *IEEE Commun. Surv. Tutorials*, vol. 22, no. 4, pp. 2408–2446, 2020, 10.1109/COMST.2020.3022751
11. 3GPP TS 36.104, "Evolved Universal Terrestrial Radio Access (EUTRA); Base Station (BS) radio transmission and reception" V13.4.0, June 2016.
12. R. Ratasuk, J. Tan, N. Mangalvedhe, M. H. Ng, and A. Ghosh, "Analysis of NB-IoTDeployment in LTE Guard-Band," In IEEE Veh. Technol. Conf., vol. 2017-June, 2017, 10.1109/VTCSpring.2017.8108184.
13. R1-157248, NB IoT – Capacity evaluation, NokiaNetworks, RAN1#83, Anaheim, USA, 2015.
14. Cellular system support for ultra-low complexity and low throughput cellular internet of things, 3GPP TR 45.820, 2015.
15. E-UTRA Physical channels and modulation – Chapter 10 Narrowband IoT, 3GPP TS 36.211, 2016.
16. G. Naddafzadeh-Shirazi, L. Lampe, G. Vos, and S. Bennett, "Coverage enhancement techniques for machine-to-machine communications over LTE," *Communications Magazine IEEE*, vol. 53, no. 7, pp. 192–200, 2015.
17. L. bin, "Discussion on the internet of things coverage enhancement technology of nb-iot," *Mobile Communication*, vol. 40, no. 19, pp. 55–59, 2016.
18. A. Goldsmith, "Wireless communications," vol. 4, no. 5, pp. 25–55, 2005.
19. J. Y. fan, Z. xiaoping, J. Xin, Z. jihua, R. dingliang, and G. yiwen, "An adaptive modulation algorithm for non-data aided error vector magnitude in fast time-varying channels," *Journal of Communications*, vol. 38, no. 3, pp. 73–82, 2017.
20. 3GPP TS 36.321, "Medium access control (MAC) protocol specification," V13.2.0, June 2016.
21. Harwahyu, R., Cheng, R. G., Liu, D. H., and Sari, R. F. "Fair Configuration Scheme for Random Access in NB-IoT with Multiple Coverage Enhancement Levels," IEEE, 2019.

10 Design of Printed Dipole Antenna Array for Directional Sensing Applications

T. Anilkumar and M. Rajan Babu
Department of ECE, Lendi Institute of Engineering and Technology, Jonnada, Vizianagaram District, Andhra Pradesh, India

B. T. P. Madhav
Antennas & Liquid Crystals Research Center, Department of ECE, Koneru Lakshmaiah Education Foundation, Vaddeswaram, Andhra Pradesh, India

M. Venkateswara Rao
Department of ECE, Dhanekula Institute of Engineering & Technology, Vijayawada, Andhra Pradesh, India

CONTENTS

DOI: 10.1201/9781003230526-10

10.1 INTRODUCTION

An array of antennas is one of the emerging domains for enthusiasts in antenna design since they offer advantages compared to the single element antennas, such as enhanced-directive gain, narrow beamwidths [1]. These are further being developed into phased arrays or scanning arrays in which the phase of the individual elements is being varied according to the required beam pattern. The tremendous growth in microstrip antennas has given a good scope for the development of antenna arrays when compared to the traditional antennas like wire, dipole, loop, parabolic, horn, etc., due to their large sizes. Attempts are still being made to develop scanning systems using antennas and their array configuration [2–9]. The feeding method is an essential parameter in designing the arrays. Mainly, the microstrip antenna arrays can be fed through serial feeding [1,8], parallel feeding [1,9], corporate feeding [1,10], etc. This chapter gives insight into the authors' simulation-based experiments to explore the radiation characteristics of the array system concerning various combinations of phase angles during the excitation of antenna elements. The experiments were done on a simple printed dipole antenna and initiated a single element design discussion.

10.2 DESIGN OF SINGLE ELEMENT PRINTED DIPOLE ANTENNA

The printed dipole antenna is designed on flame-retardant grade-4 (FR-4) substrate with 47.4 × 10.5 × 1.6 mm^3 (L_s × W_s × h) dimensions. The antenna's geometry is inspired from [6] and, as shown in Figure 10.1, consists of a pair of dipole arms. Each arm consists of two segments: a straight dipole arm and an L-shaped strip connected with it. The design evolution of this printed dipole antenna is presented in Figure 10.2.

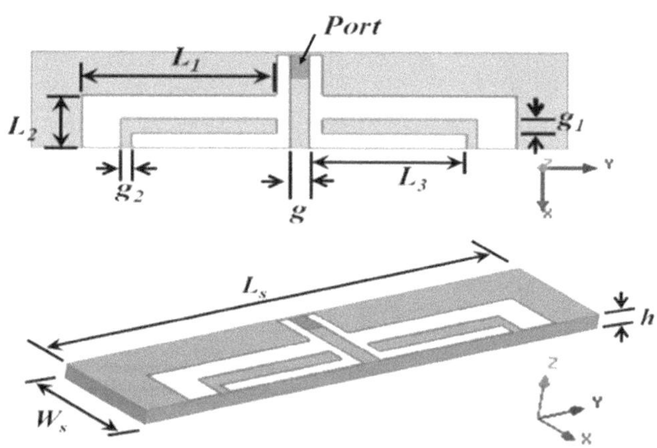

FIGURE 10.1 Geometry of single element printed dipole antenna (Dimensions: L_s = 47.2 mm, W_s = 10.5 mm, h = 1.6 mm, L_1 = 31.44 mm, L_2 = 17.97 mm, L_3 = 0.8 mm, g = 25.12 mm, g_1 = 2.76 mm, g_2 = 13.4 mm).

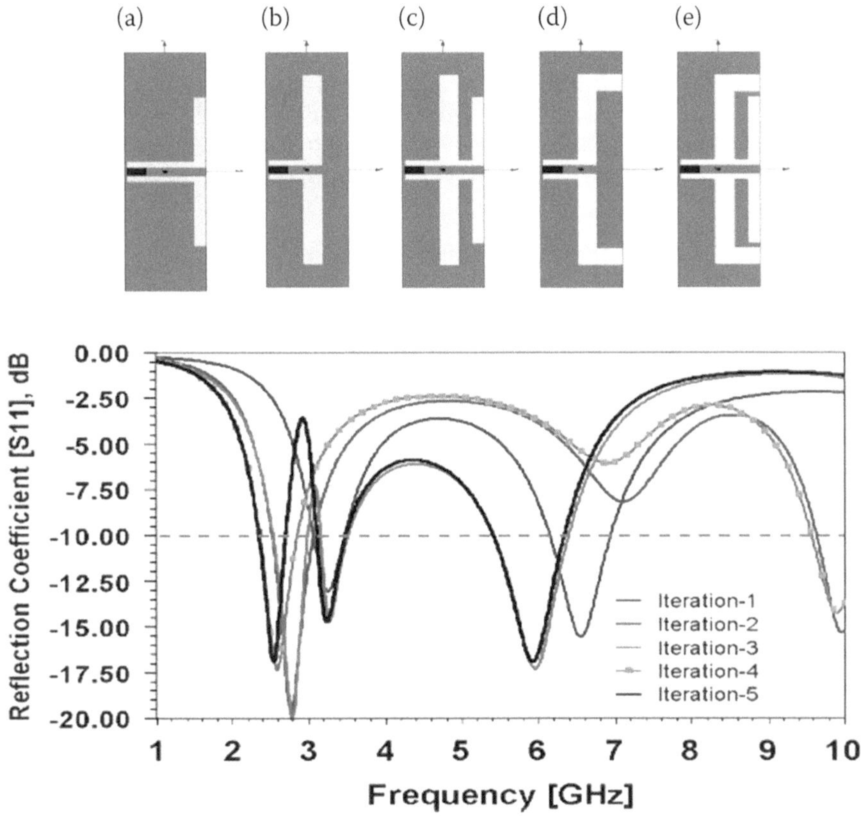

FIGURE 10.2 Iteration-wise evolution of proposed single-element printed dipole antenna (a) Iteration-1, (b) Iteration-2, (c) Iteration-3, (d) Iteration-4, (e) Proposed single-element design.

Initially, the printed dipole antenna is designed with a pair of narrow strips of length L_3. These narrow strips are fed through a pair of straight, orthogonal strips to the narrow strips, as shown in Figure 10.2(a), known as Iteration-1. Figure 10.2(b) makes the Iteration-2 in which a pair of strips are laid back compared to earlier strips in Iteration-1. Here, the length of the strips is increased to the length 'L_1'. In Iteration-3, the combination of Iteration-1 and Iteration-2 is considered, resulting in a structure with an interconnection of two pairs of strips, making a printed dipole, as shown in Figure 10.2(c). In Iteration-4 (as shown in Figure 10.2(d)), the extended version of Iteration-2 is considered and a combination of Iteration-3 and Iteration-4 produces Iteration-5, as shown in Figure 10.2(e).

The operating bands of these antenna iterations were obtained by considering the −10 dB reflection-coefficient criteria. The Iteration-1 operates in dual-band

3.2 GHz and 6.6 GHz. The iteration-2 with increased-length arms operates at 2.8 GHz and 10 GHz. The composite structure of Iteration-1 and two leads Iteration-3 is operated at 2.8 GHz and 5.9 GHz bands. The arms of Iteration-2 were made right-angled, and this causes the antenna to operate at 2.5 GHz and 9.8 GHz, respectively. The final structure, a combination of Iteration-3 and 4, make the antenna operate at 2.5 GHz, 3.3 GHz and 5.9 GHz. Thus, the printed dipole antenna operate in triple-band mode.

10.2.1 Operating Band Characteristics of Single Element Printed Dipole Antenna

The single-element printed dipole antenna operates in three bands observed from Figure 10.3(a). The resonant frequencies are 2.55 GHz, 3.25 GHz, and 5.96 GHz for the 2.35–2.71 GHz, 3.12–3.5 GHz, and 5.43–6.36 GHz. The antenna offers the reflection loss [S_{11} (dB)] of −16.92 dB, −14.58 dB and −17.27 dB, respectively, at the resonant frequencies. The antenna port offers an input impedance of 60.64–j13.5 ohms at 2.55 GHz, 42.41+j14.18 ohms, and 37.25–j4.35 ohms at 5.9 GHz with a total magnitude of 62.12 ohms, 44.72 ohms, and 37.5 ohms, respectively.

10.2.2 The Surface Current Distribution of the Single Element Printed Dipole Antenna

The surface current distribution plots are obtained at resonant frequencies of the antenna through simulations shown in Figure 10.4. At 2.55 GHz, the more significant element has more accumulated with current elements.

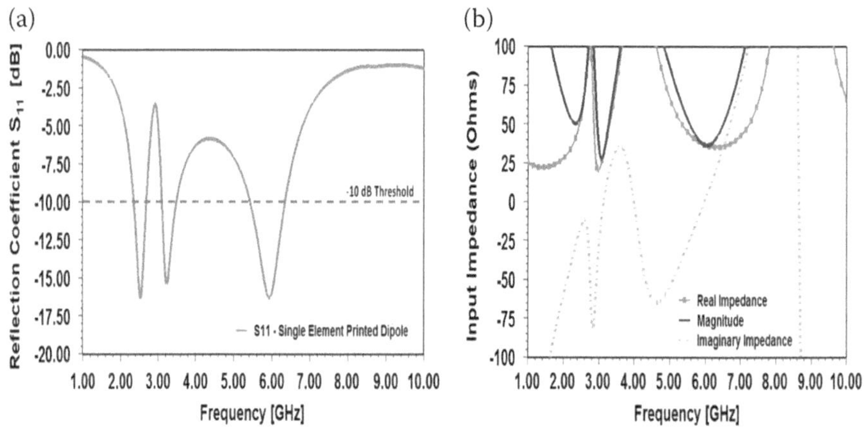

FIGURE 10.3 (a) Reflection coefficient vs frequency characteristics (b) Input impedance vs frequency characteristics of printed dipole antenna.

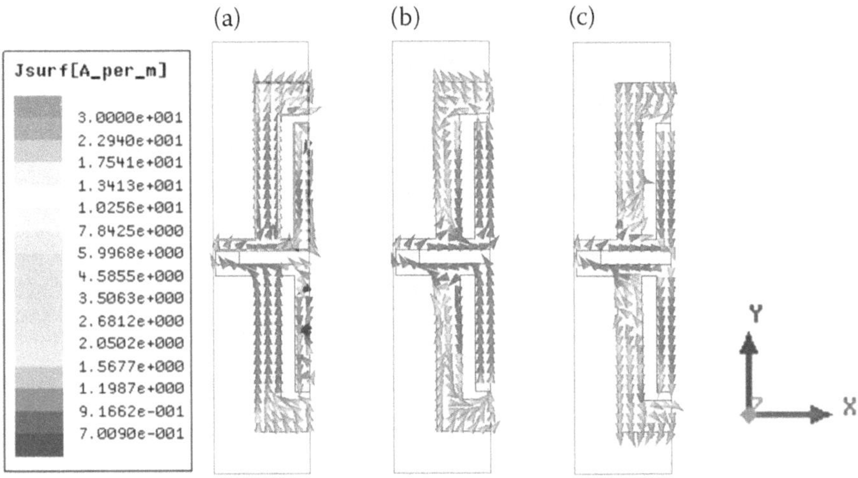

FIGURE 10.4 Surface current distribution at (a) 2.55 GHz, (b) 3.25 GHz and (c) 5.9 GHz.

The current elements are concentrated on the arms of the dipole, which are at the edge of the substrate, and this phenomenon is observed at 3.25 GHz (as in Figure 10.4(a)). Also, some current elements are concentrated on the bigger arms, which can be noticed from Figure 10.4(b). Whereas, at 5.9 GHz, the more petite strips are affected more than the larger arms of the dipole. This shows the effectiveness of the equivalent-resonant electrical lengths on the conducting portion of the printed dipole antenna.

10.2.3 Far-Field Radiation Characteristics of Single Element Printed Dipole Antenna

The far-field radiation characteristics depict the actual radiation from the antenna, which travels farther than $2D^2/\lambda$. The radiation patterns are obtained at 2.55 GHz, 3.25 GHz, and 5.9 GHz bands, respectively, shown in Figure 10.5 and Figure 10.6.

The patterns are omnidirectional in the azimuth plane, and the nulls are formed along the axial length of the dipole. In the third band, the beamwidth became wider in one orientation and narrower in the other orientation.

From Table 10.1, maximum peak gain at 5.96 GHz is observed than at 2.55 GHz and 3.25 GHz resonant frequencies. The radiation efficiency is maximum at 5.36–6.3 GHz operating band of the single-element printed dipole antenna. However, the peak gain in the lower bands has to be increased for the effective utilization of the antenna.

10.3 DESIGN OF LINEAR ARRAY (1×2) OF PRINTED DIPOLE ANTENNA

The obtained radiation characteristics are relatively wide and providing low values of gain. However, for long-distance communication requirements, antennas with lower

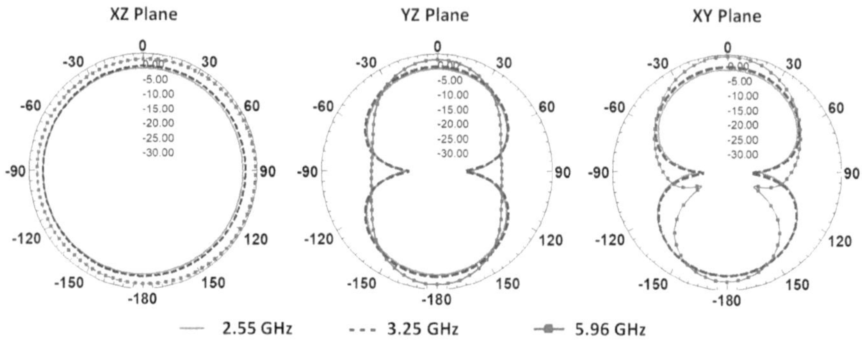

FIGURE 10.5 Radiating field characteristics of the single-element dipole antenna in XZ, YZ, XY planes.

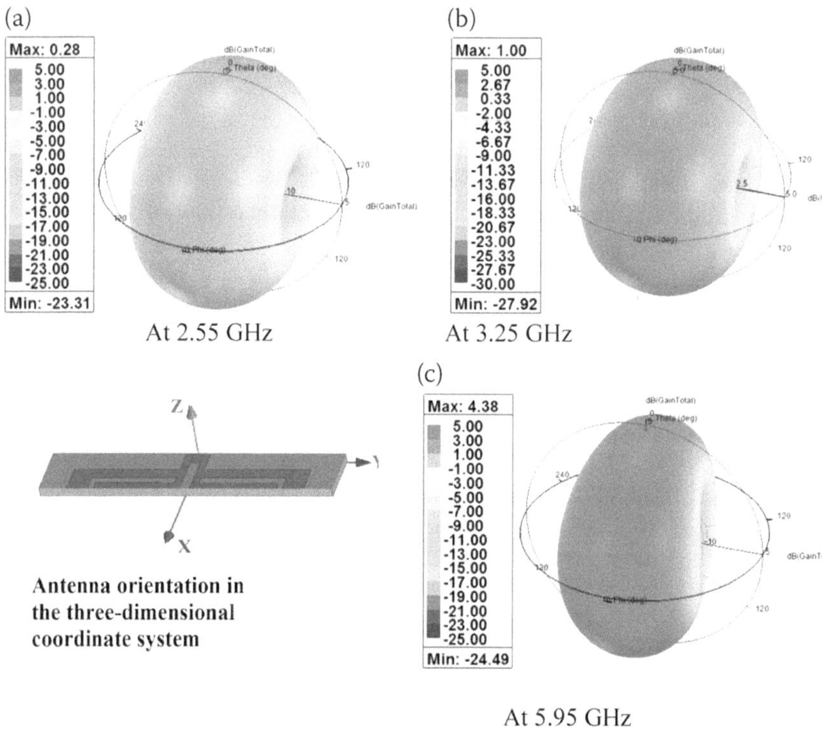

FIGURE 10.6 3D far-field characteristics of single-element printed dipole antenna. (a) Far Field Characterstics ate 2.55 Ghz. (b) Far Field Characterstics ate 3.25 Ghz. (c) Far Field Characterstics ate 5.95 Ghz.

TABLE 10.1
The overall characteristics of single element printed dipole antenna

Antenna Parameters	At 2.55 GHz	At 3.25 GHz	At 5.96 GHz
Max Radiation Intensity	0.0795906(W/sr)	0.0936143(W/sr)	0.215822(W/sr)
Peak Directivity [dB]	0.336	1.3681	4.6102
Peak Gain [dB]	0.0871	0.8744	4.3850
Radiated Power	0.925583(W)	0.858511(W)	0.93819(W)
Accepted Power	0.980313(W)	0.961882(W)	0.988122(W)
Incident Power	1(W)	1(W)	1(W)
Radiation Efficiency (%)	94.41	89.25	94.94
Front to Back Ratio	1.03985	1.16575	1.66899

gain are not preferable. Instead of increasing the electrical size of the antenna, the array configurations can be considered. Hence, the printed dipole antenna designed is converted here as a two-element antenna array. The two elements are separated by an inter-element distance of 45 mm and the consecutive edge gap 'g' = 6.6 mm, as shown in Figure 10.7. The antenna is designed and simulated to obtain the simulated radiation performance.

10.3.1 Operating Band Characteristics of Linear Printed Dipole Array (1×2) Antenna

The operating band characteristics are observed from the reflection coefficient versus frequency characteristics plotted and shown in Figure 10.8. This two-element array antenna operates in 2.55 GHz, 3.25 GHz, and 5.90 GHz, respectively. The operating band is considered based on the -10 dB reflection co-efficient criteria. The characteristics are the same when compared with the single-element design with a fraction of variations.

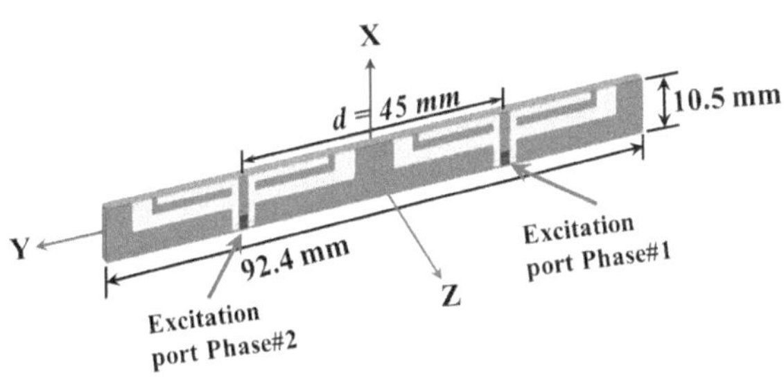

FIGURE 10.7 Geometry of 1×2 array of printed dipole antenna.

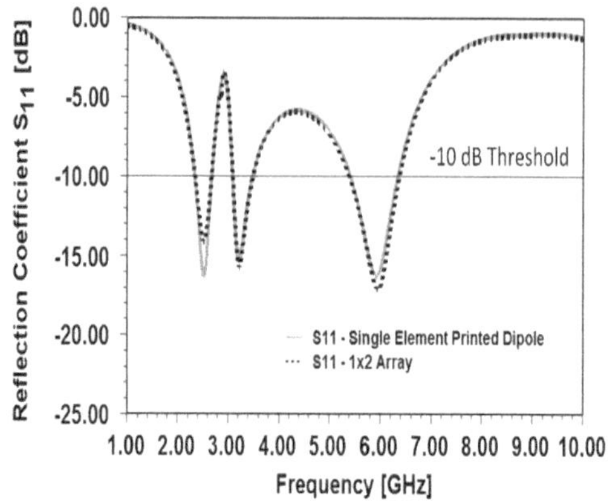

FIGURE 10.8 Operating band characteristics of single and 1×2 element printed dipole antenna.

10.3.2 Far-Field Radiation Characteristics of 1×2 Element Printed Dipole Array Antenna

The 1×2 element-printed dipole array antenna consists of three excitation ports for exciting the printed dipole radiating elements. The excitation signal consists of amplitude, frequency, and phase. As the antenna is usually designed to operate at a particular band, the frequency is mentioned to be constant and, hence, two variables exist: one is the amplitude and the other one is a phase of the excitation signal. In this experimentation, the variation of radiation characteristics concerning the change in the phase angle is concentrated. As there are two elements in this design, few of the possible combinations were considered. The phase excitation for the first and second elements are denoted by ϕ_1 and ϕ_2, respectively.

Case-1. *One of the elements is given zero phase:*
$\phi_1 = 0°$, $\phi_2 = \{0°, 15°, 30°, 45°, 60°, 75°, 90°, 105°, 120°, 135°, 150°, 165°, 180°\}$.

Case-2. *Phase angle with opposite polarity*:
$\{\phi_1, \phi_2\} = \{-15°, 15°\}, \{-30°, 30°\}, \{-45°, 45°\}, \{-60°, 60°\}, \{-75°, 75°\}, \{-90°, 90°\}, \{-105°, 105°\}, \{-120°, 120°\}, \{-135°, 135°\}, \{-150°, 150°\}, \{-165°, 165°\}, \{-180°, 180°\}$.

These cases mentioned above are evaluated and presented in Figure 10.9 for Case-1.

The radiation patterns obtained at three resonant frequencies depict much information of the scanning ability based on its beam coverage of the antenna array. In Case-1, set of phase excitations, the second lobe started appearing at 0°–90°, and an asymmetrical dual-beam pattern is obtained at 0°–180°. This is observed at 2.55 GHz. At 3.25 GHz, the occurrence of the second lobe is started at 0°–45° phase excitation. At the same time, the third operating frequency, i.e., 5.9 GHz excitation, the pattern initially (at 0°–0°) has a narrow main lobe and two symmetrical side

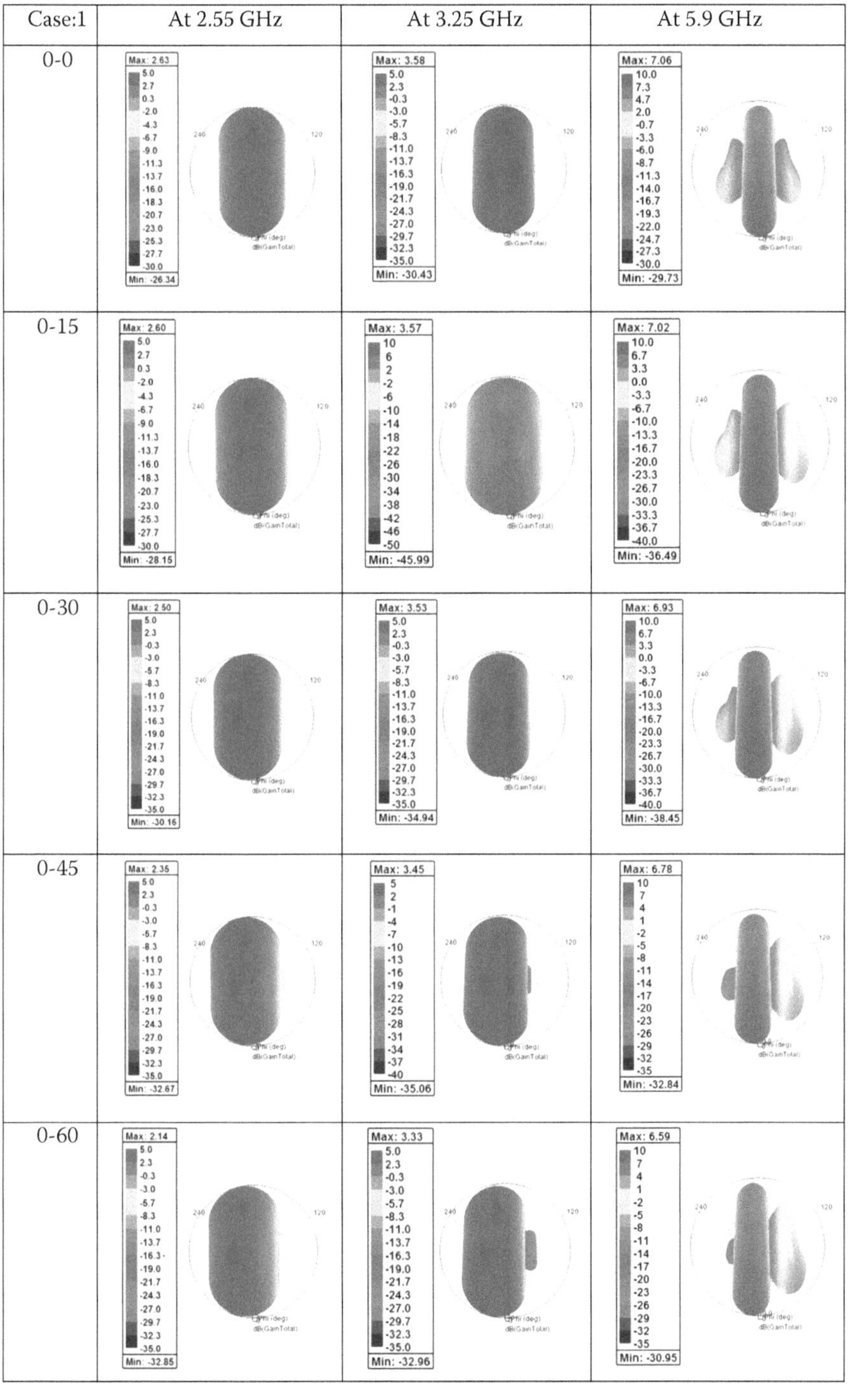

FIGURE 10.9 Far-field radiation characteristics of 1×2 array antenna with Case-1 set of phase-excitation variations (One of the elements is given zero phases).

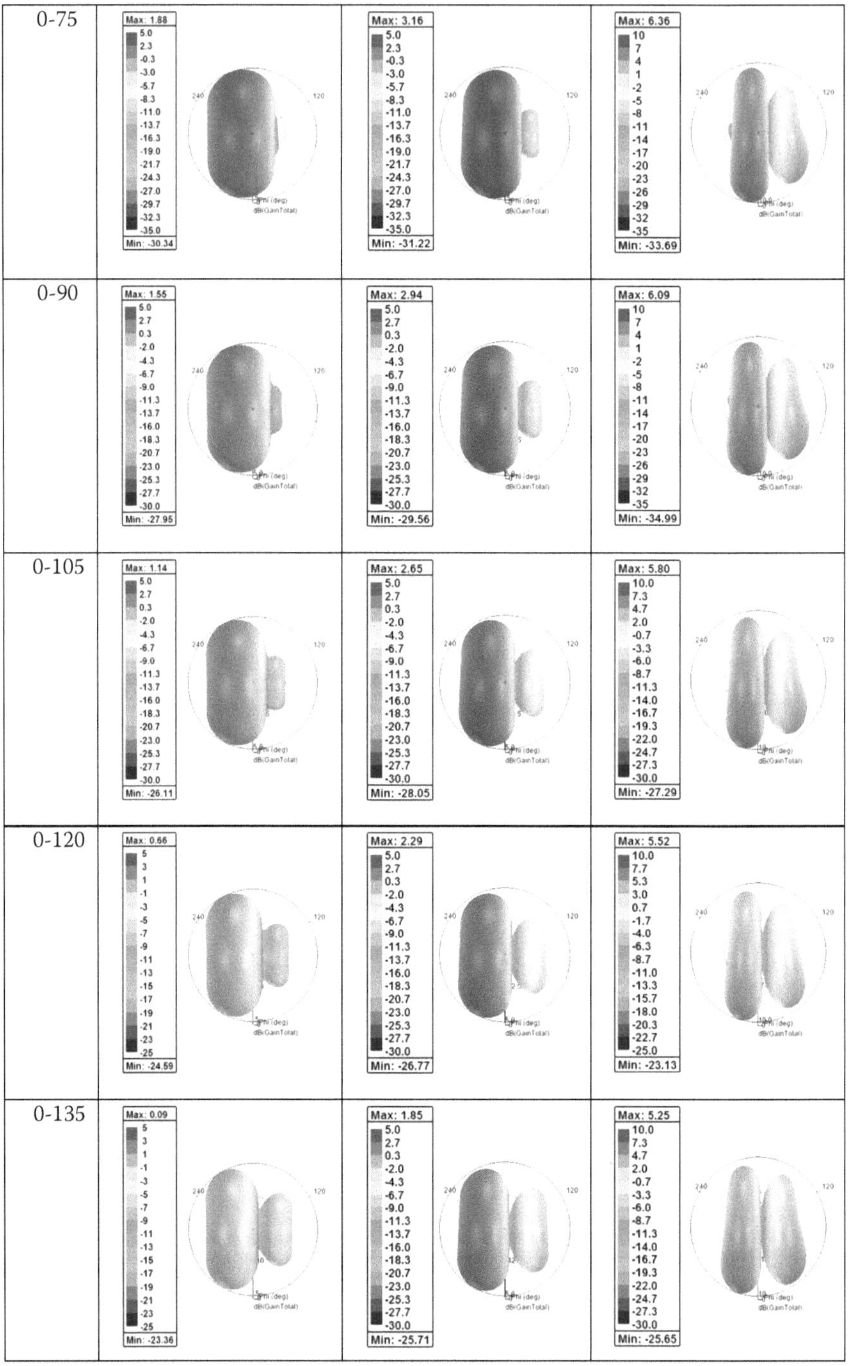

FIGURE 10.9 (*continued*)

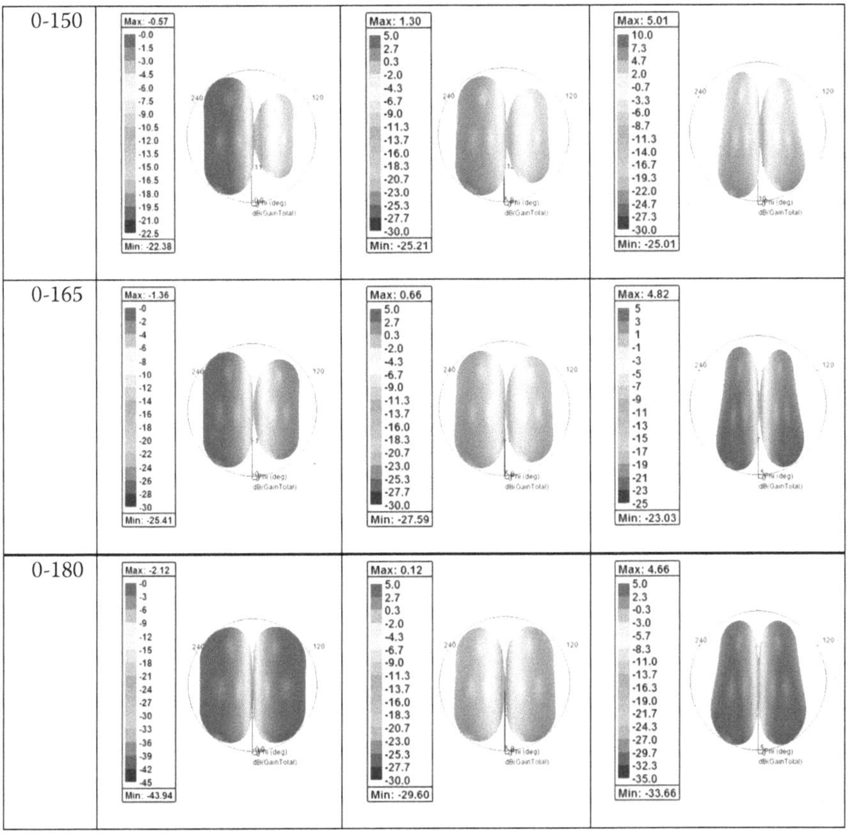

FIGURE 10.9 *(continued)*

lobes. The side lobe aligned toward Element-2 becomes dominant with the phase-angle increment at Element-2 and continues till 0°–165° condition. There, the central lobe and the side lobe will be at a similar level and will produce a pattern similar to the shape of *a pair of the foot* (Figure 10.10).

In a Case-2 set of excitations, the following things are observed. At 2.55 GHz, the sidelobe comes into existence around –45° & 45° excitation phase angles and at 3.25 GHz operating frequency, and the same thing can be observed at –30° & 30° excitation-phase angles. Whereas for 5.9 GHz operating frequency, two side lobes exist from 0deg phase-excitation itself. The critical observation is the dominance of one of the two side lobes. From 0° & 0° to –90° & 90° incremental-phase excitations, the side lobe toward the element-2 becoming dominant and at –90° & 90°the pattern consists of only two lobes instead of three (one main lobe and two side lobes). The other sidelobe is found coming into dominance from –105° & 105° till –180° & 180° phase excitation. The beam steering can be observed.

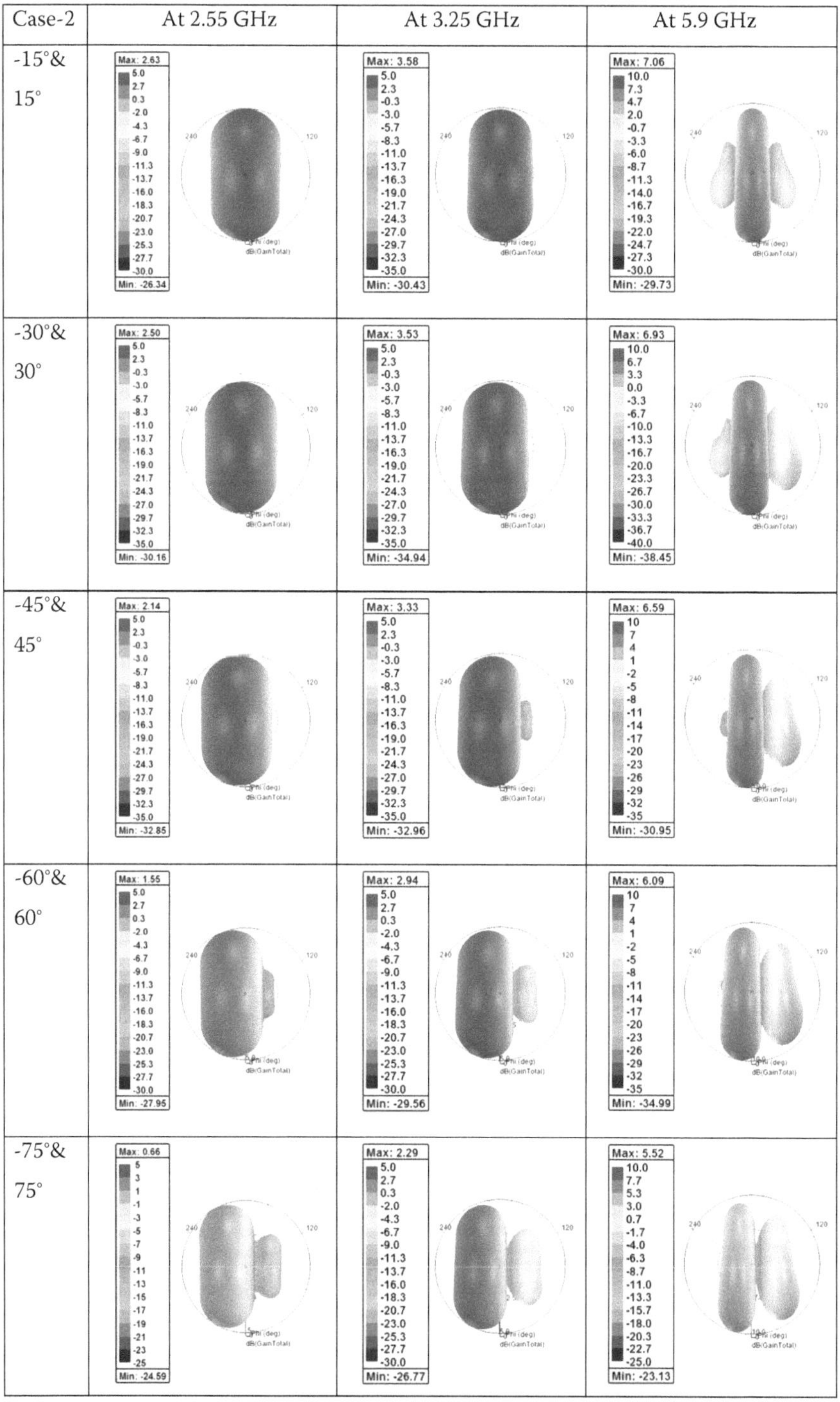

FIGURE 10.10 Far-field radiation characteristics of 1×2 array antenna with Case-2 set of phase-excitation variations (Phase angle with opposite polarity).

-90°& 90°

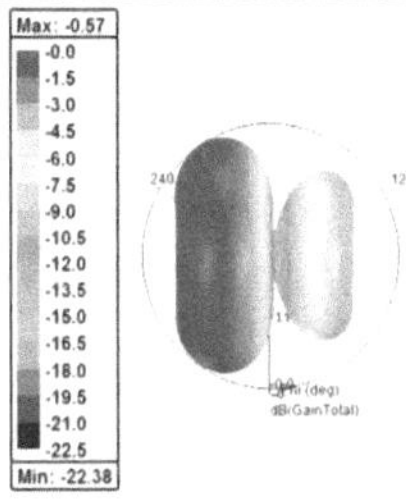

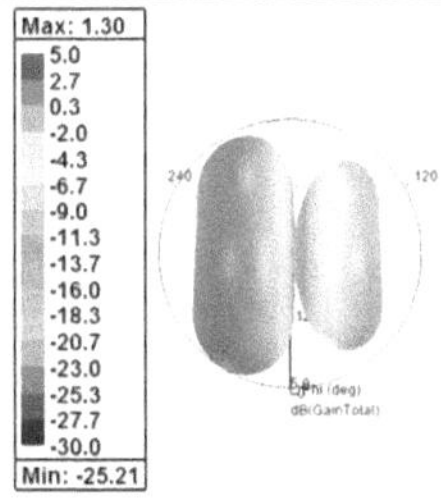

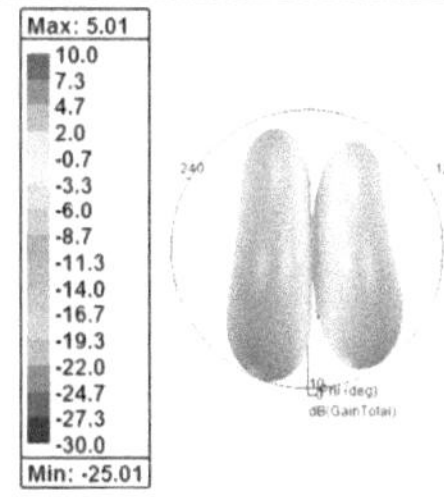

-105°& 105°

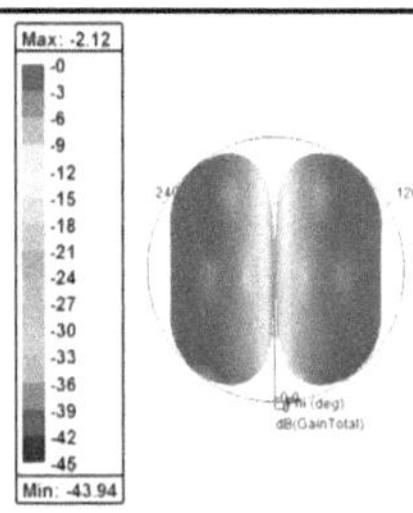

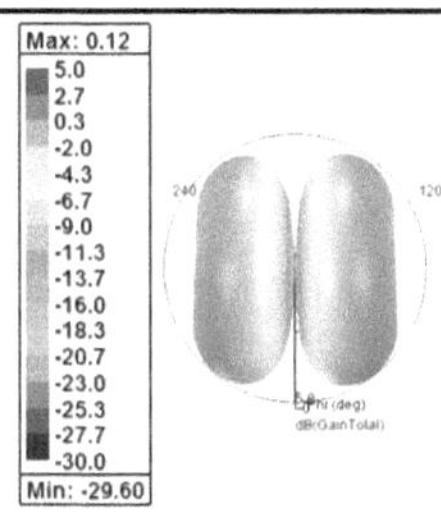

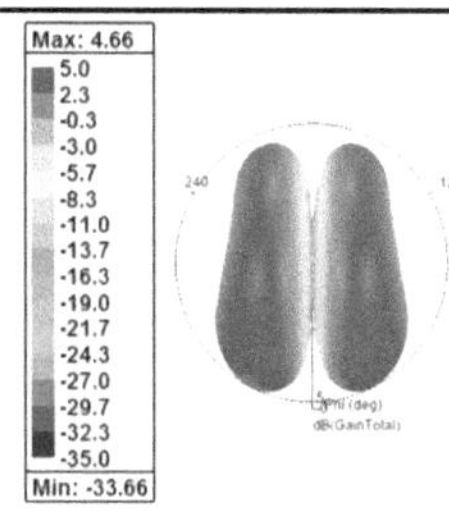

-120°& 120°

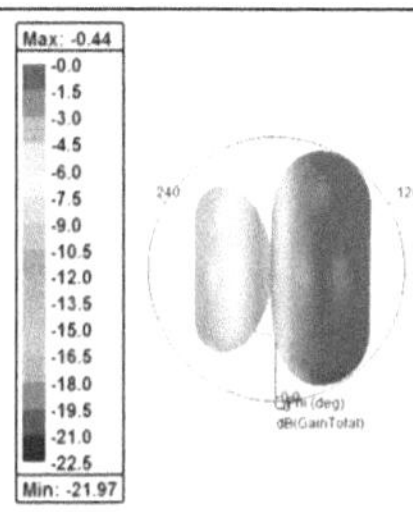

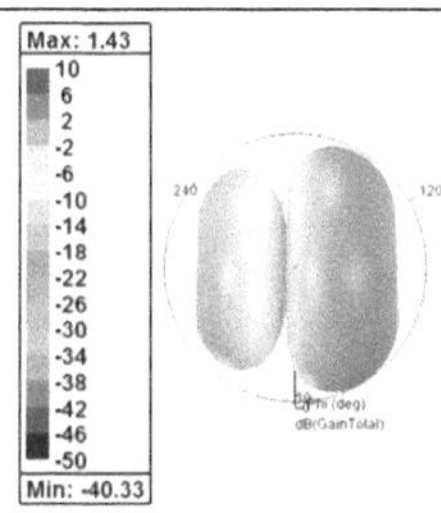

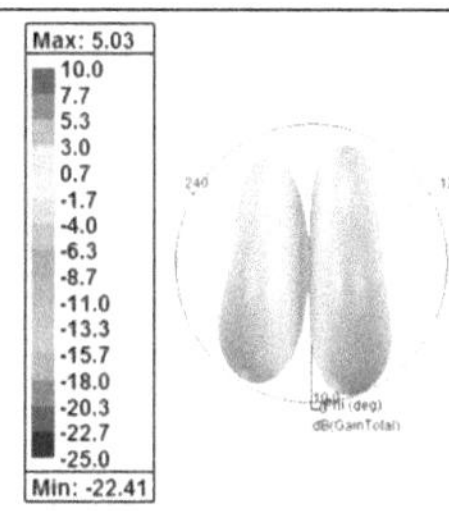

-135°& 135°

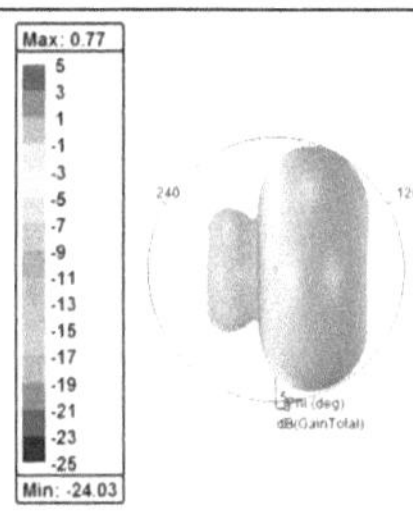

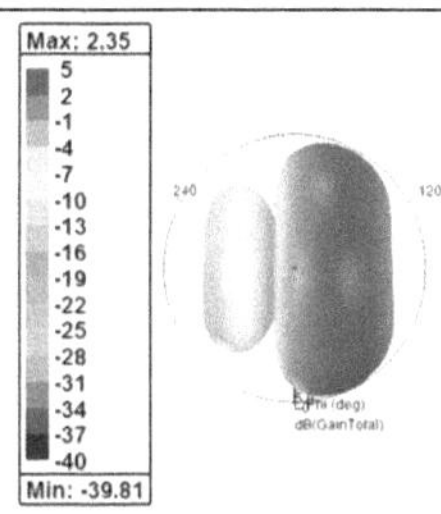

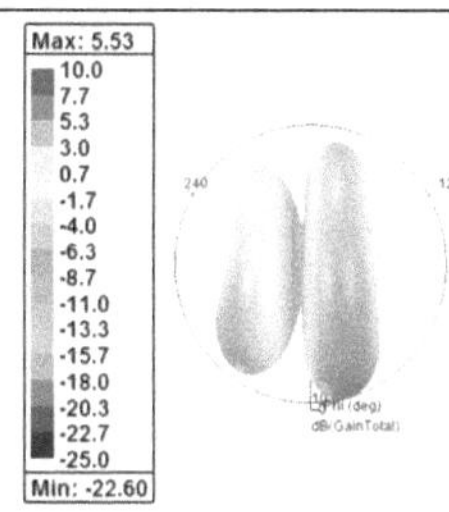

-150°& 150°

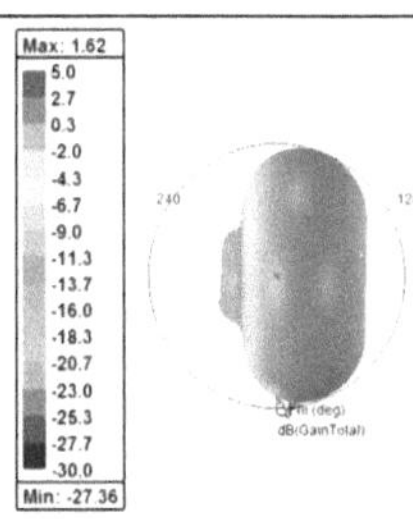

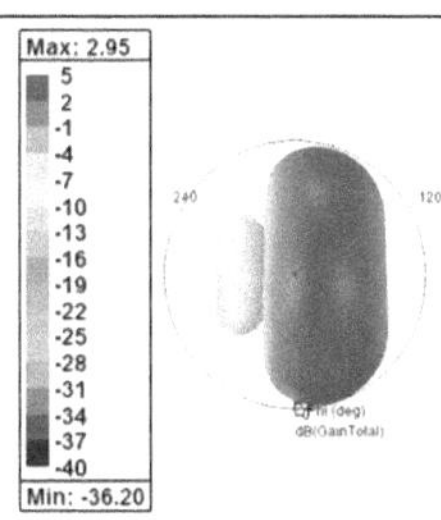

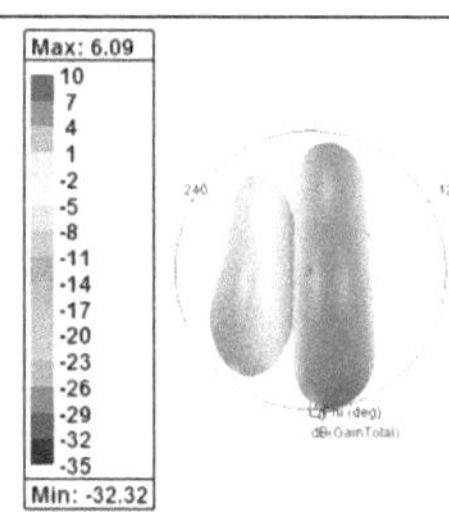

FIGURE 10.10 *(continued)*

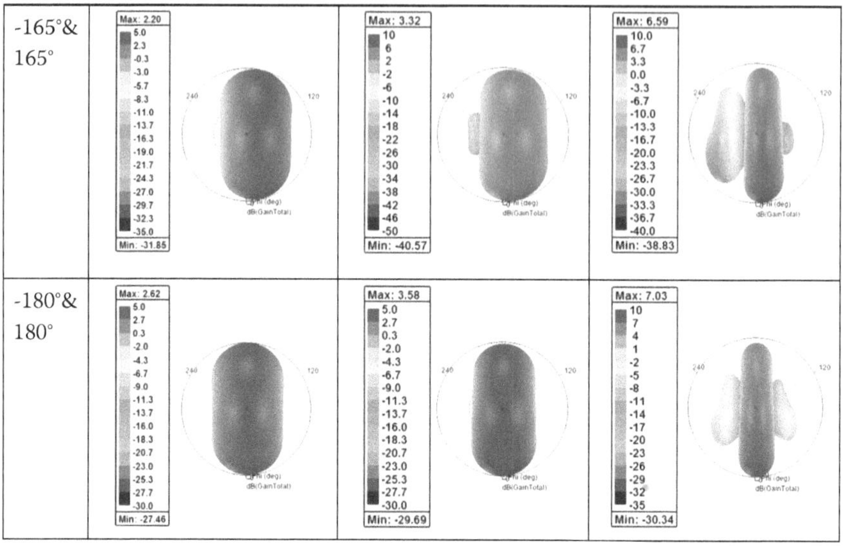

FIGURE 10.10 *(continued)*

10.4 CONCLUSION

In this research work, endeavors have been made for designing a triple-band antenna array with directive characteristics. This antenna array also provides scannable far-field radiation with the variation in the phase excitation of the signal given to each element. The side lobe occurrence and its dominance are identified and mentioned for the possible tuning of the radiation pattern for obstacle detection and scanning purpose. The incremental phase shift by keeping one of the elements at zero phase gives the beam scanning. The reversal of this is possible with phase angle swapping.

REFERENCES

1. C. A. Balanis, *Antenna Theory: Analysis and Design*. John Wiley & Sons, 2015.
2. Y. Yu, J. Xiong, and R. Wang, "A wideband omnidirectional antenna array with a low gain variation," *IEEE Antennas Wireless Propagation Lett.*, vol. 15, pp. 386–389, 2016.
3. H. Wang, S. He, and Y. Yu, "A wideband omnidirectional antenna array," In 2016 11th International Symposium on Antennas, Propagation and EM Theory (ISAPE), pp. 112–114, 2016.
4. G. Gao, B. Hu, and J. Zhang, "Design miniaturization printed circular-slot UWB antenna by the half-cutting method," *IEEE Antennas Wireless Propagation Lett.*, vol. 12, pp. 567–570, 2013.
5. L. Wang, Z. du, H. Yang, et al., "Compact UWB MIMO antenna with high isolation using fence-type decoupling structure," *IEEE Antennas Wireless Propagation Lett.*, vol. 18, no. 8, pp. 1641–1645, 2019.
6. Y. Liu, S. Liu, Y. Jia, Z. Chen, and Y. Zhang, "A miniaturized high gain dipole array antenna for 2.4/5 GHz wireless local area network applications," *Microwave Optical Technology Lett*vol. 63, no. 2, pp. 1–6, 2020. 10.1002/mop.32618

7. M. L. S. N. S. Lakshmi, H. Khan, and B. T. P. Madhav, "Novel sequential rotated 2×2 array notched circular patch antenna." *Journal of Engineering Science and Technology Review*, vol. 8, no. 4, pp. 73–77, 2015.
8. V. R. Tumati and A. K. Tirunagari, "Design and study of serially fed array antenna for ultra-wideband applications." In Proceedings of 2nd International Conference on Micro-Electronics, Electromagnetics and Telecommunications, pp. 465–473. Springer, Singapore, 2018.
9. K. Reddy, T. Venkata, V. Rao, and T. Anil Kumar, "Design of flag-shaped 3-element array antenna for directive radiation in Ku-multiband applications." In 2018 Conference on Signal Processing and Communication Engineering Systems (SPACES), pp. 95–98. IEEE, 2018.
10. B. T. P. Madhav, T. Venkateswara Rao, and T. Anil Kumar, "Design of 4-element printed array antenna for ultra-wideband applications." *International Journal of Microwave and Optical Technology*, vol. 13, no. 1, pp. 8–17, 2018.

11 Analysis of Energy Efficient Narrowband Internet of Things (NB-IoT): LPWAN Comparison, Challenges, and Opportunities

Rasveen and Shilpy Agrawal
Department of Electronics and Communication Engineering, G. D. Goenka University Gurugram, Haryana, India

Khvati Chopra
Department of Computer Science of Engineering, Jamia Hamdard, New Delhi, India

CONTENTS

DOI: 10.1201/9781003230526-11

11.1 INTRODUCTION

Internet of Things is a new wireless communication perspective, where everything is connected to the internet. We can see there has been a remarkable change in IoT devices and applications from the last few years. IoT applications are divided into two classes: one is critical IoT applications in which the volume of IoT devices is less and deals with time-sensitive remote operation and requires precision. The second is massive IoT applications, which deal with a massive number of IoT devices. The application of massive IoT is related to commercial requirements like transport logistics, agriculture, smart cities, smart homes, etc.

Low-power wide area (LPWA) technologies were introduced in 2013 as a different class of wireless technologies to fulfill IoT data transfer demand. Many LPWAN solutions are accessible, but choosing the right technology for the IoT device is not an easy task. Some of the LPWAN technologies are working in an unlicensed band (LoRa, Sigfox, Weightless, Symphony link, ZigBee, Z-Wave), and other newly introduced technologies work in the licensed band (MTC, LTE-M, NB-IoT) [1]. NB-IoT and LTE-M are newly emerged cellular technologies and will become a promising addition to LPWAN space. These cellular technologies hold many features for massive IoT devices like long battery life, less device complexity, extension in coverage, high reliability, high coupling, and deep penetration.

We mainly discuss the narrowband Internet of Things (NB-IoT) among the newly introduced cellular wireless technologies. This technology is proposed and standardized by 3GPP in release 13 (Std, 2018). It perfectly matches the LPWAN market requirement, and the main attraction of NB-IoT is that its architecture is like existing long-term evolution (LTE) technology. That is why NB-IoT can easily coexist with LTE with minor changes. Due to this, the initial network design cost is reduced and assists the network operator in quickly entering a new business. If compared with the LTE-M, NB-IoT will provide deep penetration. On the other hand, LTE -M is best suited for roaming applications, and NB-IoT has benefits for static devices. Both the technologies are highly secured, standardized, managed by the operator, and work in a licensed spectrum. Here, Figure 11.1 shows the coverage provided by the different wireless access networks (licensed and unlicensed) for IoT applications.

In this paper, we discuss the current and future IoT technologies. The paper is arranged as follows. Section 11.2 shows the comparison and technical specifications of IoT connecting technologies. Section 11.3 discussed the NB-IoT standard and release. Section 11.4 analyzed the different channel-coding schemes used by NB-IoT. Section 11.5 identifies the latest research open issues and challenges in NB-IoT.

11.2 COMPARISON OF IOT CONNECTING TECHNOLOGIES

IoT is divergent and has different features, so it is challenging for a single IoT technology to fill all the requirements fully. The network must be robust standards because the same company will not build the network itself and IoT devices. For different applications, technology requirements are different, broadly divided into short-range technologies (Bluetooth, Wi-Fi, ZigBee, Z-Wave) and long-range technologies (LoRa, INGENU, LTE-M, NB-IoT). Here in Table 11.1, we have

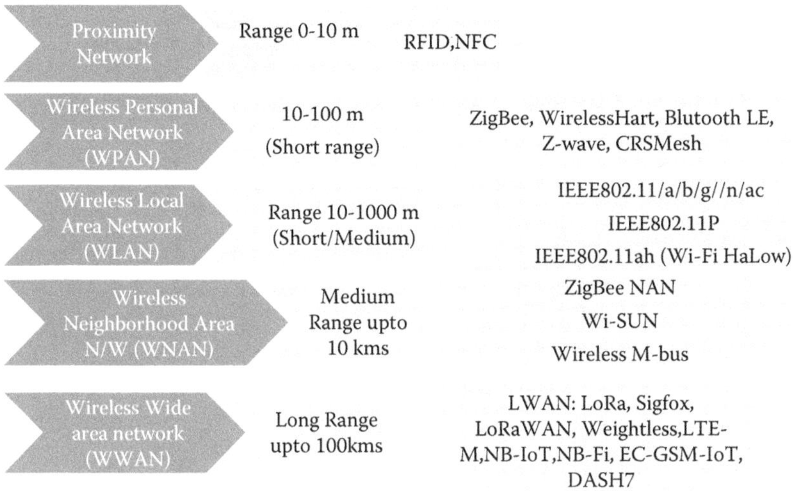

FIGURE 11.1 Wireless access network and coverage for IoT applications.

compared the different IoT technologies on different parameters. Choose the technology according to the application requirement. Each of the solutions has been introduced with its own strength and weaknesses in various network criteria. Selecting a particular technology out of many depends on many factors (privacy issues, cost, coverage requirement, battery consumption, latency, and so forth) [2]. Each IoT connecting technology is briefly discussed below, better to understand its specifications and usefulness for IoT applications.

Bluetooth – is a short-range technology defined under the wireless personal area networks category. It shares the data between point-to-point or point-to-multipoint. Bluetooth is designed as a low-powered consumer IoT application. It is used in conjunction with electronic devices, like smartphones, that behave like a hub for transferring data to the cloud. It is a perfect solution for entertainment, smart home, HVAC system, remote controls, medical wearables devices, and easily communicated data through smartphones. The weak points of BLE are that it consumes a lot of battery power and has short-range connectivity, so it is not preferred for industrial applications.

Wi-Fi – is an 802.11 a/b/g/n/ac standard it uses radio-wave for communication. It is a technology that helps devices (mobile, tablet, laptop) to connect to the internet. IEEE introduces two new protocols that consume less power and support thousands of devices first is 802.11ah – HaLow, which can be easily penetrated and provide a range of 1 km with data ranging from 150 kbps to 40 Mbps. The second is 802.11ax provides four-time better-improved capacity, data rate. It is an extremely demanding technology in the M2M market. For different application solutions, the choice of Wi-Fi and deployment options vary. There are four types of deployment: cisco mobility express, converged deployment, centralized deployment, and cloud-based deployment. Wi-Fi uses 2.4 GHz; it is best suited for line-of-sight. Wi-Fi is

TABLE 11.1
Summary of current IoT wireless technologies [3,4]

Attribute	Short Range Technologies				Long Range technologies (LPWAN)				
	Bluetooth	Wi-Fi	Zigbee	Z-Wave	Sigfox	Lora	INGENU	LTE-M	NB-IoT
Range	Less than 100 m	70–250 m	75–100 m	~100 m	3–10 km for urban and 30–50 km for rural.	2–5 km for urban area and 15km for a suburban area	15 km	~11 kms	10–15 km
Throughput	0.27 Mbps	802.11n 600 Mbps	250 kbps	40 kbps	~100 bps ultra-low	~50 kpbs	30 Mbps	<1 Mbps	~150 kbps
Bandwidth	2.4 GHz	22 MHz	2 MHz	–	100 Hz	<500 kHz	1 MHz	<1.08 MHz	180 KHz
Data Rate	1 Mbps	600 Mbps	250 kbps	40–100 kbps	>100 bps	50 kbps with FSK	156 kbps for UL and 624 kbps for DL	300/375 kbps	200 kbps
Frequency	2.4 GHz	2.4 GHz/ 5GHz	2.4 GHz, 915 MHz, 868 MHz	908.42 MHz	915–928 MHz	868 MHz and 915 MHz Sub 1 GHz	2.4 GHz	Work in the cellular band	700,800,900 MHz
Latency	~3 ms less than 10 ms	>20 ms	~15 ms	1 s	1–30 ms	1–10 ms	-10 s	10 s–15 ms	<10 s
Modulation Type	GFSK	QPSK	O-QPSK	GFSK	GFSK (for downlink) DBPSK (for uplink)	FSK technique	Patent RPMA	QAM, QPSK	BPSK, QPSK
Ongoing Cost	Low cost	Moderate	Low	Low	Low	Low	Low	Low	Low
Battery life	–	–	Many Years	10 years	10 years	>10years	More than 10–20 years	>10 years	>10 years
Advantages	Best for sending a small chunk of data, more privacy	Easy access and deployment	Exceptionally reliable	Less vulnerable, signal strength up to 100 feet	Best for infrequent data transmission, consume significantly less amount of power.	The adaptive data rate, low range, immune to interference, low power consumption	Reliable, excellent coverage, High capacity	Low latency, Higher speed compared to other LPWAN, Bidirectional communication.	Wider deployment, Reliable, Global reach, cost-saving
Applications	Automobiles, smartphones, watches	Use to connect network	Healthcare, retail, smart home	Home automation system, energy management	Perfect for Agro environment, Best for low data transfer, pet tracking, smoke detector	Animal tracking, fleet tracking, home security	Smart city, Smart home, oil field automation	Asset training. Intelligent cities, wearable, E-health solutions.	Agriculture, manufacturing, resource management, healthcare.

more vulnerable to privacy attacks compared to wired technologies. That is why this technology uses password protection and secures the data from eavesdrop.

ZigBee – is another short-range technology that addresses the unique requirement of low-cost, low-power wireless IoT networks. It is IEEE 802.15.4 based specifications used to create a personal area network. Transmission distance is limited to 10–100 m (line of sight). In this mesh technology, repeaters enhance the range, but this will enormously raise the deployment cost [5]. The ZigBee network is scalable, and we can easily add a new end device to the network. It provides extended battery life and is less complex than Bluetooth. However, it is not as secure as a Wi-Fi-based system, but it is exceptionally reliable, easy to install, and has a flexible network structure.

Z-Wave – is another addition to the short-range group. It is primarily designed for home and office automation and is controlled by smartphones and computers with internet facilities. It wirelessly manages and controls home appliances and other devices like doors, windows, locks, security systems, swimming pools, and many more. Z-wave interoperability at the application layer allows all hardware and software to work together and share the information. It is a mesh networking technology. All the nodes can communicate directly to one another, but if the node is not in the range, it first links with another node and then exchanges the information. Z-wave provides low latency for small data packets and is highly reliable. Z-wave has a unique beaming technology that enables low latency (1 sec). Z-Wave 700 series extends the range to 300 ft; hence, it is suitable for IoT application [6,7].

In the above paragraphs, we have seen that short-range technologies provide many advantages. Still, due to a few limitations, we cannot prefer these technologies for IoT applications. They offer limited mobility facilities and require repeaters to extend the range, which will increase the deployment cost. To overcome these limitations and to offer new features, LPWAN wireless technology is introduced. It is divided into two categories, unlicensed and licensed [8]

Sigfox – is an LPWAN technology working in an unlicensed band, operating on a cloud-based approach, and using ALOHA. It builds in a wireless network to communicate with low-power devices. These devices continuously transfer the data and remain in one state. The network is based on a one-hop star topology to communicate in the long-range, and it uses ultra-narrow-band modulation. Each one of the messages is 100 Hz wide and data rate of 100 or 600 bps. The devices send a random frequency message, and along with it, two more replicas are transferred on different frequencies and times; this is known as time and frequency diversity. The main advantages of Sigfox are that it is deployed with extended coverage, requires fewer base stations, and has unique anti-jamming capabilities). Table 11.2 gives the technological summary of Sigfox for a better understanding of the technology.

Lora – is an abbreviation of long-range, and it is a wireless technology where a low-powered sender transmits small data packets (0.3 kbps to 5.5 kbps) to a receiver. Chirp-spread spectrum modulation technique is used by LoRa. A gateway can handle hundreds of devices simultaneously, and gateways are powered and connected to the internet. Also, the gateways can handle multiple frequencies simultaneously. LoRa-WAN works asynchronously because this is attracted by the applications that are in extended sleep mode. Lora cannot handle dense networks and provide quality service (QoS) because it works in an unlicensed spectrum.

TABLE 11.2
Technical characteristics summary of Sigfox

Characteristics	Sigfox
Multiple access technique	Uplink and downlink: Proprietary FHSS/UNB
Medium access control (MAC)	Unslotted ALOHA/Asynchronous
Modulation type	DBPSK for uplink and GFSK for downlink
Frequency	Europe uses 868 MHz ISM band North America uses 900 MHz ISM band
Range	3–10 km (URBAN), 30–50 km (RURAL)
Link budget	~146–162 dB
Bandwidth	Channel bandwidth: uplink 100 Hz Downlink 600 Hz Listening bandwidth 200 kHz
Number of UNB channels	360
Roaming	yes
Handover	Not required
Overhear/ Signaling	No Signalling
Battery lifetime	3.6 V AA battery 1 msg/day:10 years 10 msg/day: 6 years
Payload length	8B for downlink, 12B for uplink
Output power	Max.25 W (14 dBm)
Node per base station	variable

Table 11.3 shows the technical characteristics of LoRaWAN according to SEMTECH, LoRa alliance and other sources.

eMTC/NB-IoT – 3GPP proposed a licensed LTE-M/eMTC for enabling global IoT connectivity. It is a type of LPWAN standard technology designed to enable a wide range of cellular devices. LTE-M has many advantages like mobility, more data rate, and voice capability over the network, but it requires more bandwidth than NB-IoT. The specification introduced for LTE Cat-M1 was frozen in release 13. After this, the 3GPP proposed IoT technologies like EC-GSM-IoT and NB-IoT.3GPP, in its release 13, introduced a new narrowband cellular technology named narrowband Internet of Things (NB-IoT). It uses a single bandwidth of 200 kHz with a maximum 26 kbps downloading speed and is increased up to 127 kbps in release 14. This technology is loaded with several advantages for IoT devices if compared with other existing ones. It provides deep penetration (deep indoor, remote areas), extended coverage, low cost, increased battery life, deployment mode (stand-alone, guard band, in-band), simplified network to reduce UE complexity, and the CIoT user plane is optimized for small data transfer. Table 11.4 shows the technical characteristics of eMTC/NB-IoT.

To conclude, all the LPWAN technologies that support IoT devices have their benefits individually. The long back existing technology like LoRa, Sigfox has its own hold on the IoT market, offers adequate link budgets, provides QoS, is easy to

TABLE 11.3
Technical characteristics of LoRa

Characteristics	Lora/LoRaWAN
Modulation	Lora, Standard FSK
Frequency	Used in Europe: 863–870 MHz ISM band Utilized in the US: 902–928 MHz ISM band
Range	2–5 km for urban, 15 km for suburban Variable range is used
Multiple access method	Proprietary chirp spread spectrum (CSS) Pseudo-random channel selection Optional FHSS
Link budget	~150–157 dB
MAC	Unslotted ALOHA/Asynchronous
Forward error correction	yes
Signaling/Overhear	No signaling The coding rate ranges from 1.25 to 2
Roaming	No method for cross-network nomadicity, The future specification will include nomadic roaming
Output power	Max 25 W (14 dBm)
Handover	Not required
Bandwidth	Use 125 kHz Also, use 200 kHz and 500 kHz
Adaptive data rate	Yes
Number of spreading channels	Europe use 8–10
Bidirectional	Yes
Authenticity	AES-encrypted authentication AES encrypted payload
Peak data rate	0.3 to 50 kbps
Number of messages per day	variable
Battery lifetime	Ten years or more
OTA updates	Yes
Nodes per base station	Variable 20000–60000 (less than one message per day)
SIM from factor	Soft sim equivalent, eUICC
Localization	Yes
Payload length	Variable 51–222 B and up to 250 B

deploy, and gives tough competition to EC-GSM NB-IoT. However, these are the proprietary technologies working in an unlicensed band, due to which the number of connections is limited, less secure, and suffering from duty cycle constraints.

These drawbacks allow NB-IoT to become an attractive option. NB-IoT can support many connections as it is working in a licensed spectrum. It provides numerous advantages like long battery life up to ten years, low and infrequent data

TABLE 11.4
Technical characteristics summary of eMTC/NB-IoT

Characteristics	eMTC	NB-IoT
Multiple access	Frequency hopping/multi-subframe repetition UL-SC-FDMA DL-OFDMA	UL-SC-FDMA (single tone, multitone) DL-OFDMA
Modulation	UL-QPSK DL-QPSK, 16 QAM,64 QAM	UL-BPSK, QPSK DL-BPSK, QPSK, optional 16 QAM
Frequency	450 MHz–3.5 GHz	450 MHz–3.5 GHz
Link budget	155.7 dB (MCL)	164 dB (MCL)
Latency	10–15 ms	1.4–10 sec
Bandwidth	1.4 MHz	200 MHz
Peak data rate	10 kbps to 1 Mbps	DL-250 kbps UL-single tone 20 kbps UL-multitone 250 kbps
FEC	Turbo code	UL-turbo code
Bidirectional	Yes, Full/half-duplex	Yes, Half-duplex
SIM	UICC or eSIM/eUICC	UICC or eSIM/eUICC
Node per BS	Unknown	Minimum 50 k per cell
Battery life	Ten years of operation with a 5Wh battery. Rest depends upon the traffic load	Ten years of operation with a 5 Wh battery. Rest depends upon the traffic load
Output power	20/23 dB	20/23 dB
Peak data rate	10 kpbs to 1 Mbps	DL-250 kbps UL-Single-tone 20 kbps -multitone 250 kbps

transmission, and 20 dB more coverage enhancement compared to existing technologies, as shown in Figure 11.2. Hence, the future IoT applications are more inclined to NB-IoT technology. Analysts predict that the number of LPWA connections will likely grow amazingly fast in the coming years. Moreover, it is essential to know about NB-IoT standards and release offered by 3GPP. The 3GPP standards-based solutions are working in both licensed and unlicensed spectrum. The section ahead will discuss each 3GPP Release (8 to 17) in brief.

11.3 NB-IOT STANDARD AND RELEASE

NB-IoT is a freshly introduced cellular technology for the IoT market and works in conjunction with LTE. NB-IoT's main objective is to enable low-power, low cost, low data-rate communication and to support M2M communication. This new way of communication increases the demand for studying cellular technology to provide a solution to cellular IoT. The low-power vast area network (LPWAN) market overgrows in early 2014, and due to this, the demand for IoT is also increasing

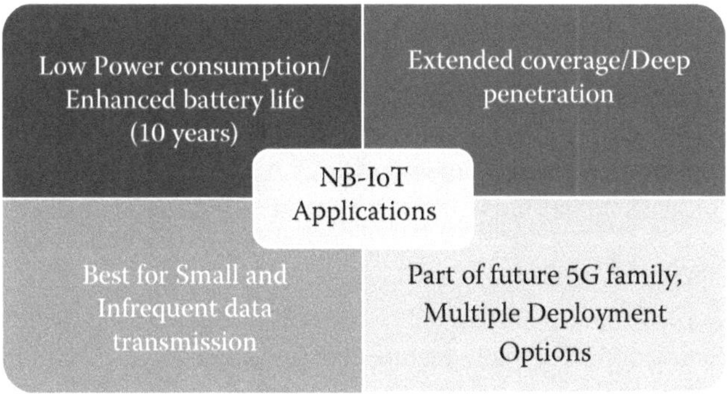

FIGURE 11.2 Advantages of NB-IoT.

rapidly. The same year proposed two technologies: Narrowband Machine to Machine (NM-M2M) and Narrowband Orthogonal Frequency Division Multiplexing (NB-OFDM). In 2015, the existing two proposals were merged, originating a new proposal, known as narrowband cellular IoT (NB-IoT). After few months, one more proposal was introduced, named Long-Term Narrowband Evolution. Moreover, by the end of 2015, all the 3GPP proposals are combined and named Release 13. In 2016, NB-IoT was acknowledged as a novel clean-slate RAT. Below will discuss the releases (Rel-13 to Rel-17) and standards in brief:

11.3.1 Release 13

3GPP, in its release 13, introduced a new NB-IoT technology to fulfill the requirement of cellular IoT devices for diverse applications like low power consumption, low cost, less complicated system.

- **Multitone Transmission** – NB-IoT Resource Unit (RU) is allocated to multiple User Equipment (UE) to achieve the objectives of massive IoT device deployment. However, in LTE, the resource unit is allocated to only a single UE. For the downlink transmission, 15 kHz bandwidth is allocated for both single-tone and multi-tone. On the other side, uplinks either occupy 3.75 kHz or 15 kHz of transmission bandwidth.
- **Deployment** – There are three modes in which NB-IoT can deploy: standalone, in-band, guard band. NB-IoT's primary resource block (PRBs) occupies 180 kHz of bandwidth in both downlink and uplink transmission. If the deployment option is standalone, it occupies 200 kHz bandwidth by "reframing" the global system for the mobile communication spectrum.
- **Energy Efficiency** – One of the vital features of NB-IoT devices is to provide a long battery life of up to ten years. It supports two features, i.e., power-saving mode (PSM) and extended discontinuous reception (eDRx). If the PSM mode is utilized, then the device life is around 413 days. Moreover, if the device uses eDRx, it remains in inactive mode (from few minutes to a few hours).

- **Extended Coverage** – NB-IoT is designed to provide 20 dB more coverage extension than other legacy technologies (GPRS, LTE). It provides extended coverage to the applications, which are inside deep doors and basements. To extend the coverage, NB-IoT repeats the transmission 128 times in uplink and 2048 times downlink.

11.3.2 Release 14

After Release 13, further enhancement is needed to upgrade the quality of service. For the enhancement of QoS, new features are introduced by 3GPP in its release 14. The enhancements include mobility, improved non-anchor carrier operation, system throughput, and service continuity [9].

- **Multicast Service** – This mechanism aims to optimize the resources and transmission latency. Therefore, in release 14, single-cell point to multipoint (SC-PTM) transmission mode is used to support Multimedia Broadcast Multicast Service (MBMS) to reduce the latency and improve the radio efficiency. This release also enhanced the supports for both stationary and mobile television services over eMBMS and unicast.
- **Improved Positioning** – In NB-IoT, to extend UE position measurement of cell identity (CID), release 14 proposes an indoor advanced positioning method for observed time difference of arrival (OTDOA). The UE measures the arrival time (ToAs) of positioning reference signals (PRSs) in this method. The enhanced cell identity measurement requirement is reference signal received quality (RSRQ), base station receiving and transmitting time difference, and reference signal received power (RSRP).
- **Transport Block Size** – In release 14, the data rate is improved by enhancing the (TBS) size to 2536 bits. However, if compared with release 13, NB-IoT supports a low data rate. The downlink uses 250 kbps, and the uplink uses 226.7 kbps.
- **UE Mobility Increment** – In release 14, Radio Resource Control (RRC) is re-established for the mobile use cases. This proposed feature (RRC re-establishment) hides if any radio interference occurs in the upper layer and hides the temporary losses.
- **Multiple-connectivity** – In this feature, multiple links are offered to UE, and it has two options. UE offered multiple links in the first option, and they are limited in number; hence, only a particular set of radio links are active. In the second option, all the available radio links are active.

11.3.3 Release 15

Release 15 came forward with many enhancements like improved quality of service and rapid adoption of massive deployment.

- **Small Cell Support** – Small cell deployment introduced by release 15 has improved the capacity and coverage. Whatever the maximum power is decided, the NB-IoT UE cannot transmit more power than that. Furthermore, to enhance the IoT connectivity for remote and rural areas, NB-IoT now supports up to a 100 km range.

- **Latency Reduction** – Novel features like reducing power consumption and transmission latency are introduced in release 15. Whenever the UE wants to decode the physical downlink control channel on a paging occasion, a wake-up signal is transmitted in the ideal mode and reduces power consumption. Also, power is reduced by sending a standard wake-up signal to the group of UEs.
- **Semi-Persistent Scheduling** – This feature is proposed to support voice messages for applications. To provide the voice message facility, a fixed resource unit is used for NB-IoT UE, which uses a specified number of intervals, reducing the control plane overhead. The base station pre-configured the UE with the Radio Network Temporary Identifier (SPS-RNTI) to differentiate one radio channel from another. SPS enables the NB-IoT data reception at a regular configured periodicity.
- **Time Division Duplex Support** – One more feature introduced in Release 15 is TDD support, a new TDD frame structure (type 2). NB-IoT supports a standard cyclic prefix for both (3.75 kHz and 15 kHz) spacing. To support some TDD configurations with few downlink subframes, few systems information can be transmitted on non-anchor carriers. With this feature, UE will have to reduce the search system and information acquisition [9,10].

11.3.4 Release 16

The enhancement in release 16 of 5G New Radio includes an extension in the existing release 15 feature. The main highlighted feature of release 16 is in the area of multi-input, multi-output (MIMO), dynamic spectrum sharing (DSS), carrier aggregation (CA), beamforming enhancement, power-saving, and user equipment.

- **Beamforming and Multi-user Enhancement** – Release 16 enhancing the multi-user (MU-MIMO) support, beam handling, favor single UE for multiple transmission and reception point (multi-TRP), and channel state information (CSI) feedback. Further, in uplink transmission mode, the full-power transmission is provided from multiple UE antennas. The enhancement in these features reduces overhead, provides more robustness, and increases throughput.
- **Carrier Aggregation and Dual Connectivity** – This release minimizes the latency rate for the set and activation of CA/DC, due to which system capacity is improved and data rate is increased. Release 15 cannot configure and report until the UE does not enter the fully connected state. However, in release 16, without extensive signaling for reporting and configuration, the connection can be resumed after periods of inactivity. Also introduced aperiodic triggering of CSI reference signal transmission [11].

- **Integrated Access and Backhauling (IAB)** – IAB provides an alternative to fiber backhaul by extending NR to support wireless backhaul [12]. This NR can be used for a wireless link from one central location to distributed cell sites also used between the cell sites. This is advantageous to deploy small cells and use them for emergencies and temporary deployment. ABI can use any frequency band, and NR can operate in the same.
- **IoT for Industries** – New Radio (NR) release 16 majorly focuses on IIoT to enhance the potential of IIoT applications like electrical power distribution, transport industry, and factory automation.
- **Network Management Tool Enhancement for the Improvement of UE Differentiation** – In Release 15, the UE can be differentiated according to traffic model and battery usage indication. However, in release 16, NB-IoT UE is differentiated based on how much radio resource is used and how much delay can be tolerated.

Release 16 notices significant changes in other areas also:

UE positioning, transport management service, vehicle to vehicle, vehicle to stationary and advanced (V2X) communication, new radio communication in the unlicensed band.

11.3.5 Release 17

3GPP Release 17 is the next remarkable chapter in New Radio standardization. It will introduce more new applications where mobile communication can be used [13]. It enables the new feature for the three prominent use cases: Ultra-Reliable Low-Latency Communication URLLC, massive machine-type communication (mMTC), enhanced mobile broadband [13]. These new features aim to support the enormous growth in mobile- data traffic, customizing NR for automotive, media and manufacturing, logistic, and public safety. Table 11.5 shows the different applications introduced in Release 17.

After discussing the difference between different LPWA technologies, technical specifications, and their releases, we know that the IoT devices will increase massively in the future, which introduces a problem. The problem is that the NB-IoT devices that are placed remotely need a long battery life because, frequently, replacing the battery is impossible. Two mechanisms are used to resolve this issue and make NB-IoT more energy-efficient: power-saving mode (PSM) and extended discontinue reception (eDRX). PSM minimizes the energy consumption by taking the device into deep sleep mode during RRC Idle state. During the whole process, the device is always connected to the network but cannot be reached. The device will connect again only when there is data available for the transmission in uplink mode. Also, to maximize the NB-IoT energy efficiency, choosing the appropriate channel-coding technique is crucial, reducing the number of retransmissions. In a communication system, channel coding is frequently used to ensure reliable link transmission. Many researchers have given their thought to understand this (channel-coding scheme), which is explained in the next section, which further assists the researchers in making innovative changes.

TABLE 11.5
Applications introduced in Release 17

	eMTC attribute
Support of reduced capability NR devices	• Battery life enhancement • UE power saving • Addresses issues including complexity reduction • Targeted applications such as video surveillance (MTC), Provide data rates between Narrowband IoT/LTE-M to the wearables, industrial sensors, and complete new radio data rate.
	URLLC attribute
Anything reality (XR) evaluation	• It fulfills the reliability and delay requirements of critical MTC (cMTC) applications like the manufacturing and robotics field. • Concurrently, exceptionally high data rates • support many augmented virtual reality and reality forms collectively referred to as XR. • A new NR device type (NR Light) is specially designed to support an industrial wireless sensor network.
	eMBB attributes
Multicast and broadcast service	• The main target is public safety, software delivery, IoT use case, and V2X
Side link relaying	• UE-to- network relaying and UE-To-UE • L2 versus L3 relaying • Side link standardized in ITS use case can be leveraged as broadcast.
Supporting NR from 52.6 GHz to 7.1 GHz	• If NR operates higher than 52.6 GHz, it enhances the capacity of applications like indoor mobile broadband service and dense urban scenarios. • A new channel access mechanism and OFDM notation are introduced for unlicensed spectrum. • The Extended NR frequency range provides more spectrum, including the 60 GHz unlicensed band.

11.4 DIFFERENT CHANNEL CODING SCHEMES USED BY NB-IOT

Most of the research work is concentrated on how to increase the network coverage of NB-IoT in comparison to other legacy technologies. However, there are other areas to focus on. One more important feature of the NB-IoT standard is to provide long battery life. Thus, the devices placed in remote areas do not need frequent battery replacement and run up to ten years. To achieve this, appropriate channel coding schemes are required to enhance NB-IoT energy efficiency [14]. NB-IoT reuses the technological numerologies (rate matching, channel coding, orthogonal frequency-division multiple-access (OFDMA), interleaving, etc.) of LTE. Thus, it reduces the time to develop an NB-IoT architecture and specification [15,16]. Energy efficiency is the leading research problem, and to solve this issue, many researchers have given the thought. To further assist the researcher here in Table 11.6, we concisely explain the channel coding schemes proposed by different authors.

TABLE 11.6
Survey of literature for modulation and channel coding schemes

Channel Coding Schemes Proposed by Different Authors	
• Automatic Repeat Request – is an error-control method for data transmission and received an acknowledgment which indicated that data packet is correctly received; if not, receiver requested to retransmit the data packet. Different authors have proposed different schemes for error control. • Forward error correction – As the name implies, data always flow in the forward direction from the encoder to a decoder. It assists the decoder in detecting and correct transmission. Instead of retransmitting the data in this technique, some redundant bits are added to the transmitting bitstream.	1. In this proposed study [17], the NB-IoT network supports half-duplex transmission and the Hybrid Automatic Repeat Request (HARQ) approach, which reduces processing time at the IoT node. The results show that by using the HARQ method, scalability in the NB-IoT network will not affect the energy efficiency performance and save up to 20% of the energy consumption. 2. In this approach [18], the author proposed an open-loop forward error correction technique. The channel-coding approach is simplified, and the energy consumption is reduced because there is no need to check particular received packet data units at every transmission. The signaling only checks that downlink data is transferred entirely or not. Moreover, every time in the downlink transmission, the data rate performance is enhanced at the transceiver IoT node, reducing energy consumption. 3. The author [19] proposed an NB-IoT energy-efficient adaptive channel coding scheme. This two-dimensional approach works to select the appropriate channel-coding scheme and reduce the number of repetitions. This approach can easily select the higher and lower modulation coding scheme (MCS) by creating a single mixed gradient. By dynamically selecting the MCS, the overall energy efficiency is enhanced, and transmission number is reduced. 4. In the proposed research [20] author introduced a hybrid channel-coding approach. It utilized a repetition code for error correction and, acknowledging the narrowband physical downlink shared channel (NPDSCH), it uses signaling hybrid automatic repeat request (HARQ). In this approach, user equipment (UE) works on three subcarriers (12,6,3). However, only the lower tones are used because if, the higher tones are used, then the energy consumption will be increased.

TABLE 11.6 (Continued)
Survey of literature for modulation and channel coding schemes

Channel Coding Schemes Proposed by Different Authors	
	5. The author [21] proposed a turbo channel-coding algorithm for the uplink transport channel for error correction and detection. In the noisy channel (AWGN) to transfer the NB-IoT data packet from MAC to the physical layer, the author used turbo code with a tail bit convolutional encoder and 24-bit cyclic redundancy check (CRC) syndrome calculator. The proposed algorithm reduces the bit error rate and provides CRC error detection FRC error correction computation.
	6. In this work [22] author proposed an error correction algorithm for multicast transmission (for the LTE network), further extend the work for the NB-IoT network. The model works in different NB-IoT downlink transmission scenarios and includes different error distributions and varying loss conditions. Instead of using the individual (HARQ and FEC) method, the results show that combining a hybrid approach and forward error control improves energy efficiency.
	7. The author in [23] proposed an adaptive forward error correction (AFEC) algorithm for wireless sensor networks and compared the performance with static FEC. The author compared the performance of two AFEC and forward error control. The result shows that using FEC coding in uncoded IEEE 802.15.4 transmission, gain in PDR is achieved and enhances the throughput and, on the other hand, minimizes the energy consumption. Also, this introduced solution does not need any additional feedback channel.
	8. The author [24] used a forward error correction scheme, which is a quasi-cyclic code. This proposed scheme avoids retransmission and conserves energy. It corrects up to the 2-bit error. When the error increases by more than 2, then the retransmission takes place. This scheme can effectively handle Burst Error (8 bit) occur.

(Continued)

TABLE 11.6 (Continued)
Survey of literature for modulation and channel coding schemes

Channel Coding Schemes Proposed by Different Authors	
	9. In the proposed [25]work for saving power, a low area interleaved is implemented for error correction and provides 5% area saving. The turbo encoder initially uses two blocks of RAM, but the author proposed a design in which only one block with dual-port RAM is used. This design saved 128 bytes of RAM. The design reduces the switching activity (sub-blocks) and minimizes the dynamic power. 10. In this [26]work a CRC code is used in BLE, and IEEE 802.15.4 and two iterative decoding technique is utilized for error correction. The natural environment is used for the investigation. Achieves up to 2.5 dB of SNR gain, 35% of corrupted data packets are corrected, which finally reduces the retransmission and increases battery life. No additional signal processing is used, which reduces the energy consumption.

If we compare LTE devices with NB-IoT devices, most of them are battery-driven, and to preserve the energy for a long time; we need to apply different approaches as proposed above by different authors. The other required and current research topic is a modulation scheme for IoT applications to enhance the high data rate demand in licensed and unlicensed bands. The following important part is a selection of modulation schemes. Here in NB-IoT, binary phase-shift keying (BPSK) and quadrature phase-shift keying (QPSK) are the only two modulation schemes to which it is restricted. For downlink communication, it uses only QPSK, and in the uplink, for single tone ($\pi/4$-QPSK, $\pi/2$-BPSK) is used, and for multitone, only QPSK is used. QPSK is frequently used in conjunction with orthogonal frequency-division multiplexing for base-band modulation and works as a complex number (I+JQ). NB-IoT works on cellular technology it means the NB-IoT device can directly communicate with the base station. When the device directly communicates, there are many transmission parameters on which the device must be configured. Like bandwidth, coding rate, transmission power, and for meeting this requirement, lots of transmission energy is consumed, which is cost-effective. Moreover, the challenge is to maintain communication performance by using the required parameters to minimize transmission energy. The researcher has proposed many designs and models to deal with this issue, essential for LPWAN. The throughput of the system is directly dependent on the modulation-coding scheme (MCS) level. High power and low MCS selection can upgrade transmission reliability and increase network coverage [27]. If the MCS is not selected efficiently, it results in a lower data rate. In [28], the author proposed an enhanced NB-IoT (eNB-IoT) framework that uses non-orthogonal spectral efficient frequency division multiplexing. This method speeds up the uplink signal transmission in NB-IoT. In addition to this, a sphere decoding (SD) detector is also used. This eNB-IoT improves the data rate and BRE up to 50% and requires 3dB less transmission power than typical NB-IoT. Less power consumption means long battery life and enhanced coverage for IoT devices. Also, to simplify the base signal processing, the author proposed an overlapped SD (OSD) detector due to which the small size SD kernels can speed up the signal detection. Hence, the gain remains the same, and complexity is reduced. If the SINR is poor, then QPSK is used, it is robust against transmission error but has lower spectral efficiency. If the signal-to-noise ratio is high, it is required to use high-level modulation schemes, higher spectral efficiency, which increases the bit rate for which 64 quadrature amplitude modulation is used. If the code rate is high, it enhances the channel quality and consequently higher the data rate. For data channel in NB-IoT system turbo codes is used [27]. To enhance the code, rate repetition and puncturing are used and combined in the Rate Matching module.

11.5 OPEN ISSUES AND CHALLENGES IN NB-IOT

NB-IoT is an upcoming growing technology. Lots of research work has been done, and still, many grey areas need work on the technical performance issues. Some of the significant challenges in the NB-IoT path are latency, security, network architecture, energy efficiency, system-level integration, network reliability and scalability, and data rate enhancement. These challenges are discussed below for better understanding.

1. Latency – The existing NB-IoT architecture design will not bear the load of billions of IoT devices. That is why it is challenging to provide millisecond latency. The traffic of IoT is massively significant, due to which the latency rate for the real-time application is increased. The latency depends on the number of technical factors like the type of deployment option used (each having a different link budget), transport block size (TBS is directly linked to several resource units), number of data-transmission repetitions, channel schedule.
2. Energy efficiency – NB-IoT devices are designed for long battery life and are placed in a remote areas where it is difficult to change the battery frequently. So, conserving the battery or energy is a significant issue. Due to the worst channel condition, the signal is repeated to reduce the bit error rate, draining the energy and reducing battery life. Many authors proposed the algorithms, but all consume lots of energy. Other alternatives are solar, vibration, etc, instead of using PSM and eDRx to resolve the energy harvesting problem.

3. Coverage – To enhance the coverage in NB-IoT, most preferably transmission repetition techniques are used. We have to increase the resource unit and lower the modulation coding scheme (MCS) to increase the data rate. It creates an opportunity in the field of research to select which MCS technique and link-adaptation method are more appropriate for NB-IoT technology to extend the coverage.
4. Subcarrier – In NB-IoT, uplink transmission utilizes single-tone as well as and multi-tone transmission. However, the maximum work is done on single-tone transmission. There is still minimal work done in the direction of NB-IoT multi-tone transmission. NB-IoT uses three deployment options out of these three options, which give better performance results for the massive IoT device is still not clear.
5. Network design/ Data rate – Many network designs are proposed for NB-IoT, but the main issue to study is network designs that can handle more data rates. Those interested in network design should have a piece of explicit knowledge about its architecture and deployment options. It creates lots of research questions on how to build an NB-IoT network at the different network layers. NB-IoT and other legacy technologies (LoRa, Sigfox) all work to enhance the data rate. An opportunity is developed to choose the best modulation scheme to enhance the physical layer's data rate. On the other side, network layer NB-IoT design considers single-hop data links within the NB-IoT cell. This design sends a lower data rate to the base station. The opportunity arises for which network/communication protocol is more suitable for processing and storing data at the NB-IoT node.
6. Interference – NB-IoT rehashes the LTE design, and it shares the LTE spectrum, so cancellation, estimation, and interference prediction become a challenge to handle. Many solutions are introduced to avoid interference problems in NB-IoT (better scheduling schemes, time and frequency synchronization, resource blanking). However, for the better performance of NB-IoT, it is still challenging to deploy effective schemes [28].
7. Random Access – Sometime in NB-IoT, the transmission of uplink data has an issue of random-access preamble collision. Much work is done in this direction to upgrade the NPRACH success rate and for the fine time-of-arrival estimation. However, many of the proposed schemes do not think about channel estimation impairment, actual channel condition. Many algorithms and mathematical models are proposed, but which method or scheme is more productive for massive deployment is still vague.
8. Testing tools – NB-IoT technology is growing amazingly fast, but there is still no standard tool to validates whether the produced product matches the standards or not. If there is a proper testing tool, then significant NB-IoT features like mobility selection, modulation coding scheme, number of repetitions required can be easily identified and develop a different model for different scenarios.

9. Scalability – Future NB-IoT is a vast market and connecting billions of devices simultaneously is a big challenge. NB-IoT works with an existing LTE cellular network, and in LTE, there is a centralized server to connect different nodes in a network. As the number of devices increases, it is difficult for the centralized system to handle all the devices, and, in turn, the investment in maintaining the particular cloud server also increases. A decentralized system is used based on artificial intelligence, network topology, fog computing, and peer-to-peer communication. In peer-to-peer communication, security challenge is the main issue, for this, Blockchain is the solution. In Blockchain, each block is linked with the previous block and contains a cryptographic hash. With this technology end to end, secure communication is possible [29]. Decentralized Blockchain keeps a copy of the Blockchain, ensuring computational trust, reducing network latency, and enhancing network energy efficiency. By applying these methods in the future, we enhance the NB-IoT network scalability.
10. Mobility – All the applications are not stationary; some of them are involved moment. Moreover, most of the proposed work ignored NB-IoT's mobility impact on channel modeling if we see the literature. If the Doppler shift is considered during channel estimation, the device complexity is increased because it supports mobility and handover attributes [10]. As the device mobility increases due to this, channel conditions will change frequently. Hence, the designed algorithms could involve low power, frequent CSI reporting, and the optimal number of repetitions.
11. Synchronization Signal – For the frequency and time synchronization of the base sequence, NB-IoT uses two signals named NPSS and NSSS. Initial synchronization in NB-IoT depends on NPSS only. If NPSS is detected, the other channel can easily be detected. The challenging part is when the channel condition is imperfect; it severely affects the cell camping procedure. If the carrier frequency offset (CFO) is slight, it results in phase shift and degrades the performance of synchronization and cell search. The frequency diversity technique enhances synchronization performance for the reception improvement of both NPSS and NSSS.
12. System Integration – The IoT sector is growing extremely fast but lacking in standards, this complicating commercial success. IoT framework also lacking in system understanding, resource requirement, costly budget, limited testing tools. This is all happening because the system integrates computational part, communication, sensors, different protocols, lots of applications needs different programming led to technological risk.

11.6 CONCLUSION

This paper has explored, analyzed, and compared the IoT connecting technology. It discussed the short-range and long-range wireless technology like, Wi-fi, Bluetooth, ZigBee, Z-Wave, Sig-fox, LoRa, INGENU, LTE-M, and NB-IoT. The rest of the article is revolved around the NB-IoT technology. This article has given an exhaustive outline of NB-IoT standards From Releases 13 to Release 17. All the 3GPP releases are discussed according to their proposed enhancement and advantages for the future IoT requirement. Lastly, we discussed the research challenges and the future standard research focus on different NB-IoT issues like conserving battery life, scheduling, data rate, handover mobility support, latency, and optimal resource usage. This article tries to provide clear and detailed information about the different LPWAN technology, 3GPP Releases, and future open research issues to the best of the author's knowledge. This article will help the researcher get crucial and relevant information about the NB-IoT research concept and suggest a possible solution.

REFERENCE

1. U. Raza, P. Kulkarni, and M. Sooriyabandara, "Low Power Wide Area Networks: An Overview," *IEEE Commun. Surv. Tutorials*, vol. 19, no. 2, pp. 855–873, 2016, 10.1109/COMST.2017.2652320
2. M. Elkhodr, S. Shahrestani, and H. Cheung, "Emerging Wireless Technologies in the Internet of Things: A Comparative Study," *Int. J. Wirel. Mob. Networks*, vol. 8, no. 5, pp. 67–82, 2016, 10.5121/ijwmn.2016.8505
3. T. Kura and M. Pūkaha, "Attachment' Main/TechnicalReportSeries/IoT_Technologies_embfonts. pdf' Does Not Exist," p. 2021, 2021, [Online]. Available: https://ecs.victoria.ac.nz/foswiki/pub/Main/TechnicalReportSeries/IoT technologies_embfonts.pdf
4. G. A. Akpakwu, B. J. Silva, G. P. Hancke, and A. M. Abu-Mahfouz, "A Survey on 5G Networks for the Internet of Things: Communication Technologies and Challenges," *IEEE Access*, vol. 6, no. December, pp. 3619–3647, 2017, 10.1109/ACCESS.2017.2779844
5. C. Wang, T. Jiang, and Q. Zhang, *ZigBee® Network Protocols and Applications*. Taylor and Francis, 2016.
6. Musewerx, Z-Wave Wireless Control Technology, System and Applications, White paper, pp. 1–14.
7. S. Popli, R. K. Jha, and S. Jain, "A Survey on Energy Efficient Narrowband Internet of Things (NBIoT): Architecture, Application and Challenges," *IEEE Access*, vol. 7, no. c, pp. 16739–16776, 2019, 10.1109/ACCESS.2018.2881533.
8. H. Wang and A. O. Fapojuwo, "A Survey of Enabling Technologies of Low Power and Long Range Machine-to-Machine Communications," *IEEE Commun. Surv. Tutorials*, vol. 19, no. 4, pp. 2621–2639, 2017, 10.1109/COMST.2017.2721379.
9. R. Ratasuk, N. Mangalvedhe, Z. Xiong, M. Robert, and D. Bhatoolaul, "Enhancements of Narrowband IoT in 3GPP Rel-14 and Rel-15," in *2017 IEEE Conference on Standards for Communications and Networking, CSCN 2017*, 2017, pp. 60–65, 10.1109/CSCN.2017.8088599.
10. A. Karandikar, N. Akhtar, and M. Mehta, *Mobility management in LTE heterogeneous networks*. Springer. 2017.
11. M. Trends and M. Networks, "3GPP Release-16: Further LTE and 5G NR Enhancements Further Enhancements to LTE," pp. 1–10, 2019, [Online]. Available: https://www.grandmetric.com/2018/10/29/3gpp-release-16-further-lte-and-5g-nrenhancements/
12. C. Madapatha *et al.*, "On Integrated Access and Backhaul Networks: Current Status and Potentials," *arXiv*, vol. 1, no. September 2020, 10.1109/ojcoms.2020.3022529.
13. Y. Osman and S. Riikka, "A Look at Key Innovation Areas of 3GPP Rel-17," *Ericsson*, 2019. https://www.ericsson.com/en/blog/2019/12/3gpp-rel-17.

14. E. Migabo, K. Djouani, A. Kurien, and T. Olwal, "A Comparative Survey Study on LPWA Networks: LoRa and NB–IoT," in *Proceedings of the Future Technologies Conference (FTC)* , vol. 2017, no. November, pp. 29–30, 2017.
15. E. M. Migabo, K. D. Djouani, and A. M. Kurien, "The Narrowband Internet of Things (NB-IoT) Resources Management Performance State of Art, Challenges, and Opportunities," *IEEE Access*, vol. 8, pp. 97658–97675, 2020, 10.1109/ACCESS.2020.2995938.
16. V. S. V. Mobile and S. Belgrade, "Narrowband Internet of Things," vol. 16, no. March, pp. 913–923, 2020, 10.4018/978-1-7998-3479-3.ch063.
17. R. Ratasuk, B. Vejlgaard, N. Mangalvedhe, and A. Ghosh, "NB-IoT System for M2M Communication," in *2016 IEEE Wirel. Commun. Netw. Conf. Work. WCNCW 2016*, no. Wd5g, pp. 428–432, 2016, 10.1109/WCNCW.2016.7552737.
18. Sami Tabbane, "IoT Long Range Technologies: Standards," *Powerpoint Present.*, no. December, pp. 1–98, 2017.
19. E. Migabo, K. Djouani, and A. Kurien, "An Energy-efficient and Adaptive Channel Coding Approach for Narrowband Internet of Things (NB-IoT) Systems," *Sensors (Switzerland)*, vol. 20, no. 12, pp. 1–29, 2020, 10.3390/s20123465.
20. T. Inoue, D. Vye, C. AWR Group, and NI, El Segundo "Simulation Speeds NB-IoT Product Development," Microwave Journal, Article no. 29478, 2017.
21. C. Yu, L. Yu, Y. Wu, Y. He, and Q. Lu, "Uplink Scheduling and Link Adaptation for Narrowband Internet of Things Systems," *IEEE Access*, vol. 5, pp. 1724–1734, 2017, 10.1109/ACCESS.2017.2664418.
22. J. M. Cornelius, A. S. J. Helberg, and A. J. Hoffman, "An Improved Error Correction Algorithm for Multicasting over LTE Networks," Ph.D. Thesis, University of North West, Potchefstroom, South Africa, 2014.
23. K. Yu, F. Barac, M. Gidlund, and J. Åkerberg, "Adaptive Forward Error Correction for Best Effort Wireless Sensor Networks," in *IEEE International Conference on Communications*, 2012, pp. 7104–7109, 10.1109/ICC.2012.6364798.
24. M. P. Singh and P. Kumar, "An Efficient Forward Error Correction Scheme for Wireless Sensor Network," *Procedia Technol.*, vol. 4, pp. 737–742, 2012, 10.1016/j.protcy.2012.05.120.
25. A. Abdelbaky and H. Mostafa, "New Low Area NB-IoT Turbo Encoder Interleaver by Sharing Resources," *Proc. Int. Conf. Microelectron. ICM*, vol. 2017, no. December, pp. 1–4, 2018, 10.1109/ICM.2017.8268892.
26. M. Bor and U. Roedig, "LoRa Transmission Parameter Selection," in *Proc. – 2017 13th Int. Conf. Distrib. Comput. Sens. Syst. DCOSS 2017*, vol. 2018, no. January, pp. 27–34, 2018, 10.1109/DCOSS.2017.10.
27. Y. P. E. Wang *et al.*, "A Primer on 3GPP Narrowband Internet of Things," in *IEEE Communications Magazine*, vol. 55, no. 3, pp. 117–123, 2017, 10.1109/MCOM.2017.1600510CM.
28. T. Z. Oo, N. H. Tran, W. Saad, D. Niyato, Z. Han, and C. S. Hong, "Offloading in HetNet: A Coordination of Interference Mitigation, User Association, and Resource Allocation," *IEEE Trans. Mob. Comput.*, vol. 16, no. 8, pp. 2276–2291, 2017, 10.1109/TMC.2016.2613864.
29. M. A. Khan and K. Salah, "IoT Security: Review, Blockchain Solutions, and Open Challenges," *Futur. Gener. Comput. Syst.*, vol. 82, pp. 395–411, 2018, 10.1016/j.future.2017.11.022.

12 Comprehensive Analysis of S-Parameters of 2-D MODFET for Microwave Applications

Ramnish Kumar
E.C.E. Department, GJU S&T, Hisar, Haryana, India

CONTENTS

12.1 INTRODUCTION

AlGaN/GaN Pseudomorphic High Electron Mobility Transistors (pHEMTs) are very large common microwave devices due to its several merits like high voltage, high power, large drift velocity, high temperature, high thermal stability, large breakdown electric field, and high-conduction band discontinuity [1,2]. Pseudomorphic HEMTs

DOI: 10.1201/9781003230526-12

are quickly replacing to conventional MESFET technology in both military as well as commercial applications. pHEMTs are promising device for millimeter wave and optical communication systems due to their low noise performance and excellent high frequency for wireless communication systems [3–6]. At the component level, pHEMTs are widely used in monolithic microwave integrated circuits (MMICs) because of its superior electron mobility [7]. Various methods have been reported for extracting equivalent circuit parameters using analytical methods [8–11].

Here, the RF pHEMT performance i.e., the impact of gate-bias on the S-parameters, has been analyzed. The present analysis starts with finding the intrinsic device Y-parameters in relations of the several elements of its two-port network. The extraction of the s-parameters of the device plays a vital part for the comprehensive analysis for predicting the high-frequency performance of the device. The final aim of this paper is the modeling of such device used to design a low-noise amplifier.

12.2 MODEL FORMULATION

The elementary AlGaN/GaN pHEMT structure considered in the current analysis is shown in Figure 12.1. The given diagram is having similarities with MOSFET due to its gate, source, drain and substrate terminals. But in case of pHEMT, the substrate is a combination of different band gap materials only to exploit the motion of electrons to achieve high electron mobility. A commonly used material combination is GaN with AlGaN, which is having very wide band gap variations.

The threshold voltage, V_{th} of the MODFET is strongly dependent on polarization charge density and is given as [12]:

$$V_{th}(m) = \phi_m(T_0) + m\left[\varepsilon_{gap}^{AlGaN}(T) - \varepsilon_{gap}^{AlGaN}(T_0)\right] - \frac{\Delta E_g}{2} - \frac{ET}{2}\ln\left[\frac{N_c^{AlGaN} N_V^{GaN}}{N_v^{AlGaN} N_c^{GaN}}\right] + \Delta E_{exc}$$
$$- \Delta E_{im} + \Delta E_b - \frac{qN_d d_d^2}{2\varepsilon(m)} - \frac{qN_b}{c_b} + \frac{E_f(m)}{q} \qquad (12.1)$$

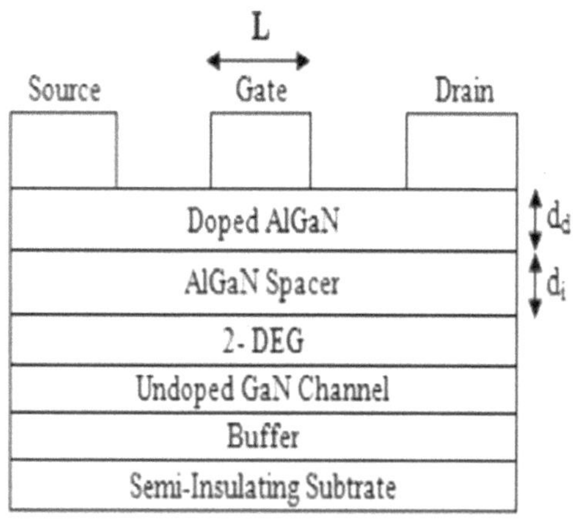

FIGURE 12.1 Schematic of AlGaN/GaN pHEMT.

Where, N_d = doping-concentration, ΔE_b = interface dipole-bond energy, $\varphi_m(T_0)$ = Schottky barrier-height, $E_f(m)$ = Fermi-potential, N_C and N_V are the concentration of electron in the conduction and valence band, $\Delta E_c(m)$= discontinuity in conduction-band, σ(m) = net polarization, ΔE_{im} = dielectric image force, ΔE_{exc} = exchange energy correlation of electron,

The 2-DEG charge density of sheet can be found from Poisson equation [13] as

$$n_s(m, x) = \frac{\varepsilon(m)}{q.\, D}\left[V_{gs} - V_c(x) - V_{th}(m)\right] \tag{12.2}$$

Where, q = electronic charge, D = space between the channel and gate, m = mole fraction, $\varepsilon(m)$ = dielectric constant.

12.2.1 Current-Voltage Characteristics

The channel drain source current is found from the below equation of current density [14] as:

$$\mathrm{I_{ds}}(\mathrm{m}, \mathrm{x}) = \mathrm{z}q\mu(x, m)\left(n_s(m, x)\frac{dv_c(x)}{dx} + \frac{K_B T}{q}\frac{dn_s(m, x)}{dx}\right) \tag{12.3}$$

Where, z = width of gate, K_B = Boltzmann's constant, T = temperature, $\mu(x)$ = electron mobility which is given as:

$$\mu(m) = \frac{\mu_0(m)}{1 + \left(\frac{\mu_0(m)E_c - V_{sat}}{E_c V_{sat}}\right)\frac{dv_{cx}}{dx}} \tag{12.4}$$

By substituting Equation (12.2) and Equation (12.4) in Equation (12.3)

$$\begin{aligned} V_c(x)|_{x=0} &= I_{ds}R_s \\ V_c(x)|_{x=L} &= V_{ds} - I_{ds}(R_s + R_d) \end{aligned} \tag{12.5}$$

Where R_s = parasitic source and R_d = drain resistances.

The channel drain to source current is given as

$$I_{ds} = \frac{-\alpha_1 + \sqrt{\alpha_1^2 - 4\alpha_2\alpha_3}}{2\alpha_2} \tag{12.6}$$

Where,

$$\alpha_1 = \mathrm{E}_1(2R_s + R_d) - \frac{E_2 R_d}{2}(R_d + 2R_s)$$

$$\alpha_2 = -\left[L + E_1 V_{ds} + E_2\left(V_{gs} - V_{th} - \frac{K_B T}{q}\right)(2R_s + R_d) - E_2 V_{ds})(R_s + R_d)\right]$$

and

$$\alpha_3 = E_2\left(V_{gs} - V_{th} - \frac{K_B T}{q}\right)V_{ds} - \frac{E_2}{2}V_{ds}^2$$

The drain saturation voltage V_{dsat} is obtained as

$$V_{dsat} = \frac{-\delta_1 + \sqrt{\delta_1^2 - 4\delta_2\delta_3}}{2\delta_2} \tag{12.7}$$

Where,

$$\delta_1 = \frac{\mu_0 Z\varepsilon(m)E_c}{d}\left[\frac{\alpha_1\mu_0 z\varepsilon(m)E_c}{d} + E_1 - E_2(R_s + R_d)\right] - \frac{E_2}{E}$$

$$\delta_2 = \frac{\mu_0 Z\varepsilon(m)E_c}{d}\left[\frac{-2Q\alpha_1\mu_0 z\varepsilon(m)E_c}{d} + E_1 Q - E_2 Q(R_s + R_d) + L + E_2 Q(2R_s + R_d)\right] + E_2 Q$$

$$\delta_3 = \frac{Q\mu_0 Z\varepsilon(m)E_c}{d}\left[\frac{\alpha s_1\mu_0 z\varepsilon(m)E_2 Q}{d} - L - \frac{E_2(2R_S + R_d)}{d}\right]$$

With

$$Q = V_{gs} - V_{th} - \frac{K_B T}{q}$$

12.2.2 Capacitance – Voltage Characteristics

The analytical equations governing capacitance-voltage characteristics are determined in the foregoing section. The channel current equation in terms of electron-velocity and sheet carrier-concentration is given as [15]

$$I_{ds} = Zqn_s(x)v_d(x) \tag{12.8}$$

Where

$$V_d(x) = \frac{\mu_0 E(x)}{1 + \frac{1}{E_3} E(x)} \tag{12.9}$$

$$E_3 = \frac{\mathrm{E_c\,V_{sat}}}{\mu_0 \mathrm{E_c} - \mathrm{V_{sat}}},\ E(x) = \frac{dV_c(x)}{dx}$$

$$Q = \int_0^t I_{ds}\,dt = \int_0^L \frac{I_{ds}\,dx}{v_d(x)} = \text{Total channel charge} \tag{12.10}$$

Rearranging Equation (12.8) and using Equation (12.2), we get

$$\frac{I_{ds}}{v_d(x)} = \frac{Z\varepsilon(m)}{D}\left\{V_{gs} - V_{th}(m) - V_c(x)\right\} \tag{12.11}$$

Now, Integrate the Equation (12.11) from 0 to L,

$$\int_0^L \frac{I_{ds}}{v_d(x)} dx = \frac{Z\varepsilon(m)}{D} \int_0^L \left\{V_{gs} - V_{th}(m) - V_c(x)\right\} dx \tag{12.12}$$

Using [16]

$$dx = \frac{Zqn_s\mu_0 - \frac{I_{ds}}{E_3}}{I_{ds}} dV_c(x) \tag{12.13}$$

By substituting Equation (12.13) into Equation (12.12) and using boundary conditions in Equation (12.5), the total channel charge is obtained as

$$Q = -\frac{A_1}{3I_{ds}}[B - \{V_{ds} - I_{ds}(R_s + R_d)\}]^3 - [B - I_{ds}R_s]^3 - A_2[B\{V_{ds} - I_{ds}(R_s + R_d)$$
$$- BI_{ds}R_s - \frac{1}{2}\{V_{ds} - 1_{ds}(R_s + R_d)\}^2 + \frac{1}{2}I_{ds}^2 R_s^2 \tag{12.14}$$

Where,

$$A_1 = \frac{z\varepsilon(m)^2 z\mu_0}{D^2},\ B = V_{gs} - V_{th},$$
$$R = R_s + R_d,\ A_2 = \frac{Z\varepsilon(m)}{E_3 D}$$

12.2.3 Gate-Drain Capacitance

It is defined as the rate of change in charge (Q) with respect to drain bias by keeping the gate and source potentials constant. So, gate-drain capacitance can be evaluated as:

$$
\begin{aligned}
C_{gd} = \frac{\partial Q}{\partial V_{ds}} &= \frac{-A_1}{3I_{ds}}[3\{B - (V_{ds} - 1_{ds}R)\}^2\{1 - g_d R\}] \\
&+ [\{B - (V_{ds} - I_{ds}R)\}]^3\left[\frac{A_1}{3I_{ds}^2}g_d\right] + \frac{A_1}{I_{ds}}\cdot(B - I_{ds}R_s)^2(-g_d R_s) \\
&+ (B - I_{ds}R_s)^3\left(\frac{-A_1}{3I_{ds}^2}g_d\right) - A_2[B(1 - g_d R) - BR_s g_d - (Vds - I_{ds}R) \\
&(1 - g_d R) + I_{ds}R_s^2 g_d
\end{aligned} \tag{12.15}
$$

12.2.4 Gate-Source Capacitance

It is defined as the variation in the total charge as change in gate voltage, which can be obtained as

$$
\begin{aligned}
C_{gs} = \frac{\partial Q}{\partial V_{gs}} &= \frac{-A_1}{3I_{ds}}[3\{B - (V_{ds} - I_{ds}R)\}^2\{1 - g_m R\}] \\
&+ \{B - (V_{ds} - I_{ds}R)\}^3\frac{A_1}{3I_{ds}^2}g_m + \frac{A_1}{I_{ds}}\cdot(B - I_{ds}R_s)^2(1 - g_m R_s) \\
&+ (B - I_{ds}R_s)^3\left(\frac{-A_1}{3I_{ds}^2}g_m\right) - A_2[B(-g_m R) + (V_{ds} - I_{ds}R)\} - \{Bg_m R_s + I_{ds}R_s\} \\
&- (V_{ds} - I_{ds}R)(-g_m R) + I_{ds}R_s^2 g_m]
\end{aligned} \tag{12.16}
$$

12.3 SMALL-SIGNAL PARAMETERS

The small-signal parameters (drain/output conductance, transconductance, cut-off frequency and transit time) are extremely important for estimating the microwave performance of the device.

12.3.1 Drain/Output Conductance

The drain conductance, which is equivalent to the maximum voltage gain achievable from a device, as shown below:

$$
g_d(m) = \left.\frac{\partial I_{ds}(m)}{\partial V_{ds}}\right|_{V_{gs}}
$$

$$g_d(m) = \frac{1}{2\alpha_1}[-E_1 + E_2(R_d + R_s) + \frac{1}{2\sqrt{\alpha_1^2 - 4\alpha_2\alpha_3}}\{2\alpha_2(-E_1 + E_2(R_d + R_s)) - 4\alpha_1 E_2(Q - V_{ds})\}] \tag{12.17}$$

12.3.2 Transconductance

It is evaluated as

$$g_m(m) = \left.\frac{\partial I_{ds}(m)}{\partial V_{gs}}\right|_{V_{ds}}$$

$$g_m(m) = \frac{1}{2\alpha_1}\left[-E_2(2R_s + R_d) + \frac{1}{2\sqrt{\alpha_1^2 - 4\alpha_2\alpha_3}}\{2\alpha_2(-E_2(2R_s + R_d)) - 4\alpha_1 E_2 V_{ds}\}\right] \tag{12.18}$$

12.3.3 Cut-Off Frequency

The cut-off frequency is given below as:

$$f_t = \frac{g_m}{2\pi(C_{gs} + C_{gd})} \tag{12.19}$$

Using Equations (12.15), (12.16) and (12.18), the cutoff frequency can be obtained from which the ultimate speed of the device can be determined.

12.3.4 Transit Time

Lesser transit times are very needed to get a high-frequency response of device which is shown as:

$$T(t) = \frac{1}{2\pi f_t} \tag{12.20}$$

Substituting Equation (12.19) into Equation (12.20), transit time can be obtained.

12.4 Y-PARAMETERS EXTRACTION

The admittance parameters can be derived from Figure 12.2 using circuit analysis technique. The input and output currents can be given as:

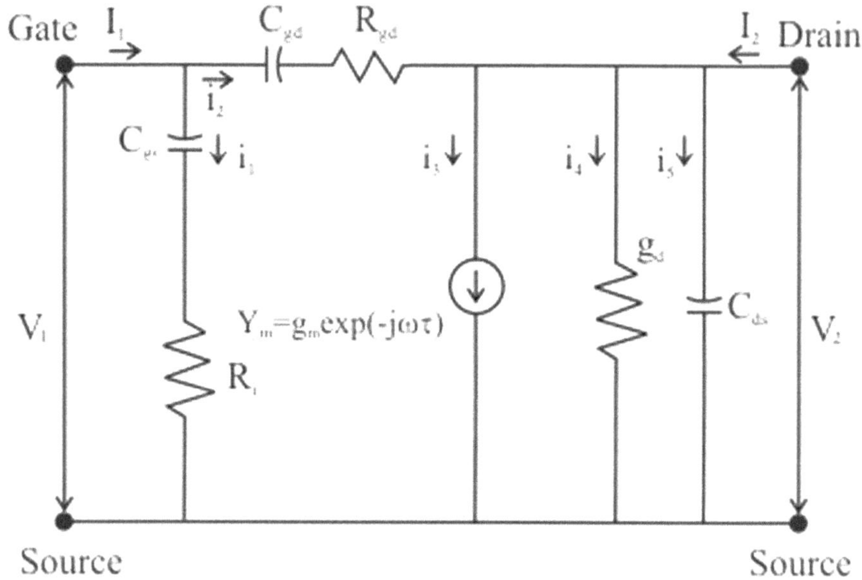

FIGURE 12.2 Small signal equivalent circuit of MODFET for network analysis.

$$I_1 = i_1 + i_2 \tag{12.21}$$

$$I_2 = i_4 + i_5 + i_3 + -i_2 \tag{12.22}$$

I_1 and I_2 can be obtained using nodal analysis

$$I_1 = \frac{V_1(j\omega C_{gs})}{(j\omega C_{gs}R_i + 1)} + \frac{(V_1 - V_2)(j\omega C_{gd}R_{gd})}{1 + (j\omega C_{gd}R_{gd})} \tag{12.23}$$

$$I_2 = V_2 j\omega C_{ds} + V_2 g_d + \frac{V_1}{(1 + j\omega C_{gs}R_i)} g_m(-j\omega\tau) - \frac{(V_1 - V_2)(j\omega C_{gd})}{1 + (j\omega C_{gd}R_{gd})} \tag{12.24}$$

The several admittance parameters may be identified as:

a. The input admittance Y_{11} is defined as:

$$\mathrm{Y}_{11} = \left.\frac{I_1}{V_1}\right|_{V_2=0} \tag{12.25}$$

Using $V_2 = 0$ from (12.23),

$$Y_{11} = \frac{\omega^2 C_{gs}^2 R_i^2}{D} + \frac{\omega^2 C_{gd}^2 R_{gd}}{D'} + \frac{j\omega C_{gs}}{D} + \frac{j\omega C_{gd}}{D'} \tag{12.26}$$

Where

$$D = 1 + \omega^2 R_i^2 C_{gs}^2$$
$$D' = 1 + \omega^2 R_{gd}^2 C_{gd}^2$$

b. The reverse transfer admittance Y_{12} is defined as:

$$Y_{12} = \left.\frac{I_1}{V_2}\right|_{V_1=0} \tag{12.27}$$

from (12.23), using $V_1 = 0$

$$Y_{12} = \frac{-j\omega C_{gd}^2}{D_2} - \frac{\omega^2 C_{gd}^2 R_{gd}^2}{D_2} \tag{12.28}$$

c. The forward transfer admittance is defined as:

$$Y_{21} = \left.\frac{I_2}{V_1}\right|_{V_2=0} \tag{12.29}$$

Putting $V_2 = 0$ in Equation (12.24)

$$Y_{21} = \frac{g_m \exp(-j\omega\tau)}{1 + j\omega C_{gs} R_i} - \frac{j\omega\ C_{gd}}{1 + j\omega R_{gd} C_{gd}} \tag{12.30}$$

d. The output admittance is defined as:

$$Y_{22} = \left.\frac{I_2}{V_2}\right|_{V_1=0} \tag{12.31}$$

Substituting $V_1 = 0$ in Equation (12.24)

$$Y_{22} = g_d + j\omega\left[C_{ds} + \frac{C_{gd}}{D'}\right] + \frac{\omega^2 C_{gd}^2 R_{gd}}{D'} \tag{12.32}$$

The Equations (12.26), (12.28), (12.30) and (12.32) can be further simplified assuming R_{gd} very small

$$Y_{11} = \frac{\omega^2 C_{gs}^2 R_i^2}{D} + j\omega\left(\frac{C_{gs}}{D} + C_{gd}\right)$$

$$Y_{12} = -j\omega C_{gd}$$

$$Y_{21} = \frac{g_m \exp(-j\omega\tau)}{1 + j\omega C_{gs} R_i} - j\omega C_{gd}$$

$$Y_{22} = g_d + j\omega\left[C_{ds} + C_{gd}\right]$$

with

$$\omega = 2\pi f$$

12.5 S- PARAMETERS EXTRACTION

The Y-parameters can be converted into s-parameters using [17]

$$S_{11} = \frac{(1 - Z_0 Y_{11})(1 + Z_0 Y_{22}) + Z_0^2 Y_{12} Y_{21}}{\nabla} \tag{12.33}$$

$$S_{12} = \frac{-2(Z_0 Y_{12})}{\nabla} \tag{12.34}$$

$$S_{21} = \frac{-2(Z_0 Y_{21})}{\nabla} \tag{12.35}$$

$$S_{22} = \frac{(1 + Z_0 Y_{11})(1 - Z_0 Y_{22}) + Z_0^2 Y_{12} Y_{21}}{\nabla} \tag{12.36}$$

where

$$\nabla = (1 + Z_0 Y_{11})(1 + Z_0 Y_{22}) - Z_0^2 Y_{12} Y_{21} \text{ and}$$

$Z_0 = 50\,\Omega$, is the characteristics impedance at each port.

12.6 GAINS

12.6.1 Unilateral Power Gain

It is the maximum possible gain that an active port can achieve. It is obtained using an expression [18]:

$$U_T = \frac{\frac{1}{2}\left|\frac{S_{21}}{S_{12}} - 1\right|^2}{k\left|\frac{S_{21}}{S_{12}}\right| - \mathrm{Re}\left[\frac{S_{21}}{S_{12}}\right]} \tag{12.37}$$

Where k is Stability factor and is defined as:

$$k = \frac{1 + |\Delta|^2 - |S_{11}|^2 - |S_{22}|^2}{2\,|S_{12}S_{21}|}$$

and

$$\Delta = S_{11}S_{22} - S_{12}S_{21}$$

Using Equations (12.33), (12.34), (12.35) and (12.36) in Equation (12.37), the Unilateral power gain can be evaluated.

12.6.2 Maximum Stable Power Gain

It is defined as the maximum possible power gain value that is attained before the occurrence of instability i.e.,

$$Gms = \frac{|S_{21}|}{|S_{12}|} \tag{12.38}$$

Using Equations (12.34) and (12.35), *Gms* can be evaluated.

12.6.3 Maximum Unilateral Transducer Power Gain

It can be found from the expression given as [19]:

$$G_{TU\,\mathrm{max}} = \frac{|S_{21}|^2}{(1 - |S_{11}|^2)(1 - |S_{22}|^2)} \tag{12.39}$$

Substituting Equations (12.33), (12.35) and (12.36) in Equation (12.39), $G_{TU\,\mathrm{max}}$ can be evaluated.

12.6.4 Unilateral Figure of Merit (FOM)

It can be calculated as:

$$U = \frac{|S_{11}||S_{12}||S_{22}||S_{21}|}{(1 - |S_{11}|^2)(1 - |S_{22}|^2)} \tag{12.40}$$

Using Equations (12.33), (12.34), (12.35) and (12.36) in Equation (12.40), the Unilateral figure of merit can be obtained.

12.7 RESULTS AND DISCUSSION

The frequency variation of the various S-parameters, both in real and imaginary form, are shown in the following figures. These results are analytically calculated using intrinsic equivalent model of circuit. Further, the obtained results are compared through the results found from the ATLAS 3-D device simulator [20,21].

Figure 12.3 shows the variation in input reflection coefficient for both real and imaginary part as a frequency function. Here, it is observed that the impact of gate bias does

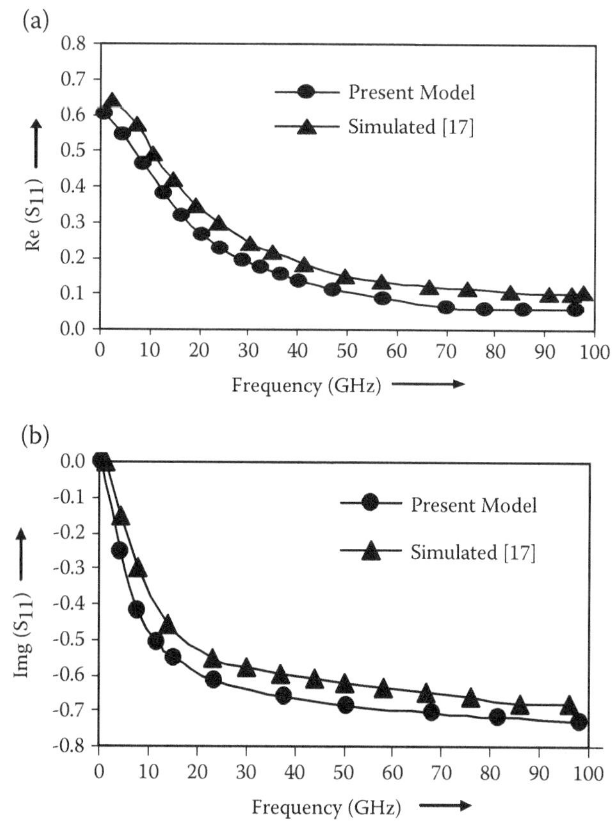

FIGURE 12.3 (a) Real part of input reflection coefficient (S_{11}) at $V_{ds} = 0.4$ V and $V_{gs} = 0.5$ V (b) Imaginary part of input reflection coefficient (S_{11}).

not have a significant impact on. S_{11} Since, S_{11} have the direct relation with the matching or mismatching level of the input amplifier circuit. Therefore, the V_{gs} does not impact on input matching of microwave amplifier as with source impedance

It can easily be observed from the figure that the V_{gs} significant impact on S_{22} is significantly less as compared with S_{21} as well as S_{12}. At higher values of V_{gs} for low frequencies, the real part of S_{22} illustrate low values. But, the imaginary part of S_{22} shows high values (Figure 12.4).

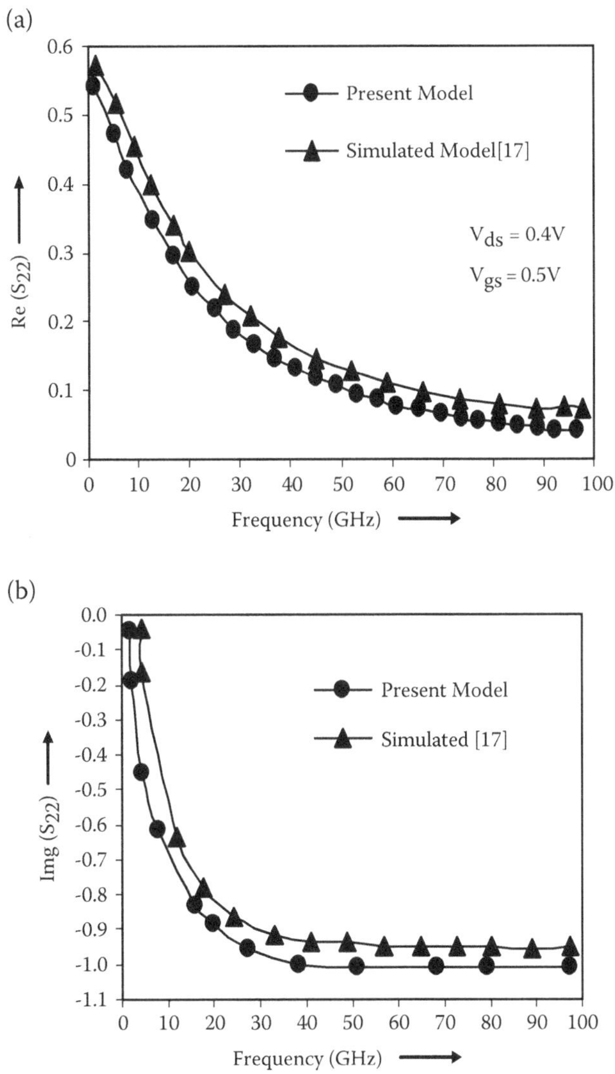

FIGURE 12.4 (a) Real part of output reflection coefficient (S_{11}) (b) Imaginary part of output reflection coefficient (S_{22}).

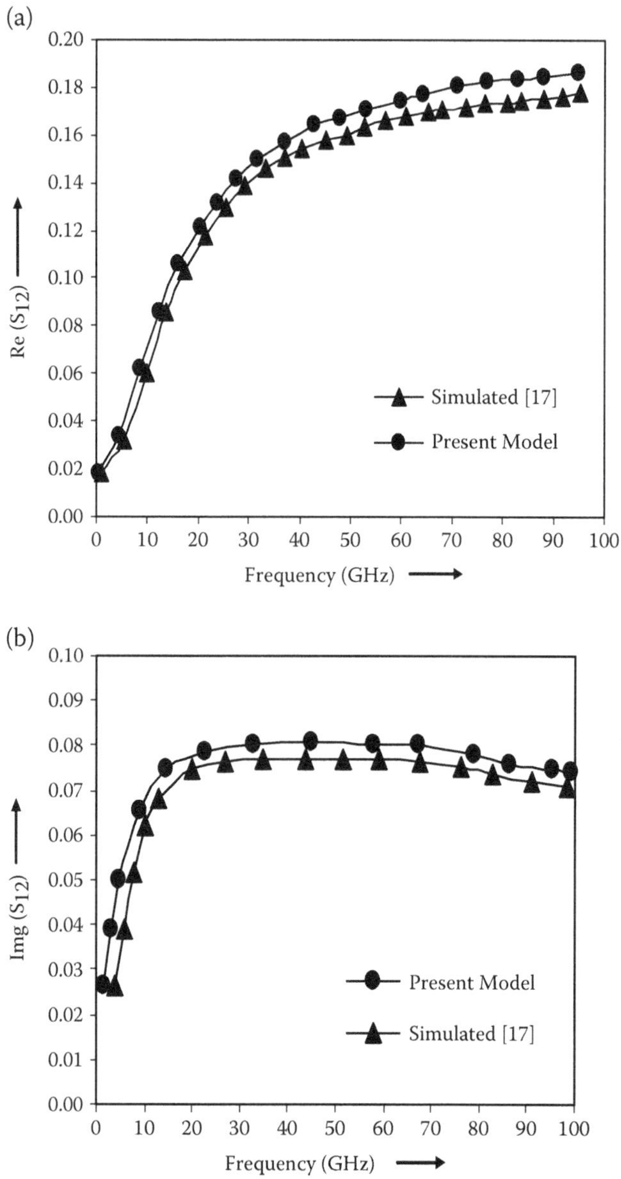

FIGURE 12.5 (a) Real part of reverse transmission coefficient (S_{12}) (b) Imaginary part of reverse transmission coefficient (S_{12}).

It is observed from the Figure 12.5, the real part of S_{12} takes a high value for high V_{gs} at frequencies range greater than 50 GHz.

It is observed from the Figure 12.6, the high value of V_{gs} gives high value of S_{21} with large impact of V_{gs} on the imaginary-part of S_{21}.

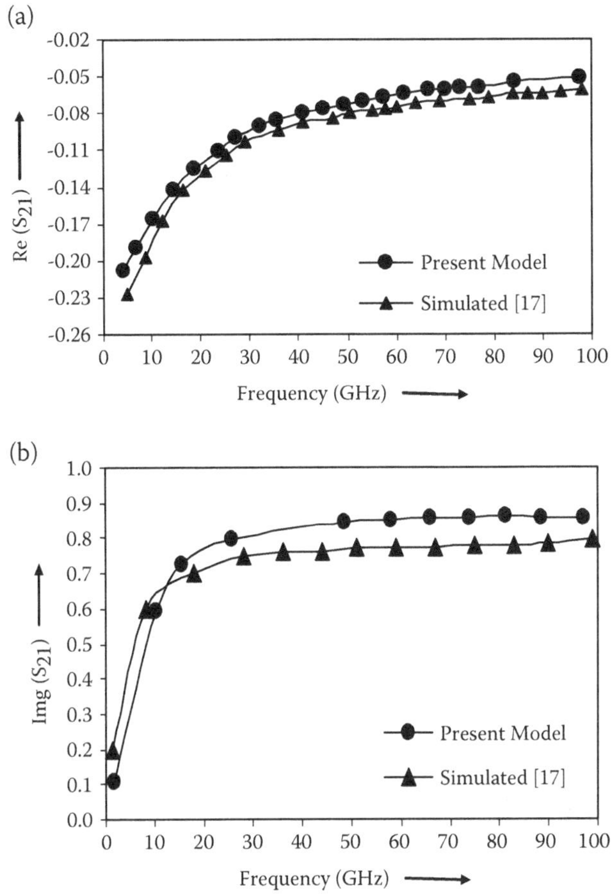

FIGURE 12.6 (a) Real part of forward transmission coefficient (S_{21}) (b) Imaginary part of forward transmission coefficient (S_{21}).

Figure 12.7 shows that an exponential unilateral power gain decreases as frequency increases. Also, the decrease in input and output mismatch observed as the frequency increases. Further, the largest unilateral transducer power gain obtained at a given frequency.

Figure 12.8 shows that the drain current increases with the increase value of drain voltage. Also, the maximum drain current present in model is 502 mA/mm at 1 V gate bias voltage. Its mean AlGaN/GaN devices can be used effectively for large power applications. The model exhibits high current driving capabilities and good pinch-off characteristics. It is seen that the current-voltage characteristics of MODFET resembles with the drain characteristics of MOSFET.

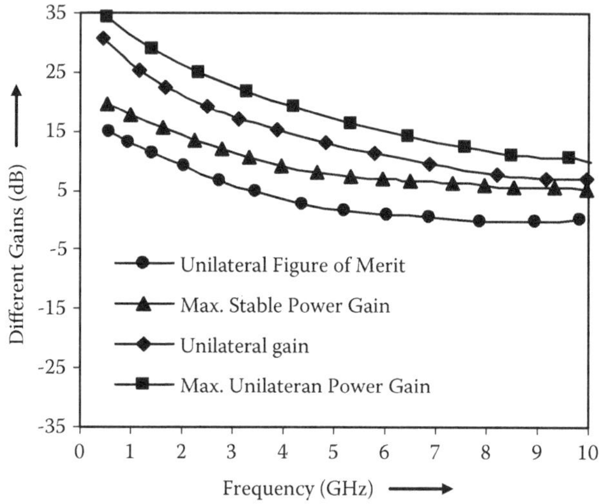

FIGURE 12.7 Variation of different gains as a function of frequency.

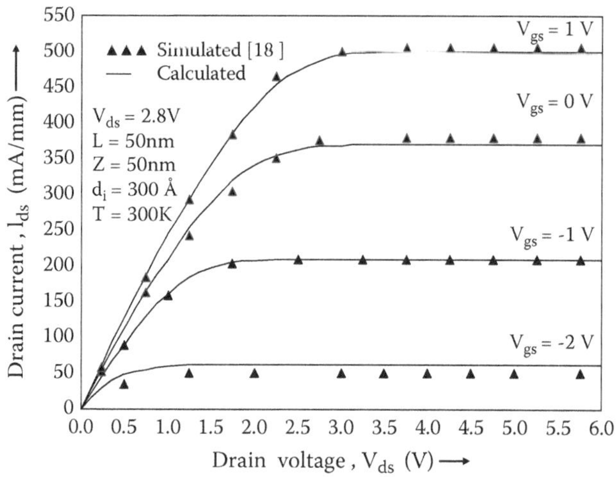

FIGURE 12.8 Variation of drain current with drain voltage.

12.8 CONCLUSION

A comprehensive analysis of s-parameters for AlGaN/GaN pHEMT has been presented in this paper to predict its performance. The V_{gs} dependence of the s-parameters of a AlGaN/GaN pHEMT were obtained by using the small-signal parameters analysis. A high value of maximum oscillation frequency and the unilateral peak value of power gain verify the device suitability for microwave applications. Further, results shows that the impact of V_{gs} is more significant for both

Transmission Coefficient. But, the dependency of reflection coefficient is significantly less on V_{gs}. The scattering -parameters calculated results are more promising with the ATLAS-3D device simulator results which validates the analytical model.

REFERENCES

1. M. N. A. Aadit, S. G. Kirtania, F. Afrin, Md. K. Alam, and Q. D. M. Khosru, "Different types of field-effect transistors – Theory and applications," *High Electron Mobility Transistors: Performance Analysis, Research Trend and Applications*, pp. 45–64, 2017.
2. Francisco E. Rangel Patino and J. Rodrigo Camacho Perez, "Modeling and simulation of Pseudomorphic HEMTs for analog circuit design and analysis," *Instituto Tecnologico de Chihuahua*, pp. 277–282, ELECTRO 2001.
3. M. B. Tayel and A. H. Yassin, "Parameters Extraction for Pseudomorphic HEMTs using Genetic Algorithms," *World Academy of Science, Engineering and Technology*, vol. 26, no. 2, pp. 757–760, 2009.
4. P. K. Chopra, S. Jain, and A. Ahlawat, "ANN modeling approach for designing low noise phemt amplifier in wireless communication systems," *Optical Memory and Neural Networks*, vol. 20, no. 4, pp. 271–278, 2011.
5. S. Sano, K. Ebihara, T. Yamamoto, T. Sato, and N. Miyazawa, "GaN HEMTs for wireless communication," *Journal of SEI Technical Review*, vol. 1, no. 86, April 2018.
6. A. Helali, M. Gassoumi, M. Gassoumi, and H. Maaref, "Design and optimization of LNA amplifier based on HEMT GaN for X-band wireless-communication and IoT applications,' *Springer*, vol. 13, pp. 2645–2653, August 2020.
7. M. G. Goswami and R.S. Gupta, "Analysis of scattering parameters and thermal noise of a MOSFET for its microwave frequency applications," *MOTL*, vol. 31, no. 2, pp. 97–105, 2001.
8. L. Sang, Y. Xu, Y. Chan, Y. Guo, and R. Xu, "Large signal equivalent circuit model for package AlGaN/GaN HEMT," *Progress in Electromagnetics Research Letters*, vol. 20, pp. 27–36, 2011.
9. M. Han, Y. Dai, J. Zhou, and C. Xu Li, "A new approach to determine parasitic elements of GaN HEMT by cold FET s-parameter," *Terahertz Science and Technology*, vol. 5, no. 4, pp. 189–195, 2012.
10. A. K. Visvkarma, R. Laishram, S. Kapoor, D. S. Rawal, S. Vinayak, and M. Saxena, "Improvement in DC and pulse characteristics of AlGaN/GaN HEMT by employing dual metal gate structure," *Semiconductor Science and Technology*, vol. 34, no. 10, 2019.
11. S. Zhu, H. Jia *, T. Li, Y. Tong, Y. Liang, X. Wang, T. Zeng, and Y. Yang, "Novel high-energy-efficiency AlGaN/GaN HEMT with high gate and multi-recessed buffer," *Micromachines*, vol. 10, no. 444, pp. 1–12, 2019.
12. S. K. A. Ramnish and A. Ahlawat, "An analytical two-dimensional C- V model of AlGaN/GaN MODFET for high speed circuit applications," *The IUP Journal of Telecommunications*, vol. V, no. 4, pp. 49–68, 2013.
13. R. K. Tyagi, A. Ahlawat, M. Pandey, and S. Pandey, "An analytical two-dimensional model for AlGaN/GaN HEMT with polarization effects for high power applications," *Microelectronics Journal*, vol. 38, pp. 877–883, 2007.
14. A. Ahlawat, M. Pandey, and S. Pandey, "Microwave analysis of 70 nm InGaAs pHEMT on InP substrate for nanoscale digital IC application," *Microwave and Optical technology letters*, vol. 49, no. 10, pp. 2462–2470, 2007.
15. A. Aggarwal, A. Goswami, S. Sen, and R. S. Gupta, "Transconductance extraction for Pseudomorphic modulation doped field effect transistor (AlGaAs/InGaAs) for microwave and millimeter wave applications," *MOTL*, vol. 22, no. 1, 1999.

16. S. Sen, M. K. Pandey, and R. S. Gupta, "Two-dimensional C-V model of AlGaAs/GaAs modulation doped field effect transistor (MODFET) for high frequency applications," *IEEE Transactions on Electron Devices*, vol. 46, no. 9, pp. 1818–1823, 1999.
17. S. Ramo, J. R. Whinnery, and T. Van Duzer, *Fields and Waves in Communication Electronics*, Third Edition, John Wiley & Sons Inc. pp. 537–541, 1993.
18. V. Guru, J. Jogi, M. Gupta, H. P. Vyas, and R. S. Gupta, "An improved intrinsic small-signal equivalent circuit model of delta-doped AlGaAs/InGaAs/GaAs HEMT for microwave frequency applications," *MOTL*, vol. 37, no. 5, pp. 376–379, 2003.
19. A. Goswami, S. Agarwal, S. Bose, M. G. Haldar, and R. S. Gupta, "Substrate effect dependent scattering parameter extraction of short gate length IGFET for microwave frequency applications," *MOTL*, vol. 24, no. 5, pp. 341–348, 2000.
20. ATLAS Device Simulator, SILVACO International, 2009.
21. A. Rashmi, S. Gupta, S. H. Sen, and R. S. Gupta, "Analytical model for dc characteristics and small-signal parameters of AlGaN/GaN modulation-doped field-effect transistor for microwave circuit applications," *Microwave and Optical Technology Letters*, vol. 27, no. 6, pp. 413–419, 2000.

13 U-Shape Planar Antenna Design Effects Using CST

Rishu Bhatia
Associate Professor, Ganga Institute of Technology and Management, Kablana, Jhajjar (Haryana), India

Renuka Arora
Assistant Professor, Ganga Institute of Technology and Management, Kablana, Jhajjar (Haryana), India

Rakesh Joon
Associate Professor, Ganga Institute of Technology and Management, Kablana, Jhajjar (Haryana), India

Sanjima
Assistant Professor, Chandigarh University, Mohali (Punjab), India

CONTENTS

13.1 INTRODUCTION

During the past decades, many applications have depended upon the event of planar transmitters [1–18]. These transmitters are required for a feast of submissions. They will be constructed at a mode for less expense than other expertise and are extra squeezed and positive-encumbrance. These attributes are critical for the various saleable submissions for which planar projections are widely utilized in base position or handset projections. The planar fauna of the probes makes them ideal for massive arrays and shortens the interface of other electronics, like loudspeakers and phase shifters, which are mandatory for satellite communications. Their planar environment also consents to the projections in submissions where aspect and profile are essential. These numerous requests have inspired a good sort of planar antenna courses. The use of a Microstrip radio wire is an accomplishment in remote

DOI: 10.1201/9781003230526-13

correspondence frameworks, and it is fulfilling the need of current remote correspondence innovation, compared to new advancements. Microstrip receiving wires are being used in all of these frameworks due to their numerous benefits [19], like being extremely lightweight, their organizer structure, and their efficient productivity. Alternatively, the restricted working transfer speed impedes, which forces limitations in its utilization in remote frameworks [20]. A broadband application performing different undertakings and small contraptions has become a principal part of our day-to-day correspondence life. Consequently, the prerequisite for low-profile wideband has been downsized [21]. Microstrip receiving wire fulfills most of the portable and satellite hardware prerequisites, and its usage fulfills various business necessities. Electronic circuits needed for remote applications are contracting unquestionably, where the microstrip is without a doubt reasonable. The size of the radio wires being utilized for a large portion of the application is likewise contracting. Microstrip radio wires fix plan satisfaction with the these necessities. A unique procedure has been considered [22–24], and it tracked down that appropriate impedance transmission capacity of the microstrip receiving wire can be one justification improvement. The cutting impacts of indents [25,26] are frequently exercised for wireless local area networks and future evolution applications. An essential type of the Microstrip radio wire can be built utilizing dielectric substrate as a base material and a leading emanating material tingled on the upper side of the substrate. The emanating directing material is of any mathematical shape as an essential structure or standard shape to improve the investigation and execution expectation. To satisfy WLAN standards, a multiband antenna that operates at 2.0 to 2.7 GHz is required. Various coupling-feed technologies often change the range of the antenna. The compact size antenna is favorable for several communication services. This structure is to be made on a 1.53 mm thick FR-4 type substrate [27–31].

13.2 ANTENNA CLASSIFICATION

Radio cables can be named omni directional, conveying similarly directional vitality, where energy emanates additional caption. An entirely even omni directional radio cable is not genuinely believable. Some receiving-line categories have a constant contamination enterprise in the smooth, yet send little energy vertical or descending. A "directional" radio wire, as a rule, is proposed to boost its coupling to the electromagnetic field toward the other station. An upward receiving wire transmits every which mode evenly. However, it sends less energy vertical. A dipole receiving wire accumulated on an equal plane sends minute energy in a heading trajectory corresponding to the conveyor; this district is the radio wire invalid. The dipole radio link, which justifies most extreme wire plans, is sensible, with identical yet converse powers and streams utilitarian at its two incurables. The vertical getting wire is a monopole radio wire, not offset concerning ground. The ground accepts the bit of the blaze anode of a dipole. Since monopole radio wires rely upon a conductive outside, they may be put with a minced plane to infer the effect of being appended on the Earth's outside. A graph of the electric fields (blue) and attractive (red) shows a dipole receiving wire during transmission. More mind-boggling radio wires increment the directivity of the receiving wire. Extra components in the radio wire structure, which need not be straightforwardly associated with the recipient or transmitter, increment its directionality. Radio wire "acquire" depicts the grouping of transmitted force into a specific strong point of the room. "Gain" is perhaps a lamentably picked term, by assessment with the speaker "procure," which derives a net development in power. Curiously, for radio wire "get," the power extended the best way is to the weakness of power diminished undesirably. As opposed to enhancers, radio wires are electrically "inert" devices that save hard-and-fast power, and there is no addition in complete power over that passed on from the power source, further creating a course of that fixed total. An arranged show contains somewhere around two direct getting wires, which are related together through an electrical association. This regularly includes various equal dipole radio wires with a specific dividing. Contingent upon the overall stage presented by the organization, a similar blend of dipole receiving wires can work as a "broadside exhibit." Radio wire clusters might utilize any fundamental receiving wire type, like a dipole, circle, or opening receiving wires. These components are frequently indistinguishable. A log-occasional dipole exhibit comprises various dipole components of various lengths to acquire a directional receiving wire with extensive data transmission. The dipole receiving wires forming it is thoroughly thought of as "dynamic components" since they are, on the whole, electrically associated together.

A Yagi–Uda radio cable has a single component through an electrical association; the extra sponging components communicate with the specific field to understand a receiving wire over a light transfer speed. There might be various purported "chiefs" before the dynamic component toward engendering and at least one "reflector" on the contrary side of the dynamic component. More important directionality can be obtained utilizing pillar framing procedures like an allegorical reflector or a horn. At low frequencies, varieties of vertical pinnacles are utilized to accomplish objectives and possess huge land spaces. For get-together, a broad infusion radio wire can have raw directivity. For non-controlling advantageous use, a short vertical radio wire or little circle getting wire works outstandingly, with the essential arrangement challenge being that of impedance organizing. With a vertical radio wire, a stacking twist at the establishment of the getting wire may be used to drop the loose fragment of impedance; tiny circle getting wires are tuned with equivalent capacitors consequently. A radio wire indication is a line that associates the receiving wire to a transmitter or collector. The "receiving wire feed" may allude to all parts interfacing the radio wire to the transmitter. In an alleged "gap receiving wire," like a horn or explanatory dish, the "feed" may likewise allude to a fundamental emanating radio wire inserted in the whole arrangement of reflecting components which could be viewed as the one dynamic component in that receiving wire framework. A microwave receiving wire may likewise be taken care of straightforwardly from a waveguide in its place of a transmission line. A receiving wire balance is a design of conductive substantial, which improves the ground effects. It very well might be associated with or protected from the regular ground. In a monopole getting wire, this guides in the capacity of the normal crushed. Such a plan is usually connected with the apparent relationship of an unfit transmission line, like a coaxial relationship. An electromagnetic propensity refractor in getting wires is a section because of its layout and spot abilities to explicitly concede or drive pieces of the electromagnetic wave going over it. The refractor adjusts the longitudinal attributes of the tendency on one side proportional with the differing side. This is what could be compared to an optical focal point. A radio wire coupling network is an uninvolved organization that coordinates impedance in the middle of the receiving wire and the transmitter. This might be utilized to further develop the standing upsurge proportion to bound misfortunes in the transmission line and to bounce the spreaderaverage resistive impedance.

13.3 ANTENNA FORMATION AND DESIGN TECHNIQUE

Antennas with half-dipole structures have small bandwidth. To extend bandwidth, the slotted antenna is used for broadband application. During this design, various sort of slotting is employed for effective result. The antenna angle is directly changed with the bandwidth of the antenna. The bandwidth of the antenna is often adjusted by changing the slot dimension on horizontal or vertical sides. The slot projection, covering a narrow-milled plane, maybe a very actual antenna. With changes in structure, the scheme is effective to waveguide and feeding schemes and has initiated a claim for facets of wireless and radar society. The slot-based transmitter was explored in the 1940s. Planar microstrip fed hole antennas were utilized in 1972.

A fixed radio wire is a category of receiving cable with a place of security, which can be based on an exterior. It comprises a planar rectangular, round, three-sided mounted over a more significant piece of metal called a ground flat. They are the first generous microstrip receiving wire depicted by Howell in 1972. The radiation system emerges from bordering fields along the emanating edges. The radiation at the controls makes the receiving wire act somewhat superior electrically than its actual dimensions, so all together, for the radio wire to be thunderous, a span of microstrip program line marginally more limited portion of the frequency. (Figure 13.1).

FIGURE 13.1 Slotted planar antenna design.

Change with the bandwidth of the antenna. The bandwidth of the antenna is often adjusted by changing the slot dimension on horizontal or vertical sides. The slot projection, covering a narrow-milled plane, maybe a very actual antenna. With changes in structure, the scheme is effective to waveguide and feeding schemes and has initiated claim altogether facets of wireless and radar society. The slot-based transmitter was explored within the 1940s. Planar microstrip fed hole antennas were utilized in 1972.

A fixed radio wire is a category of receiving cable with a place of security, which can be based on an exterior. It comprises a planar rectangular, round, three-sided mounted over a more significant piece of metal called a ground flat. They are the first generous microstrip receiving wire depicted by Howell in 1972. The radiation system emerges from bordering fields along the emanating edges. The radiation at the controls makes the receiving wire act somewhat superior electrically than its actual dimensions, so all together, for the radio wire to be thunderous, a span of microstrip program line marginally more limited portion of the frequency.

A variation of the fixed radio wire regularly utilized in cell phones is the shorted shot receiving wire. In this receiving cable, one angle of the shot is grounded with a milled pin. This difference has desirable coordinating over the typical shot. Another difference of shot receiving cable with the mostly scraped crashed planeis a highly flexible receiving cable for dual-band actions. The conclusion of varying the scope of the ground plane on the projection is investigated through replication.

The reverberating partially-wavelength space projection is of compressed scope. The transmitter has a huge input impedance, which makes it unpleasant to contest. These matching problems are often removed by expending a balance microstrip source or the creased-slot projection, which stops honestly from the creased dipole. The broad span of the projection remains around a half-wavelength but growing the total of creases resistance. (Table 13.1).

TABLE 13.1
Planar antenna characteristics

Type of Antenna	Type of Pattern	Directivity	Polarization	Bandwidth	Comments
Patch	broadside	medium	linear/circular	narrow	easiest design
Slot	broadside	low/medium	linear	medium	bidirectional
TSA (Vivaldi)	endfire	medium/high	linear	wide	feed transition
Yagi slot	endfire	medium	linear	medium	two-layer design
Quasi-Yagi	endfire	medium/high	linear	wide	uniplanar, compact
LPDA	endfire	medium	linear	wide	balun, two-layer
Leaky Wave	scannable	high	linear	medium	beam-steering, beam tilting
Ring	broadside	large	linear/circular	narrow	feeding complicated
Spiral	broadside	medium	linear/circular	wide	balun and absorber
Bow-tie	broadside	Medium/large	linear	wide	same as spiral

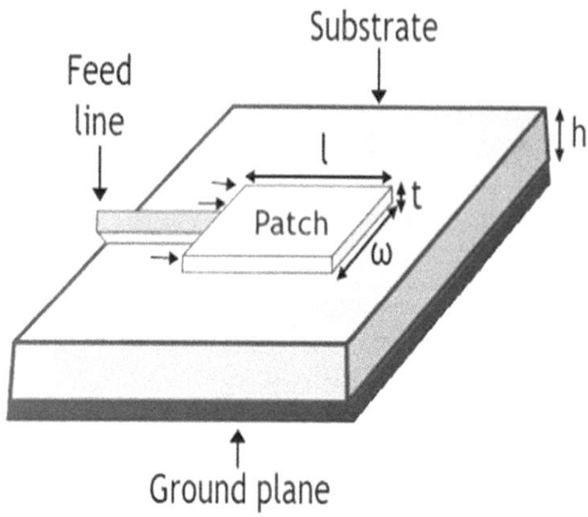

FIGURE 13.2 Top view of microstrip antenna.

The antenna fabrication is done with a 1.53 mm thick substrate of FR-4 material. First, a ground of PEC material with 0.035 mm thick is used with dimensions L = −15 to 15 mm, B = −15 to 15 mm and H = 0.0 to 0.035 mm after those five patches are considered U-shape slot cut to dimension. (Figure 13.2).

Steps for calculating the dimension of thc antenna:

Step 1. **Design of the Width (*W*)**:

The width of the Patch of the antenna is given as:

$$W = \frac{c}{2f_o\sqrt{\frac{\varepsilon_r + 1}{2}}}$$

Step 2. **Design of dielectric constant** ($\varepsilon reff$): The effective dielectric constant is:

$$\varepsilon_{reff} = \frac{\varepsilon_r + 1}{2} + \frac{\varepsilon_r - 1}{2}\left[1 + 12\frac{h}{W}\right]^{0.5}$$

Step 3. **Design of the Effective length** (*Leff*):

The effective length is:

$$L_{eff} = \frac{C}{2f_o\sqrt{\varepsilon_{reff}}}$$

Step 4. **Design of the extension length** (ΔL):

The length extension is:

$$\Delta L = 0.412h\frac{(\varepsilon_{reff} + 0.3)\left(\frac{W}{h} + 0.264\right)}{(\varepsilon_{reff} - 0.258)\left(\frac{W}{h} + 0.8\right)}$$

Step 5. **Design of actual patch length** (L):

The actual length is obtained by:

$$L = L_{eff} - 2\Delta L$$

Step 6. **Design of the dimensions of the ground plane** (Lg **and** Wg):

$$L_g = 6h + L$$

$$W_g = 6h + W$$

After the spacing feed is used and add Patch and Feed internally. Antenna feed is employed for connecting (Figures 13.3–13.7) (Table 13.2).

Antenna and radio signal. It is the location on the antenna where the feed line of transmitter and receiver connects. It converts the feed line impedance to the inherent impedance of the transmitter. Operating bands are also preset during production,

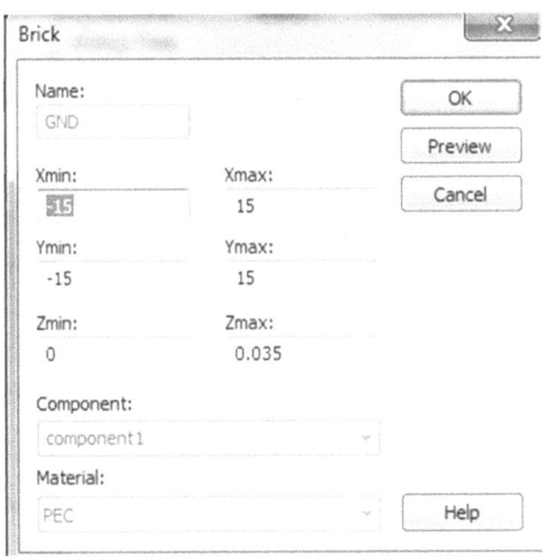

FIGURE 13.3 Ground dimension.

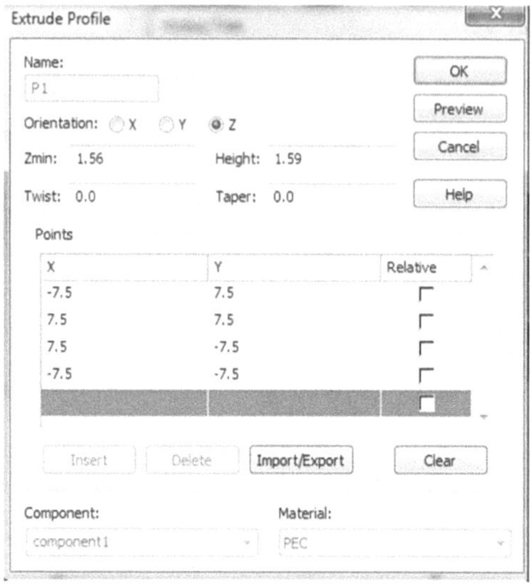

FIGURE 13.4 Patch 1st.

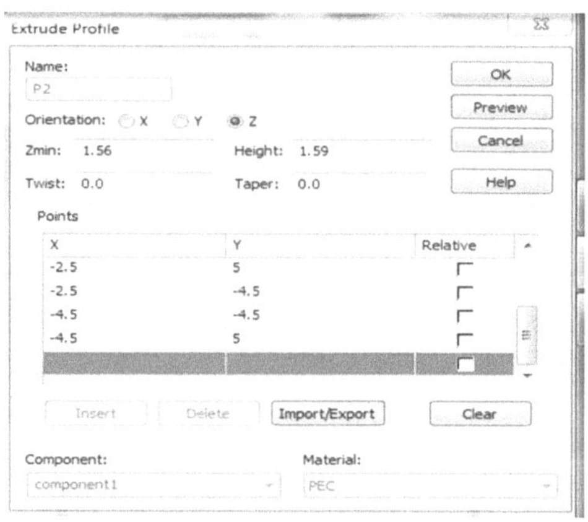

FIGURE 13.5 Patch 2nd.

which makes the antenna inflexible to put up new services. Such an antenna can recognize itself to any modification in the surroundings (Figures 13.8–13.10).

13.4 EXPERIMENTAL RESULTS

Fix radio wires come in different shapes and sizes and comprise a metal fix straight over a ground plane. The fundamental impediment of these receiving wires is their generally huge size contrasted with different sorts of radio wires. For instance, some fixed radio wires are around a large portion of a frequency on each side. The polarization can be round or direct, contingent upon the plan of the fix. In a fixed

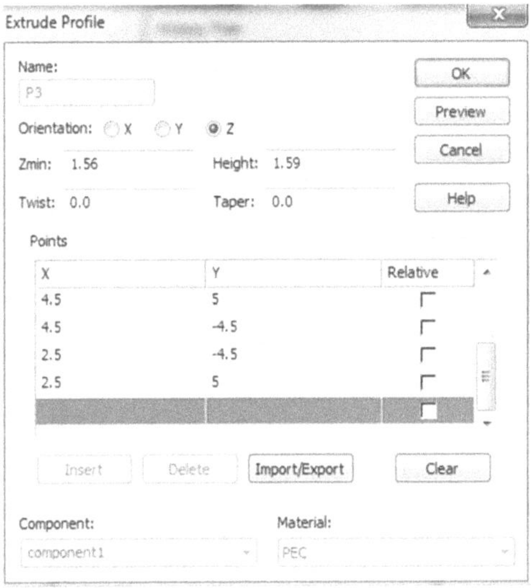

FIGURE 13.6 Patch 3rd.

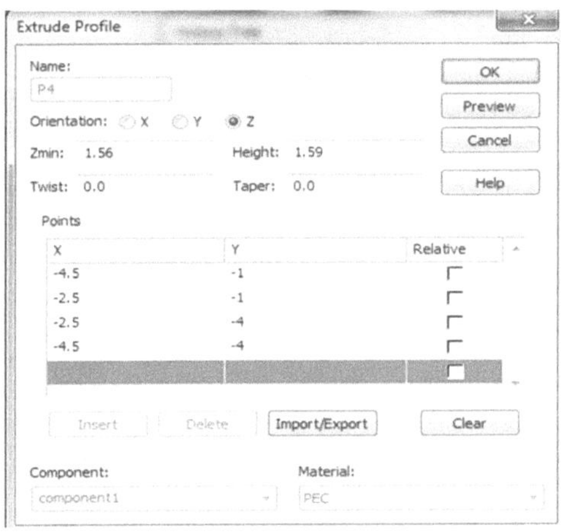

FIGURE 13.7 Patch 4th.

receiving wire, the more significant part of the spread is over the ground plane and can have a high directional increase. In this section, the slotted U-shape antenna was designed and resulted calculated with an –40 dB scale. By slotting, the planar broadcast can be formed by etching the substrate metal plane. One benefit of slotting is that impedance can be imprecise at any time by changing the width of the slot. This antenna arrangement is well-suited for 2.0–2.4 GHz. In this structure, TE mode transmission is used and their cut-off frequency range is 2.409 GHz. When experienced, the antenna sample displays a maximum gain of more than 7 dB with a

TABLE 13.2
Bandwidth and reflection coefficients at different heights

Height (h) (mm)	Bandwidth (MHz)	Reflection coefficient (-dB)
1.18	350	19.6
1.4	360	23.4
1.53	450	26
1.78	410	39.2
2.49	530	17.1

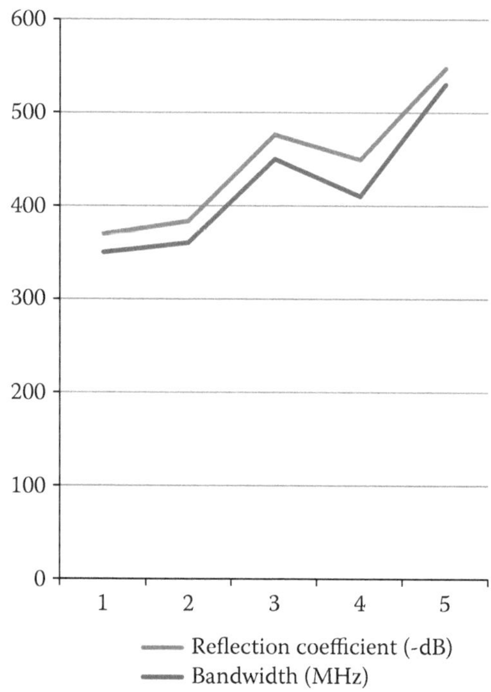

FIGURE 13.8 Bandwidth vs reflection coefficient graph.

sound wave pattern in the different frequency bands. Using PEC material contains an aerial that can be easily included on top of the course board of devices to lower the packaging cost. The radiation element is physically supported by a coaxial cable and shorting wall. The planned antenna is dense in dimension, low charge and easy to construct. These considerations are efficient for the worldwide portability of wireless communication applications. The simulated antenna concert is calculated using CST Microwave Studio Suite (Figures 13.11 and 13.12).

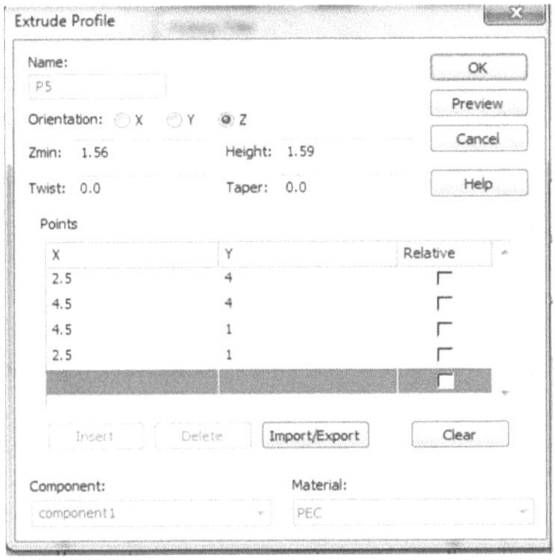

FIGURE 13.9 Patch 5th.

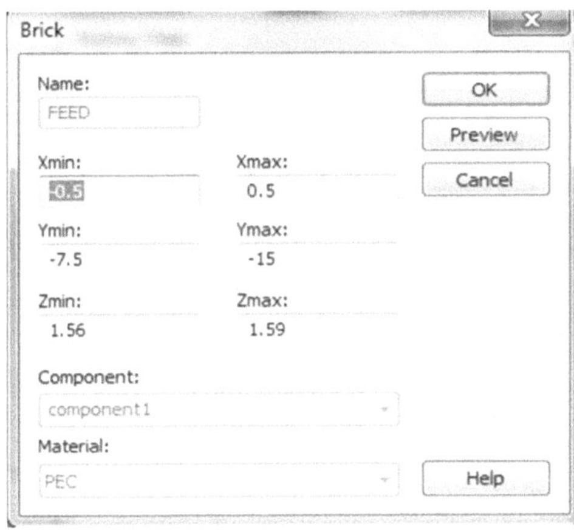

FIGURE 13.10 Feed.

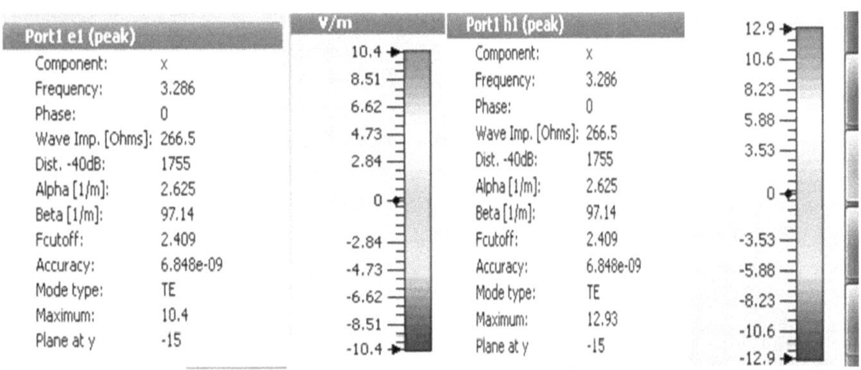

FIGURE 13.11 Concentration of field in X component with E and H field port one and port 2.

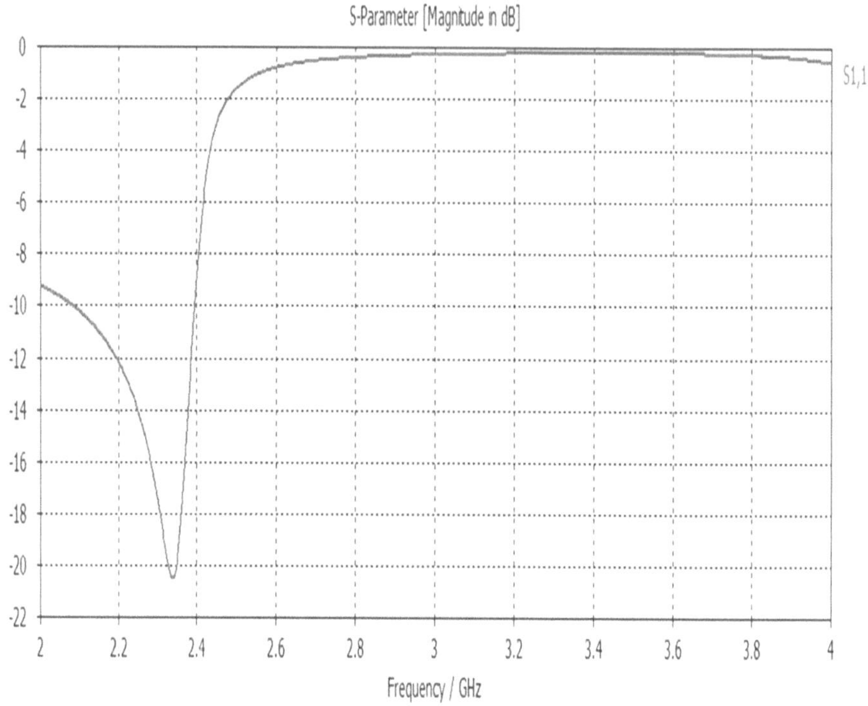

FIGURE 13.12 Simulation result.

13.5 CONCLUSION

This slotted structure is predicated on a planar design concept with an –40 dB scale. The antenna is engaged with a square patch with a microstrip line and a crushed plane with a rectangle space. The planned projection is particularly informal to be built-in through microwave functions for capable cost. Effectual simulation outputs show that the planned antenna construction is efficient in broadband characteristics and low side levels with a compact size. This paper has focused on the progress of integrated antenna structure with high bandwidth performance. Each gap is often applicable for a selected operating frequency. There is no prerequisite of repeated covering as required for multiband type antenna. The simulation result is compatible with WLAN/LTE application with the band for 2.0–2.4 GHz. This antenna has another reward as dense size and constant radiation characteristics across the practical radiation frequency.

REFERENCES

1. A. Verma, "Analysis and design of E shaped patch antenna In X band," *International Journal of Advanced Engineering Technology (Ijaet). E-Issn*, vol. 3, no. 1, pp. 223–224, 2012.
2. A. H. Ahmad and B. K. Jaralla, "Design and simulation of broadband rectangular microstrip antenna," *Eng. Tech.* vol. 26, no. 1, 2008.

3. F. Yang, X.-X. Zhang, X. Ye, and Y. Rahmat-Sami, "Wideband E-shaped patch antenna for wireless communications," *IEEE Transactions on Antennas and Propagation*, vol. 49, no. 7, July 2001.
4. S. Jain and R. Nema, "Review paper for circular microstrip patch antenna," *International Journal of Computer Technology and Electronics Engineering (IJCTEE)*, vol. 1, no. 3, 2014.
5. Nasimuddin and Z. N. Chen, "Wideband multilayered microstrip antennas fed by coplanar waveguide-loop with and without via combinations," *IET Microw. Antennas Propag.*, vol. 3, pp. 85–91, 2009.
6. S. K. Oh, H. S. Yoon, and S. O. Park, "A PIFA-type varactor – Tunable slim antenna with a PIL patch feed for multiband applications," *IEEE Antennas Wireless Propag. Lett.*, vol. 6, pp. 103–105, 2007.
7. J. Soler, C. Puente, and J. Anguera, "Advances in loading techniques to design multifrequency monopole antennas," *Microw. Opt. Technol.Lett.*, vol. 41, no. 6, pp. 434–437, June 2004.
8. T. H. Kim and D. C. Park, "A CPW-fed triple-band monopole antenna for WiMAX/WLAN applications," in *Proc. 38th EuMC*, October 2008, pp. 897–900.
9. Y. Wang, Q. Y. Zhang, and Q. X. Chu, "Triple-band monopole antenna with dual-sleeves inside the ground plane," in *Proc. APMC 2009*, December 2009, pp. 1980–1983.
10. H. W. Liu, C. H. Ku, and C. F. Yang, "Novel CPW-fed planar monopole antenna for WiMAX/WLAN applications," *IEEE AntennasWireless Propag. Lett.*, vol. 9, pp. 240–243, 2010.
11. W. S. Cheng and K. Y. Ku, "Band-rejected design of the printed open slot antenna for WLAN/WiMAX operation," *IEEE Trans. AntennasPropag.*, vol. 56, no. 4, pp. 1163–1169, April 2008.
12. L. Dang, Z. Y. Lei, Y. J. Xie, G. L. Ning, and J. Fan, "A compact microstrip slot triple-band antenna for WLAN/WiMAX applications," *IEEE Antennas Wireless Propag. Lett.*, vol. 9, pp. 1178–1181, 2010.
13. Y. Li, Z. Zhang, Z. Feng, and M. F. Iskander, "Design of penta-band omnidirectional slot antenna with slender columnar structure," *IEEETrans. Antennas Propag.*, vol. 62, no. 2, pp. 594–601, February 2014.
14. J. D. Kraus and R. J. Marhefka, *Antennas for All Applications*. New York: McGraw-Hill, 2003.
15. M. T. Wu and M. L. Chuang, "Application of transmission line model to a dual-band, stepped monopole antenna designing," *IEEE AntennasWireless Propag. Lett.*, vol. 10, pp. 1449–1452, 2011.
16. C. A. Balanis, *Antenna Theory, Analysis and Design*. Hoboken, NJ: Wiley, 2005.
17. R. Garg, P. Bhartia, I. Bahl, and A. Ittipiboon, *Microstrip Antenna Design Handbook*. London, United Kingdom and Norwood, Massachusetts, United States: Artech House, 2001.
18. R. Mishra, "An Overview of Microstrip Antenna," *HCTL Open International Journal of Technology Innovations and Research (IJTIR)*, vol.21, pp. 2–4, August 2016.
19. Coulibaly, T. A. Denidni, and H. Boutayeb, "Broadband microstrip-fed dielectric resonator antenna for X-band applications," *IEEE Antennas and Wireless Propagation Letters*, vol. 7, pp. 341–345, 2008.
20. D.-Z. Kim, W.-I. Son, W.-G. Lim, H.-L. Lee, and J. W. Yu, "Integrated planar monopole antenna with microstrip resonators having band-notched characteristics," *IEEE Trans. Antennas Propag.*, vol. 58, pp. 2837–2842, 2010.
21. R. Mishra, J. Jayasinghe, R. G. Mishra, and P. Kuchhal, "Design and performance analysis of a rectangular microstrip line feed ultra-wide band antenna," *International Journal of Signal Processing, Image Processing and Pattern Recognition*, vol. 9, no. 6, pp. 419–426, 2016.

22. R. G. Mishra, R. Mishra, P. Kuchhal, and N. Prasanthi Kumari, "Analysis of the microstrip patch antenna designed using genetic algorithm based optimization for wide-band applications," *International Journal of Pure and Applied Mathematics*, ISSN 1314-3395, vol. 118, no. 11, 2018. 10.12732/ijpam.v118i11.108
23. R. S. Kushwaha, D. K. Srivastava, J. P. Saini, and S. Dhupkariya, "Bandwidth enhancement for microstrip patch antenna with microstrip line feed," *IEEE International Conference on Computer and Communication Technology*, pp. 183–185, November 2012.
24. S. W. Su, K. L. Wong, and C. L. Tang, "Band-notched ultra-wideband planar monopole antenna," *Microwave Optical Technology, Letter*, vol. 44, pp. 217–219, 2005.
25. Garima, D. Bhatnagar, J. S. Saini, V. K. Saxena, and L. M. Joshi, "Design of broadband circular patch microstrip patch antenna with diamond shape slot," *Indian Journal of Radio and Space Physics*, vol 40, pp. 275–281, October 2011.
26. C. J. Wang and K. L. Hsiao, "CPW-fed monopole antenna for multiple system integration," *IEEE Trans. Antennas Propag.*, vol. 62, no. 2, pp. 1007–1011, February 2014.
27. R. K. Chaurasia and V. Mathur, "Enhancement of bandwidth for the square of microstrip antenna by partial ground and feedline technique," *Asia Pacific Journal of Engineering Science and Technology*, vol. 3, no. 1, pp. 49–53, March 2017.
28. R. Mishra, R. G. Mishra, and P. Kuchhal, "Analytical study on the effect of dimension and position of slot for the designing of ultra wide band (UWB) microstrip antenna," *5th IEEE International Conference on Advances in Computing, Communications and Informatics (ICACCI)*, 978-1-5090-2028-7, September 2016.
29. K. F. Lee, K. M. Luk, K. F. Tong, S. M. Shum, T. Huyn, and R. Q. Lee, "Experimental and simulation studies of the coaxially fed u–slotrectangular patch," *IEEE Proceedings of Microwave Antennapropagation*, vol. 144, no. 5, pp. 354–358, 1997.
30. R. Mishra, P. Kuchhal, and A. Kumar, "Effect of height of the substrate &width of the patch on the performance characteristics of microstrip antenna," *International Journal of Electrical and ComputerEngineering*, vol. 5, no. 6, pp. 1441–1445, 2015.
31. A. Beno and D. S. Emmanuel, "Diamond shaped symmetrical slotted miniaturized microstrip patch antenna for wireless applications," *Journal of Theoretical and Applied Information Technology(Jatit)*, vol. 47, no. 3, 31 January 2013, Issn:1992-8645E-Issn:1817-3195.

14 Smart Traffic Light Controller Using Edge Detection in Digital Signal Processing

Aayush Chibber, Anuarg, and Rohit Anand
Department of Electronics and Communication Engineering, G. B. Pant Engineering College, New Delhi, India

Jagtar Singh
Department of Electronics and Communication Engineering, N. C. College of Engineering, Israna, Panipat, India

CONTENTS

14.1 INTRODUCTION

It is essential to bring modern techniques to manage traffic regulation as traffic is continuously rising due to the restricted resources in the present infrastructure and the increasing number of vehicles [1]. A traffic light system is an automatic system

DOI: 10.1201/9781003230526-14

whose function is to provide traffic light signals at road intersections. A traffic light system, sometimes also referred to as a stop-and-go light system, is a directing or indicating device arranged at crossroads, walker crossings, or other locations to signify when it is risk-free to drive or patrol a worldwide color-code system. Traffic light signals influence the traffic flow because they can provide the right way to move the traffic on the road, but if the signals are functioning poorly, increased accidents may result.

MATLAB tools and GUI make an excellent technique for designing the different user interfaces that can also help control the traffic light. Many approaches to signal processing have emerged during the last 30 to 40 years. Digital-image processing is an application of signal processing in which an image is used as an input, while the image or any of the image characteristics is an output [2]. Several techniques are instigated for enhancing or compressing the images acquired from satellites, space probes, and many other resources. The image-compression methods benefit from the data-redundancy concept [3,4] that reduces the storage and transmission cost of the image with no compromise in image quality [5]; the enhancement techniques result in the enhancement of image contrast that in turn increases the dynamic range of the image [6]. The enhancement methods that can be categorized in signal processing are the operations such as high-pass filtering (i.e., increasing the sharpness), low-pass filtering (i.e., increasing the blurriness), increasing the intensity or brightness of the image, etc. Traffic congestion in the various lanes is determined using signal processing that is carried out with the help of video processing based on which video frames are captured using a digital camera installed beside the traffic light. The easy-to-use software and lower price of the digital camera are why the image capturing is done this way [7].

In this proposed work, image processing using MATLAB and GUI is chosen to calculate traffic density using the cameras because of better performance and reasonable rate. Using the different edge-detection methods in image processing, a technique has been proposed for better traffic control.

The chapter is organized as follows. The following section (Section 14.2) discusses the brief related work. Section 14.3 gives an idea of the proposed research work. The subsequent section discusses the implementation with the help of the flow sequence. Section 14.5 discusses the results of the proposed work with the help of MATLAB GUI outputs. Al last, a brief conclusion is drawn.

14.2 RELATED WORK

In recent years, developing a traffic surveillance system based on video technology has attracted many researchers.

The authors in [8] proposed a fuzzy logic technique based on traffic uncertainties to put forward a traffic-anomaly detection algorithm. In [9], the authors presented a novel system for traffic light detection that is computationally efficient and can eliminate the need for a global positioning system (GPS). The authors in [10] developed a scheme for detecting the traffic at night based on the vehicle's head and rear lights using the Azimuthal Blur technique. An adequate system has been presented in [11] for regulating the traffic flow based on Arduino and MATLAB. In

[12], an efficient technique has been proposed to control the traffic by defining the thresholds based on the demarcation of roads. Further, the authors in [13] presented work using a Gaussian algorithm that calculates the changes in the image by using the concept of subtraction concerning the background and detects the objects to manage the number of vehicles in all the individual lanes detected by their centroid line. The authors in [14] developed an image-processing technique and embedded systems for automatic control of the traffic and manual control using Bluetooth and wireless connection. A traffic-flow technique has been suggested in [15] based on three methodologies – temporal variation for identifying moving objects, optical flow approach for background modifications, and background extraction for extracting the moving object. The authors in [16] proposed a research work using Canny Edge detection and circular Hough transformation to detect traffic signals and send a voice signal to blind pedestrians to move across the lanes. In [17], the drawbacks of different algorithms and detectors used in image processing have been discussed. Further, in [18], the authors segregated the traffic into three regions, namely light, medium, and heavy traffic jams, and used the background subtraction and erosion operation to implement an intelligent traffic light system. The controlling of a traffic light using the histogram equalization and the Canny Edge detection to find out the traffic density of the road has been presented in [19]. In [20], a particular wavelet transform (called Gabor) and binary operating function have been utilized to separate the state-space vehicle orientation feature and divide the image into small regions for extracting the histograms and measuring the resemblance of the various automobiles in Euler space for vehicle detection. An overlapping and scheduling algorithm for finding the traffic density has been developed in [21].

The proposed work discusses the implementation of a traffic light controller with the assistance of various edge-detection operators.

14.3 PROPOSED WORK

The block diagram of the proposed system is shown in Figure 14.1, which provides an idea of the management of traffic using the concept of signal (image) processing. A reference image is taken, and an acquired image to be matched with the reference image is taken with a camera. Both the images are converted from color to grayscale and then resized to the suitable pixel resolution. The sharp content of both images is dominant in the edges of both images. This edge detection can be done using any edge-detection operator. (Sobel operator is used in the proposed work.) Both the sharpened images are matched concerning each other using pixel matching. After matching, the specific time allocation is adjusted (to be discussed in the next section).

The various components shown in Figure 14.1 are discussed in detail in the next section.

14.3.1 Image Acquisition and Formation

Generally, the images are represented by a 2D function f(x, y), where x and y indicate the locations of the plane. The intensity level that provides information

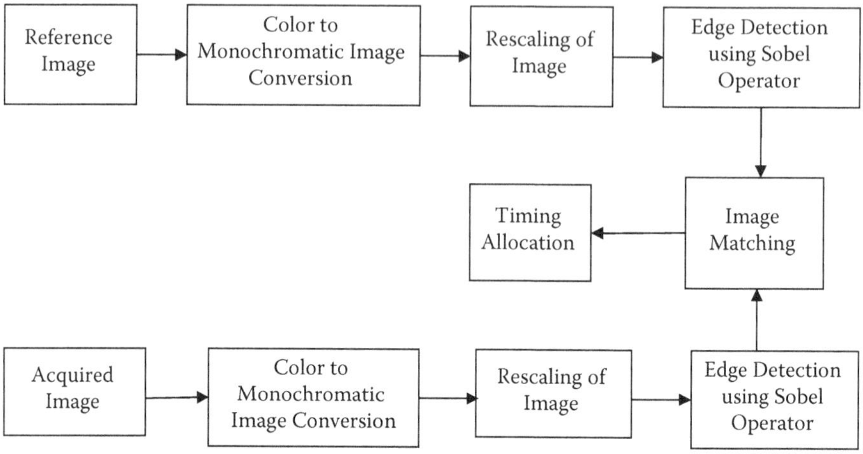

FIGURE 14.1 Block diagram for the proposed system.

about the brightness and the darkness is referred to as the amplitude of the image. The higher the value of intensity, the more the amplitude and the higher the value of that pixel. This process is also termed as the intensity or gray level of the image at the specified point. Since there is continuous variation in the intensity values of the image, there is a requirement to translate these intensity levels to the bound discrete levels to make a purely digital version of the image. The image to be given as input for the same will be provided in MATLAB. Each digital image comprises a finite number of pixels, which is the smallest resolvable part of the image (hence, the image is also called a pixel array) [22]. There are some conditions to prepare the function f(x, y) as the intensity levels of the pixels vary directly as a source transmits the optical energy. Hence, f(x, y) is not zero and must be a finite signal [22].

i.e., $0 < f(x, y) < \text{infinity}$

14.3.2 Image Pre-Processing

The pre-processing of the image can be done by color to monochromatic image conversion followed by scaling.

i. ***RGB to Gray Conversion:*** Human beings view the different colors through the sensory cells called cones present in their eyes. One cone is primarily sensitive to green light (G), one to red light (R), and one to blue light (B). All the recognizable colors can be produced by emission of a bound mixture of these primary colors (R, G and B). Hence, images are often saved in three different matrices in the Bayer Mosaic form we use in cameras. It develops one of the formats of color images, named RGB format for digital images. Grayscale images specify gray levels where we do not differentiate how many different colors are emitted. Somewhat the same amount of energy is emitted in every channel. The distinction lies in the portion of light emitted for all the pixels; a small amount of emitted light gives dark pixels, and more intensity is felt in the form of bright pixels. While converting a colored image to a monochromatic one, the red, green and blue values are considered for each pixel to generate a single value that reflects the brightness corresponding to each pixel. The most common approach is to find out the mean value of the offering from all the three channels: (R + G + B)/3. However, since the green color often contributes to the detectable brightness level, another realistic approach is to calculate a weighted average, e.g., 0.3R + 0.59G + 0.11B [23].

ii. ***Image Scaling/Resizing:*** Image scaling takes place in all the digital photographs at one stage or another. This scaling may be in the Bayer Demosaicing stage [24] or the photo-enlargement stage. It occurs when we rescale any image concerning the pixel grid. Image scaling is mandatory whenever there is a requirement to enlarge or diminish the various pixels in any image. Even if the same image rescaling is done (i.e., rescaled image is of identical dimensions as the real-time image being captured), the result can change remarkably depending upon the procedure through which the image is rescaled because of the various causes. Since the different cameras have different resolutions, it may not be valid for other camera specifications whenever a system is developed for some particular criteria. Hence, it becomes mandatory to frame the constant resolution specification for the image and hence accomplish image rescaling.

14.3.3 Edge Detection

Edges in images are regions with immediate variation in the gray level while moving from one pixel to the other. Edge detection aims to recognize those points where the image intensity varies abruptly or has noise level or discontinuities. The locations having a drastic change in the brightness will make an edge when joined together. Edge detection is a convenient technique used in signal processing, computer vision, and machine learning [25], mainly where many features are extracted. The edge-detection technique tends to reduce the volume of data and rejects the non-useful data while preserving the critical constructional characteristics of an image. Several techniques in edge detection can be divided into two categories:

a. Gradient
b. Laplacian

The gradient method [26] tends to expose the edges by finding the greatest and lowest values in the first-order differentiation of the image, whereas the laplacian method [27] looks for the points of crossings from positive to negative or vice-versa (i.e., zero-crossing points) in the second-order differentiation of the image.

14.3.3.1 Edge Detection Techniques

Different colors have different values of brightness associated with them. For example, a green image has a higher brightness than the red and blue image; it is a more blurred image than many other images, and a red image is generally a high noise image.

Following are the most commonly used edge-detection techniques:

a. Sobel Edge Detection Technique [28]
b. Roberts Edge Detection Technique [29]
c. Prewitt Edge Detection [29]
d. Zerocrossing Threshold Edge Detection Technique [30]
e. Canny Edge Detection Technique [28]

In this proposed work, Sobel Edge-Detection Technique is used because of its simplicity in calculating the gradient and ability to detect the orientations of the edges.

14.3.3.2 Sobel Edge Detection:

A handy mathematical operating function for performing the analysis can be the Sobel edge-detection operator [28], which estimates the differentiation of an image. It is calculated individually for the y and x directions. We have used a mask of 3 × 3 matrix, one for x-direction and one for y-direction. The gradient for x-direction has negative coefficients on the left side and positive coefficients on the right side, and the preservation takes place slightly on the center column pixels. Similarly, the gradient for y-direction has negative coefficients on the bottom and positive numbers on the top, and the preservation takes place slightly on the middle row pixels.

Sobel mask or filter is used to detect edges done by finding the value of gradient at all the pixels. It finds the direction of the most significant increment from light to dark and the rate of change in that direction. The results of applying the Sobel operator show how perfectly a pixel changes in brightness, and the same pixel can be counted as an edge if the pixel changes its brightness abruptly.

It also indicates how that edge is expected to be aligned. Putting the mask to a pixel in an area of the uniform gray level is a null vector. The output results in a vector, and based on that, the gradient is calculated for each pixel examined and after that, based on the threshold value, it is checked whether it is taken a valid element or not.

Mathematically, the Sobel filter uses two masks: one along with the changes in the horizontal direction and the other along with the vertical direction.

14.3.4 Image Sharpening with Laplacian

Edge points can be detected using the zero crossings of the second derivative, and the operator used for the same is Laplacian. The Laplacian uses the double derivative for image sharpening and, based on that, we will be finding out the sharpened image. This can be done by approximating the maxima and minima of the gradient magnitude and finding the points where the double derivative equals zero. We can find the zero crossings from where the maxima and minima are located. The expression given below will illustrate the image that we got after sharpening.

$$s(x, y) = f(x, y) + c[\nabla^2 f(x, y)] \tag{14.1}$$

where
s(x, y) is an image obtained after sharpening
f(x, y) refers to the input image after the Sobel edge detection
$\nabla^2 f(x, y)$is the Laplacian of the image f(x, y)
The contact c is taken as –1

Some properties of the Laplacian operator are:

1. It is economical to implement when compared with gradient because it requires only one mask.
2. It does not provide information about the edge direction.
3. It is more sensitive to noise, although it uses double derivatives for computation.
4. It is an anisotropic operator that has equal weights in all directions.

14.3.5 Image Matching

Identification methods based on coincidence indicate one class that is predefined and the other one will be improved, and then they both will be compared for matching. An uncertain pattern is allocated to one of the classes nearest to it regarding a pre-specified parameter. The usual method is the minimum-distance classifier algorithm that classifies the images based on the Euclidian distance and then finding the minimum of all the computed distances to classify them. There is one more technique related to the correlation through which the degree of similarity may be found.

The image-matching approach used in the structure is quite different. Differentiating a reference image with the actual image is done in image matching based on the pixel-by-pixel comparison. Though a few drawbacks in the matching based on pixels are there, the actual existent image is loaded in memory and translated to the required form. For both images to be exactly coincident with each other, their pixel intensity levels must be precisely the same such that they are resized on the same scale. This is a straightforward approach used in pixel matching. Finally, the extent of matching is expressed as:

$$\% \text{ age matching} = \frac{\text{Number of pixels matched successfully}}{\text{Total number of pixels}} \tag{14.2}$$

14.4 IMPLEMENTATION

The implementation is developed on the real-time working of the traffic light. In the real-time scenario, the following cases can arise:

Case no. 1. When there are no vehicles on the road, the red light is turned on.
Case no. 2. When there are very few vehicles or even one vehicle on the road, the green light is turned on for less time, typically 5 seconds or so.
Case No 3. When the number of vehicles is more (say 5 or 10 or more than that), the green light is turned on for a longer duration of 15 seconds or 20 seconds.

The frame-by-frame processing of the given video is done. After processing the given frame, the timing allocation of the traffic light signal is decided. The red light is turned on at first. After the red-light timer expires, the traffic density is calculated, the match percentage is reported, and the green light is turned on. The timing of the green light signal will be controlled, thus forming a control system.

The complete flow sequence for the proposed work consists of the various steps listed below:

1. A reference image is taken, and the image that has to be matched is captured continuously with the help of a camera located at the junction.
2. Both the above images are undergone pre-processing:
   a. The images are converted from color version (i.e., RGB) to monochromatic (i.e., gray) version.
   b. The monochromatic images are rescaled to 300 × 300 pixels.
3. The pre-processed images undergo edge detection using the Sobel technique of edge detection.
4. Both the images obtained in the last step undergo matching pixel-by-pixel.
5. The assignment of timing is performed that depends upon the degree of matching:
   a. In the case of 0% to less than 10% matching, the green light is turned on for 20 seconds.
   b. In the case of matching from 10% to less than 30%, the green light is kept on for 15 seconds.
   c. In the case of matching from 30% to less than 50%, the green light is kept on for 10 seconds.
   d. In the case of matching from 50% to less than 100%, the green light is turned on for 5 seconds.
   e. In the case of absolute matching (i.e., 100% matching), the red light is turned on for 20 seconds.

Table 14.1 shows the percentage of the match along with the timing allocation of the light.

TABLE 14.1
Percentage of match with timing allocation

S. No.	Percentage of Match	Light Signal	Timer for a Light Signal
1.	Less than 10%	Green light	20 seconds
2.	Between 10% to less than 30%	Green light	15 seconds
3.	Between 30% to less than 50%	Green light	10 seconds
4.	Between 50% to less than 100%	Green light	5 seconds
5.	100%	Red light	20 seconds

Note that the matching percentage in Table 14.1 is taken concerning the empty road, and the captured image is the image or video frame of the road, which is full of vehicles.

The flow chart illustrating the design methodology is given in Figure 14.2. It may be explained in the form of the following steps:

1. The empty road image is taken.
2. The image processing algorithm is applied to it.
3. This output image, after applying the image processing algorithm, is stored in the processor.
4. The video frame is then read by the camera.
5. The image-processing algorithm onto the video frame is now applied.
6. The percentage of the image match obtained in step 3 with that obtained in step 6 is determined.
7. If the percentage of the match is not equal to 100, then go to step eight; otherwise, go to step 10.
8. Provide timing allocation to the green light (i.e., duration for which the green light timer expires).
9. Turn on the red light for 20 seconds.
10. If there are still vehicles on the road, then go to step 4; otherwise, go to step 11.
11. Stop the program.

14.5 RESULTS AND DISCUSSION

14.5.1 Results For Reference Image

The simulation results for the intelligent traffic light controller are presented in this section.

The results will follow the following sequence for the reference image (i.e., the image of a vacant road).

1. RGB (colored) image of the reference image

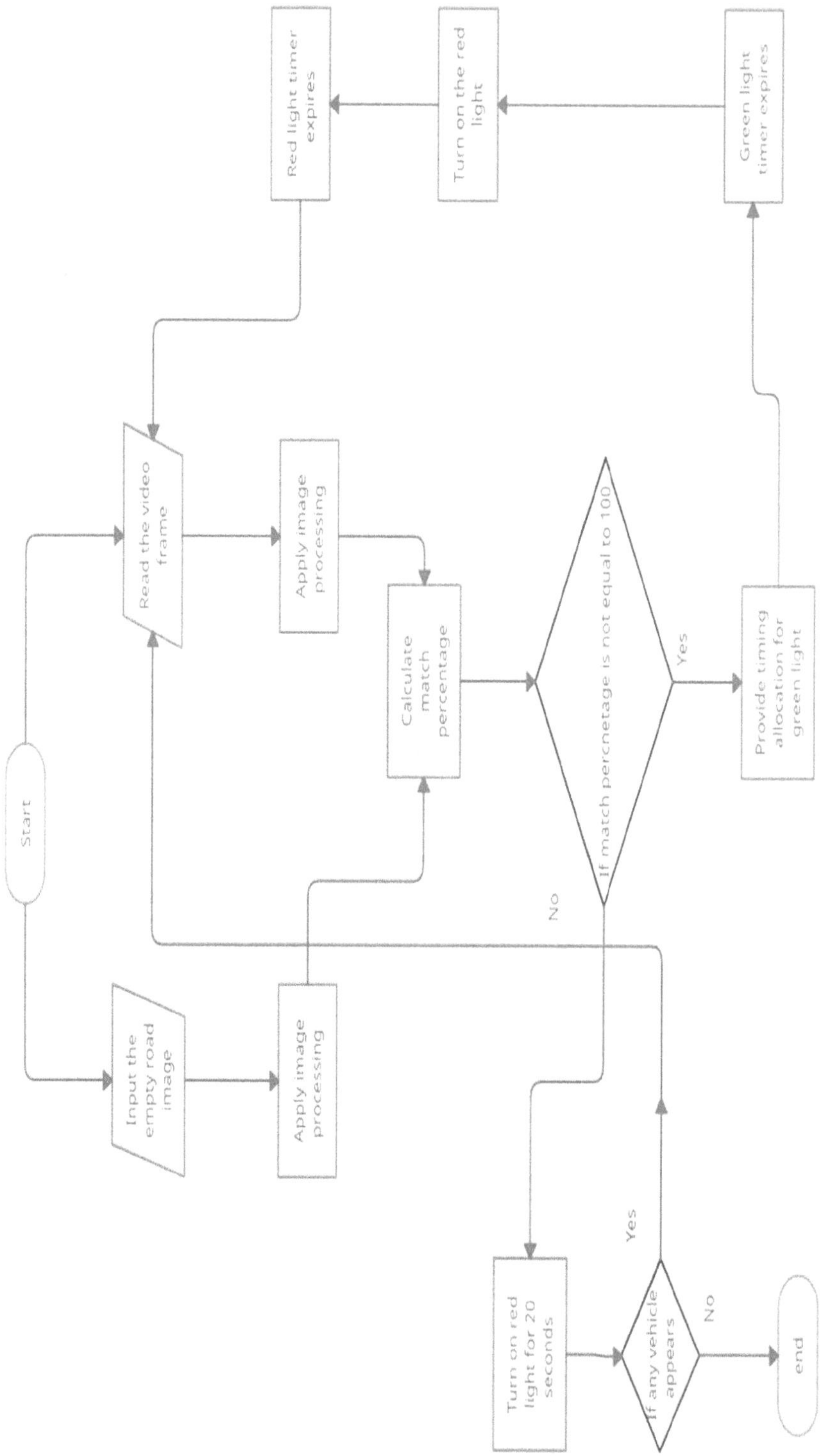

FIGURE 14.2 Flowchart for the proposed work.

2. Grayscale image of the reference image
3. The Sobel gradient image of the reference image
4. The edge-detected reference image
5. The filtered image for sharpening
6. The sharpened image, along with the horizontal and vertical directions
7. The sharpened image along with diagonal directions

All the images mentioned above are shown in Figure 14.3 to Figure 14.9.

14.5.2 Results for Captured Image

The results in this section will follow this sequence for the acquired (captured) image, which is the image of a road having the different vehicles

FIGURE 14.3 RGB reference image.

FIGURE 14.4 The grayscale reference image.

FIGURE 14.5 Sobel gradient of reference image.

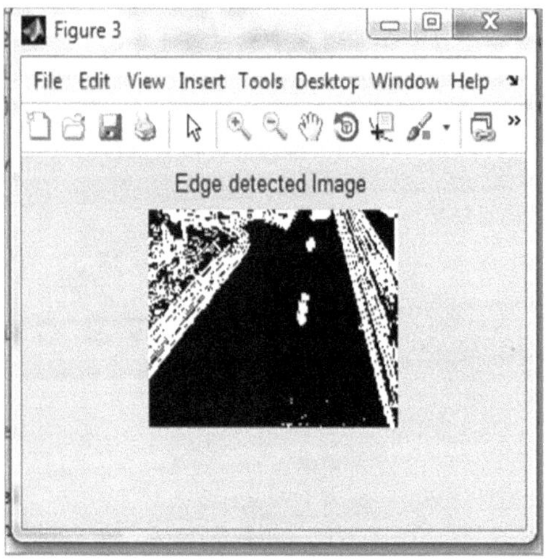

FIGURE 14.6 The edge detected reference image.

1. RGB image of the captured image
2. Grayscale image of the captured image
3. The Sobel gradient image of the captured image
4. The edge detected captured image
5. The filtered image for sharpening
6. The sharpened image, along with the horizontal and vertical directions
7. The sharpened image along with diagonal directions

All the images mentioned above are shown in Figure 14.10 to Figure 14.16.

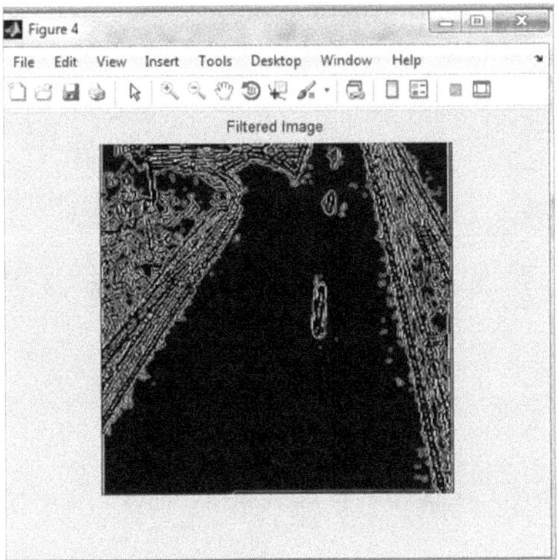

FIGURE 14.7 The filtered reference image.

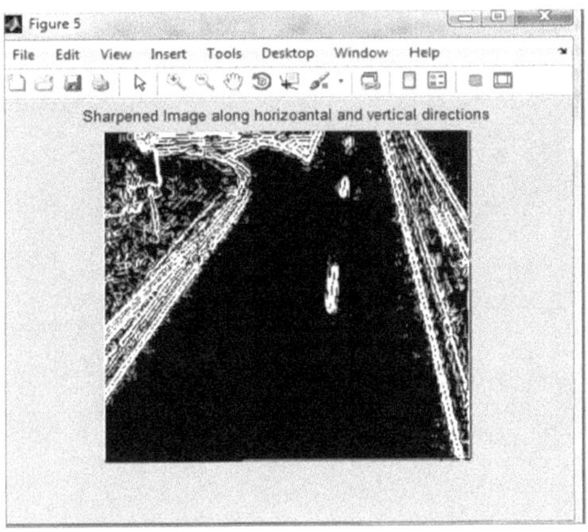

FIGURE 14.8 Sharpened image for horizontal and vertical sides.

14.5.3 MATLAB GUI OUTPUTS

The proposed methodology is based on video processing, of which the heart is image processing. The image-processing techniques have been applied extensively to get the required output. We are concentrating on the control of traffic lights using MATLAB as the core processor of the design. The MATLAB Graphical user interface (GUI) has been used to design the traffic light system.

A graphical user interface is a graphical layout in multiple windows with various sections that assist users in undergoing the various bilateral functions. The graphical user interface (GPS) person does not need to generate a script or type any command to carry out the numerous tasks. No GUI user needs to know how the various tasks

FIGURE 14.9 Sharpened image along with diagonal directions.

FIGURE 14.10 RGB captured image.

are accomplished (unlike the programs involving much coding). The various sections of graphical user interface include menus, radio buttons, push buttons, toolbars, list boxes, and sliders, and many others. These interfaces generated with MATLAB tools can do the various computations (of varying complexities), interact with other GUIs, and write and read the data files and reveal the data in the form of graphical plots or tables. The MATLAB GUI is shown in Figure 14.17.

In the GUI shown, the following points need to be noted:

a. The process button will initiate the whole algorithm, while axes 1 to 6 will show the image-processing operations for the reference image.

FIGURE 14.11 The grayscale captured image.

FIGURE 14.12 Sobel gradient of captured image.

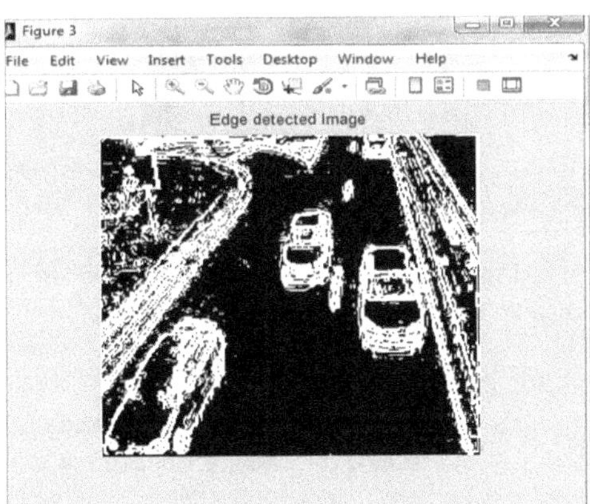

FIGURE 14.13 The edge detected captured image.

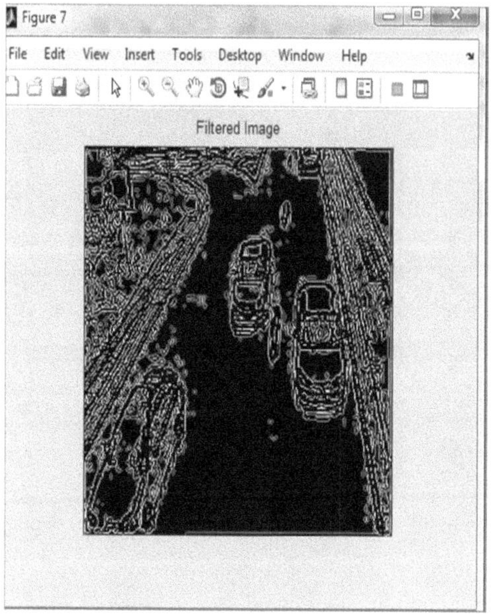

FIGURE 14.14 The filtered captured image.

FIGURE 14.15 Sharpened image for vertical and horizontal sides.

b. The lower range of the axes (i.e., axes 7 to axes 12) will show the image-processing operations for the captured image. In these axes, the whole video will be simulated and the image processing operations will be applied.

In the subsequent figures, we will show the examples of the outputs for the traffic light at different traffic densities

FIGURE 14.16 Sharpened image along with diagonal directions.

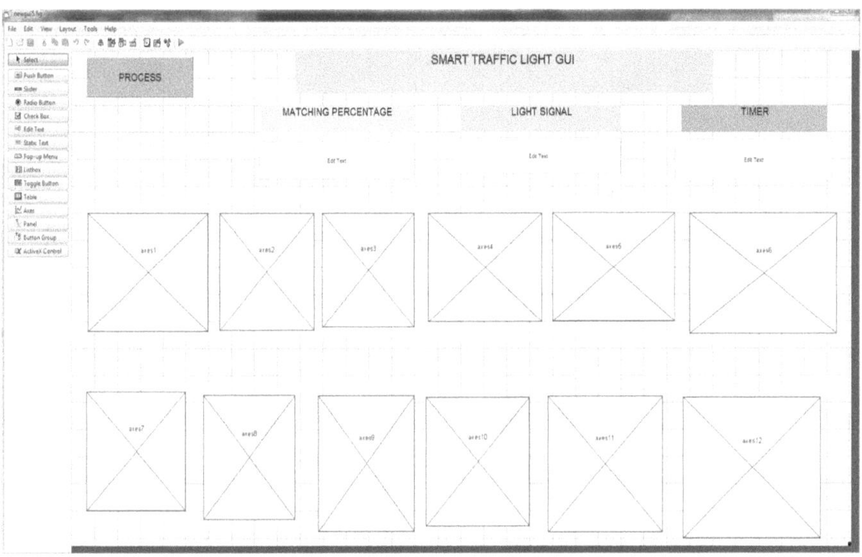

FIGURE 14.17 The primary MATLAB Graphical User InterFace.

1. Red light turned ON for 20 seconds for 100% matching (Figure 14.18)
2. Green light turned ON for 20 seconds for no matching, i.e., 0% matching (Figure 14.19)
3. Green light turned ON for 15 seconds for 25% matching (Figure 14.20)
4. Green light turned ON for 5 seconds for 50% matching (Figure 14.21)

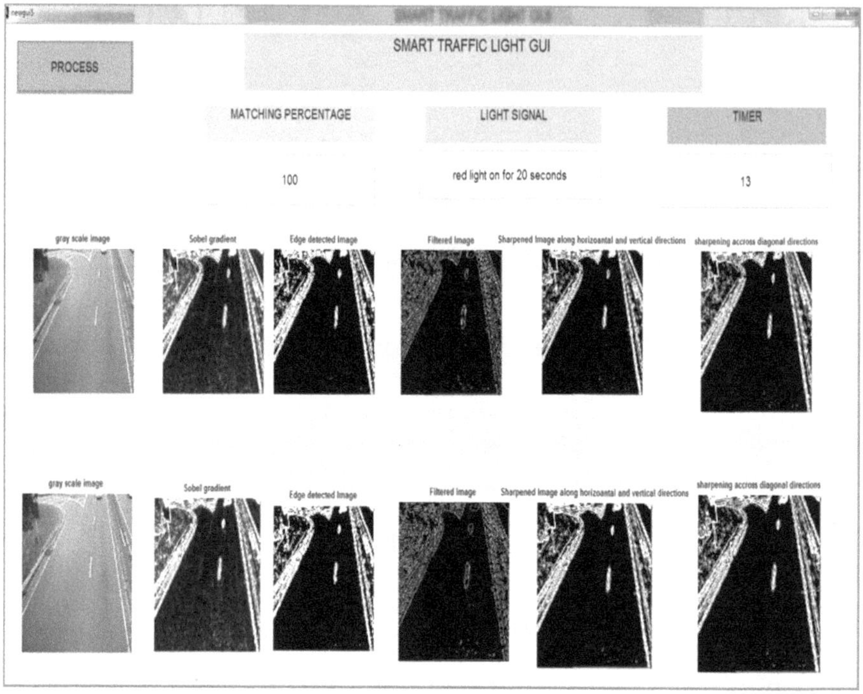

FIGURE 14.18 The MATLAB GUI output for red light ON for 20 seconds.

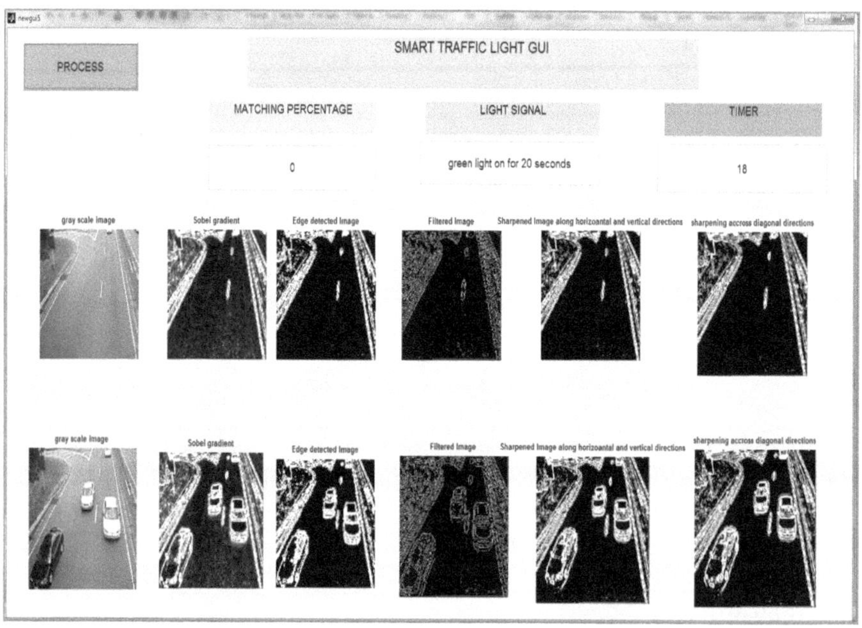

FIGURE 14.19 The MATLAB GUI output for green light ON for 20 seconds.

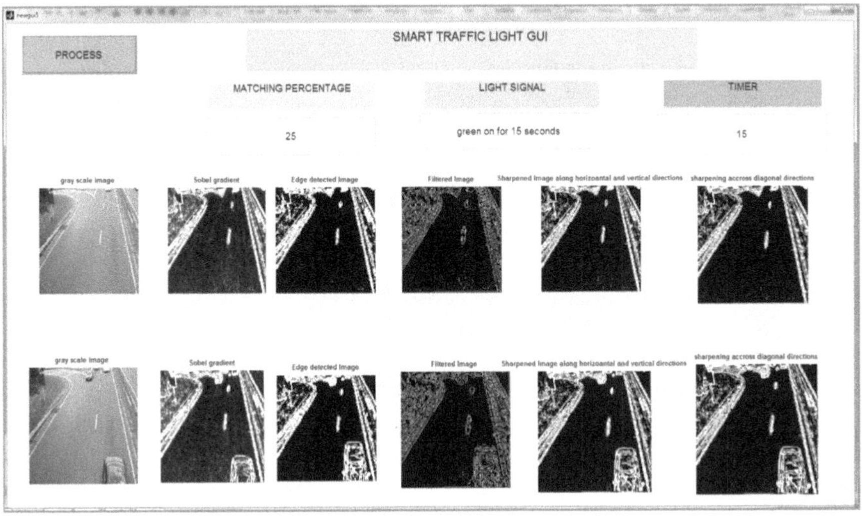

FIGURE 14.20 The MATLAB GUI output for green light ON for 15 seconds.

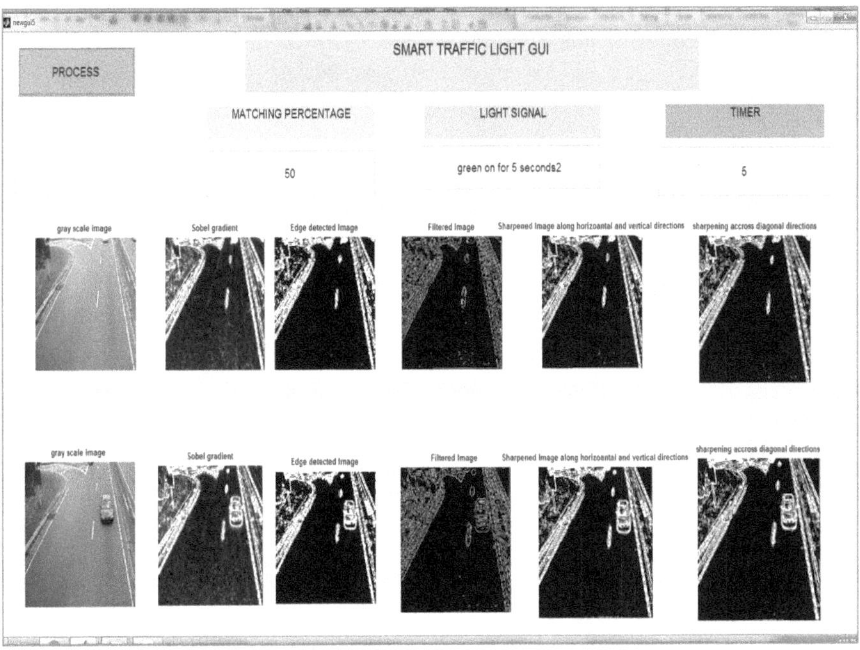

FIGURE 14.21 The MATLAB GUI output for green light ON for 5 seconds.

14.6 CONCLUSION

Traffic controls using signal (image) processing techniques intensifies most of the limitations of the older traffic systems. In automatic controlling, traffic light costs are better managed by avoiding turning on green traffic lights on vacant roads. This technique avoids the problem of having a constraint on the battery. The Sobel Edge-Detector technique is used in the proposed approach as it is a very systematic technique. Signal processing is a handy way to manage the traffic in contrast to the conventional techniques. The proposed image-processing method eliminates the requirement of extra equipment like audio sensors and electronic sensors. The primary benefit is the signal-time variance that controls the traffic compactness using matching between images. The precision in the computation of time due to a sole operational camera relies on the position while facing the road at all times. The output of MATLAB Graphical User Interface indicates presumed and trusted results. It shows the harmony in nearly every span that is indicated as boundaries like 0%, 5%, 10%, 25%, 50%, 100% etc.

The research indicates that the proposed signal-processing method is suitable to manage the changing state of the traffic light in the country. It is also more consistent in detecting the vehicles on the road. It envisions the actuality; hence, it performs much superior to those image processing systems that primarily depend upon identifying the metal content of the various automobiles. The proposed system is very efficient, but it needs to be upgraded to acquire more precision.

REFERENCES

1. P. Choudekar, S. Banerjee, and M. K. Muju, "Implementation of image processing in real-time traffic light control." In *2011 3rd International Conference on Electronics Computer Technology* (vol. 2, pp. 94–98). IEEE, 2011, April.
2. P. Choudhary and M. R. Anand, "Determination of rate of degradation of iron plates due to rust using image processing." *International Journal of Engineering Research*, vol. 4, no. 2, pp. 76–84, 2015.
3. G. Vyas, R. Anand, and K. E. Holê, "Implementation of advanced image compression using wavelet transform and SPHIT algorithm." *International Journal of Electronics and Electrical Engineering*, vol. 4, no. 3, 249–254, 2011.
4. M. Gupta and R. Anand, "Color image compression using set of selected bit planes." *IJECT*, vol. 2, no. 3, pp. 243–248, 2011.
5. S. Malik, R. Saroha, and R. Anand, "A simple algorithm for reduction of blocking artifacts using saws technique based on fuzzy logic." *International Journal of Computational Engineering Research*, vol. 2, no. 4, pp. 1097–1101, 2012.
6. S. Juneja and R. Anand, "Contrast enhancement of an image by DWT-SVD and DCT-SVD." In *Data Engineering and Intelligent Computing* (pp. 595–603). Springer, 2018.
7. R. Anand, G. Shrivastava, S. Gupta, S. L. Peng, and N. Sindhwani, "Audio watermarking with reduced number of random samples." In *Handbook of Research on Network Forensics and Analysis Techniques* (pp. 372–394). IGI Global, 2018.
8. Y. Li, T. Guo, R. Xia, and W. Xie, "Road traffic anomaly detection based on fuzzy theory." *IEEE Access*, vol. 6, pp. 40281–40288, 2018.
9. A. Almagambetov, S. Velipasalar, and A. Baitassova, "Mobile standards-based traffic light detection in assistive devices for individuals with color-vision deficiency." *IEEE Transactions on Intelligent Transportation Systems*, vol. 16, no. 3, pp. 1305–1320, 2014.

10. C. Tang and A. Hussain, "Robust vehicle surveillance in night traffic videos using an azimuthal blur technique." *IEEE Transactions on Vehicular Technology*, vol. 64, no. 10, pp. 4432–4440, 2014.
11. A. S. Rahishet, A. Indore, V. Deshmukh, and U. S. Pushpa, "Intelligent traffic light control using image processing."*International Journal of Industrial Electronics and Electrical Engineering*, vol. 3, no. 4, pp. 69–71, 2015.
12. S. Ashwin, R. A. Vasist, S. S. Hiremath, and H. R. Lakshmi, "Automatic control of road traffic using video processing." In *2017 International Conference on Smart Technologies for Smart Nation (SmartTechCon)* (pp. 1580–1584). IEEE, 2017, August.
13. M. F. Chowdhury, M. R. A. Biplob, and J. Uddin, "Real-time traffic density measurement using computer vision and dynamic traffic control." In *2018 Joint 7th International Conference on Informatics, Electronics & Vision (ICIEV) and 2018 2nd International Conference on Imaging, Vision & Pattern Recognition (icIVPR)* (pp. 353–356). IEEE, 2018, June.
14. M. M. Elkhatib, A. I. Adwan, A. S. Alsamna, and A. M. Abu-Hudrouss, "Intelligent traffic lights using image processing algorithms." In *2019 IEEE 7th Palestinian International Conference on Electrical and Computer Engineering (PICECE)* (pp. 1–6). IEEE, 2019, March.
15. V. Ellappan, B. Sindhusaranya, and T. Hailu, "Multi intelligent traffic light optimization techniques by applying modified component analysis algorithm." In *2019 Fifth International Conference on Science Technology Engineering and Mathematics (ICONSTEM)* (vol. 1, pp. 316–318). IEEE, 2019, March.
16. P. S. Swami, and P. Futane, "Traffic light detection system for low vision or visually impaired person through voice." In *2018 Fourth International Conference on Computing Communication Control and Automation (ICCUBEA)* (pp. 1–5). IEEE, 2018, August.
17. I. Gulati and R. Srinivasan, "Image processing in intelligent traffic management." *Int J Recent Techno Eng (IJRTE)*, vol. 8, no. 2S4, pp. 213–218, 2019.
18. B. Eamthanakul, M. Ketcham, and N. Chumuang, "The traffic congestion is investigating system by image processing from CCTV camera." In *2017 International Conference on Digital Arts, Media and Technology (ICDAMT)* (pp. 240–245). IEEE, 2017, March.
19. M. Puthran, S. Puthur, and R. Dharulkar, "Smart traffic signal." *Int. Journal of Comp Sc and Info Tech*, vol. 6, no. 2, pp. 1360–1363, 2015.
20. Y. Tang, C. Zhang, R. Gu, P. Li, and B. Yang, "Vehicle detection and recognition for intelligent traffic surveillance system." *Multimedia tools and applications*, vol. 76, no. 4, pp. 5817–5832, 2017.
21. M. Razavi, M. Hamidkhani, and R. Sadeghi, "Smart traffic light scheduling in the smart city using the image and video processing." In *2019 3rd International Conference on Internet of Things and Applications (IoT)* (pp. 1–4). IEEE, 2019, April.
22. R. C. Gonzalez and R. E. Woods, *Digital image processing*. 4th Edition. Pearson Education, 2018.
23. R. R. Gulati, *Monochrome and color television*. New Age International, 2005.
24. H. S. Malvar, L. W. He, and R. Cutler, "High-quality linear interpolation for demosaicing of Bayer-patterned color images." In *2004 IEEE International Conference on Acoustics, Speech, and Signal Processing* (Vol. 3, pp. 4853–488). IEEE, 2004, May.
25. P. Saini and M. R. Anand, "Identification of defects in plastic gears using image processing and computer vision: A review." *International Journal of Engineering Research*, vol. 3, no. 2, pp. 94–99, 2014.
26. P. P. Acharjya, R. Das, and D. Ghoshal, "A study on image edge detection using the gradients." *International Journal of Scientific and Research Publications*, vol. 2, no. 12, pp. 1–5, 2012.

27. M. S. M. Garcia-Verela, and E. Valencia, "Laplacian filter based on the color difference for image enhancement." In *5th Iberoamerican Meeting on Optics and 8th Latin American Meeting on Optics, Lasers, and Their Applications* (vol. 5622, pp. 1259–1264). International Society for Optics and Photonics, 2004, October.
28. S. Vijayarani, and M. Vinupriya, "Performance analysis of canny and Sobel edge detection algorithms in image mining." *International Journal of Innovative Research in Computer and Communication Engineering*, vol. 1, no. 8, pp. 1760–1767, 2013.
29. G. N. Chaple, R. D. Daruwala, and M. S. Gofane, "Comparison of Robert, Prewitt, Sobel operator-based edge detection methods for real-time uses on FPGA." In *2015 International Conference on Technologies for Sustainable Development (ICTSD)* (pp. 1–4). IEEE, 2015, February.
30. J. J. Clark, "Authenticating edges produced by zero-crossing algorithms." *IEEE Transactions on Pattern Analysis and Machine Intelligence*, vol. 11, no. 1, pp. 43–57, 1989.

15 Performance Analysis of Non-orthogonal Multiple Access over Orthogonal Multiple Access

Purnima K. Sharma
ECE Department, Srivasavi Engineering College, Tadepalligudem, Andhra Pradesh, India

Dinesh Sharma
ECE Department, Chandigarh College of Engineering & Technology, Chandigarh, India

E. Kusuma Kumari
ECE Department, Srivasavi Engineering College, Tadepalligudem, Andhra Pradesh, India

T. J. V. Subrahmanyeswara Rao
ECE Department, Sasi Institute of Engineering & Technology, Tadepalligudem, Andhra Pradesh, India

CONTENTS

DOI: 10.1201/9781003230526-15

15.1 INTRODUCTION

One significant part of cellular system design is the establishment of multiple access strategies. Its goal is to give several users the ability to share spectrum resources efficiently and cost-effectively. Many multiple-access techniques were deployed for different wireless applications [1]. Since the interference is less among users with such an orthogonal design, good system-level performance can be achieved, even with simple receivers. The multiple access technology must be reviewed because of the tremendous spread of new internet-enabled intelligent products, apps, and services, hastening the development of 5G communications. NOMA – non-orthogonal multiple access is a newly suggested 3GPP long-term evolution (LTE) technique [2,3], of which the spectrum efficiency is projected to be superior.

15.1.1 Milestones in NOMA

a. 1972: Thomas Cover takes up the problem of simultaneously achievable transmission rates and proves that superimposing the higher rate of data on the lower-rate data gives better theoretically possible rates than the time-sharing scheme of transmission. This formed the basis for Superposition Coding.
b. 1994: "Successive Interference Cancellation Scheme" for DS/CDMA appears in literature.
c. 1998: Thomas Cover publishes a more elaborate paper titled "Comments on Broadcast Channels," which summarizes some of the past research work in the field and adding new insights. The paper concludes that if information intended for each receiver is superimposed and broadcast, the receivers can peel off the information in the layers to recover the intended message.
d. 2012: Design and experimental evaluation of Superposition Coding strategies [2] for the first time.
e. 2013: Superposition Coding with SIC receiver studied under Non-Orthogonal Multiple Access by NTT Docomo [1,4].
f. 2015: 3GPP Release 13 studies about power domain.
g. 2021: Imperfect SIC can increase the efficiency of a NOMA by decreasing the interference [5].

NOMA is the best promising radio access strategy. NOMA is a protocol for reducing user traffic and interference for future wireless communications [2–4,6–16].

15.1.2 Advantages of NOMA

While NOMA has numerous advantages, many practical issues prevented its adoption. Interest in this field rose again in the early 2010s mainly due to the near exhaustion of traditional resources and the need for schemes with greater spectral efficiency. The following are the main advantages of NOMA:

a. Improved Capacity: As multiplexing is over the power domain, individual users get greater bandwidth and hence a higher capacity. The loss of capacity due to the fractional allocation of transmit power does not impact the system much as bandwidth plays a more crucial role in incapacity. NOMA achieves the capacity bound in both uplink and downlink.
b. Frequent Scheduling: The UEs are scheduled more often as the transmission is done to two UEs simultaneously. This reduces latency.
c. Improved Fairness: Cell edge users get a fairer share of throughput as more power is allocated.
d. Improved Spectral Efficiency: The same spectrum is used for simultaneous transmission, improving spectral efficiency.
e. Massive Connectivity: As the allocation of users is non-orthogonal, the number of supported devices is no longer strictly bound by the resources available. Thus, NOMA can accommodate more users and can hence advance connectivity.

15.1.3 Applications of NOMA

a. The following are the main applications of NOMA:
b. Visible Light Communication: Similar performance gains as seen in the RF case can be expected if NOMA is implemented in VLC. As the channel generally does not change most of the time, the decoding becomes simpler.
c. MIMO-NOMA: The combination of MU-MIMO, which allows multiple beams and NOMA within a single beam, can lead to a greater capacity of the system.
d. Internet of Things: The scenario in IoT is massive connectivity. The exploitation of non-orthogonal resources as a means to enhance connectivity is subject to research.
e. SoDeMA: Software-Defined Multiple Access is an active research area where the best multiple access schemes are chosen based on the conditions of the system. For example, OMA would prefer NOMA if we only have a small number of users and do not have significant SNR variance. Similarly, different scenarios call for different multiple access.

15.1.4 Different Schemes of NOMA

NOMA uses non-orthogonal transmission by purposely introducing inter-cell and intra-cell interference. The required data is decoded at the receiver using the sequential-interference cancellation (SIC) [12] approach. The receiver's complexity is higher than OMA's, but it provides better spectral efficiency, lower latency,

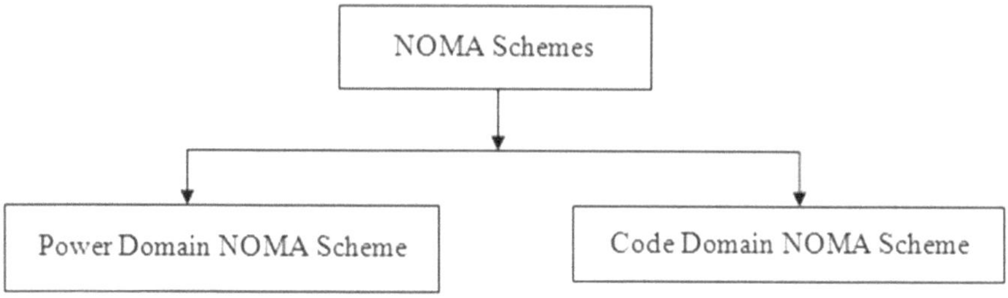

FIGURE 15.1 Different categories of NOMA schemes.

excellent dependability, and colossal connection. NOMA is founded on giving far users more power and giving near users less power [17]. The active research areas are power allocation, sum-rate maximization, optimal user pairing, performance evaluation under various scenarios, and energy efficient resource allocation. Figure 15.1 depicts different categories of NOMA schemes.

15.1.4.1 Power Domain NOMA

Different power levels are allotted for consumers according to their distance with a base station or related channel information in power-domain NOMA to achieve higher gain. Also, multi-user interferences are decreased with the use of SIC [18]. Figure 15.2 depicts the Downlink NOMA system model of power-domain multiple access. Multi-user detection (MUD) is identified at the receiver using SIC, measured with the rise of the SINR-signal to interference noise ratio. Figure 15.3 depicts the Uplink NOMA system model of power-domain multiple access.

15.1.4.2 CODE Domain NOMA

Multiplexing by code domain is the same as multiplexing by code-division multiple access – CDMA. It takes advantage of all of the time and frequency slots available to it. It can deliver good spreading and shaping gains while increasing signal

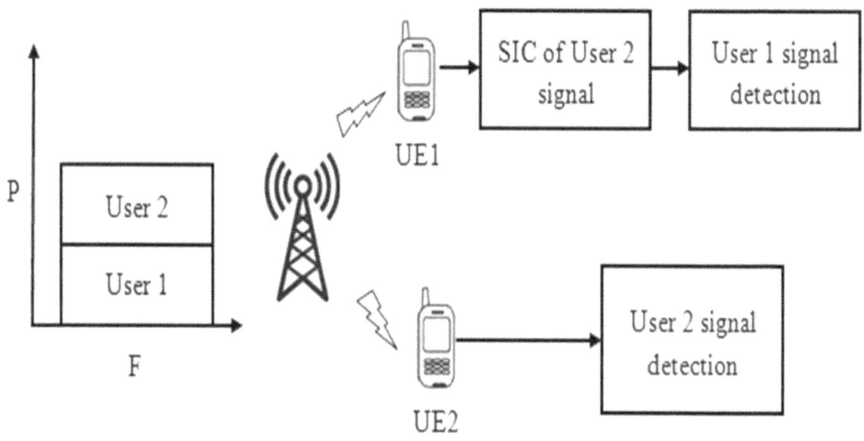

FIGURE 15.2 Simple downlink NOMA system model for two users.

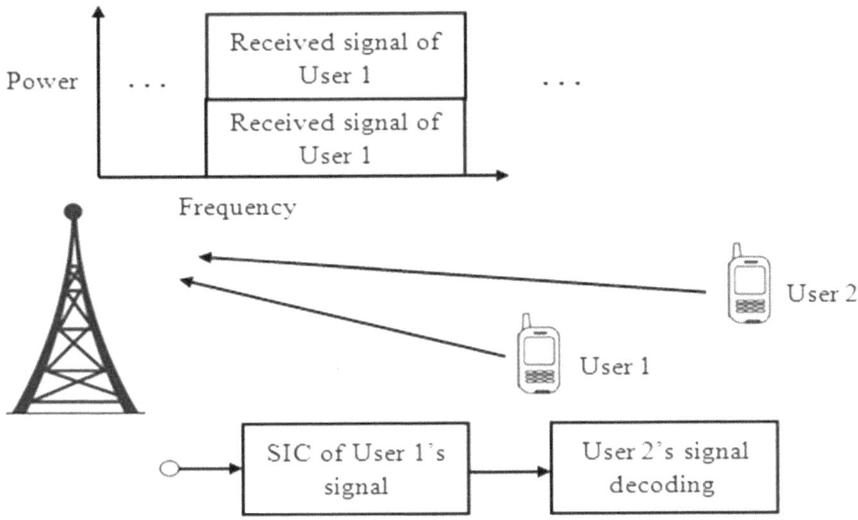

FIGURE 15.3 Simple uplink NOMA system model for two users.

bandwidth [18]. Xie et al. in [19] proposed random code according to the on and off random-access channel for designing a communication method in which a vector represents the information of N users where each corresponds to a particular user. A non-zero value represents that the user is active, and an inactive user is assigned zero. Instead of collecting the signal vector's actual value, the receiver aims to detect unknown active nodes.

In this, each node n is described with a complex Gaussian vector C_n which encodes the data S_n. And the encoded message is given as $X_n = C_n S_n$

Now the received signal is given as

$$Y = \sqrt{\rho o} \sum_{n=1}^{Na} \mathbf{h_n}(C_n \mathbf{S_n})^T + Z \tag{15.1}$$

The NOMA was first developed for downlink multi-user communications, also known as a broadcast channel, in which SC and SIC are the two main building parts. However, because the EM waves are overlaid at the receiver base station (BS), the installation of successive-interference cancellation circuits is more economical at the BS than at the consumer station; NOMA arises automatically in uplink communications. As a result, the NOMA concept has recently been extended to uplink multi-user communications. At the transmitter, multiple users' information is coded with varying power levels. Users' channel gains are considered for a specific ordering without sacrificing generality. In the literature, the user with the best channel quality is referred to as a strong user, whereas the user with less quality is referred to as a weak user. Strong and weak users' transmit powers are divided in order of channel gain. The weak user is generally allotted higher power at the transmitter to ensure equality and aid SIC decoding.

To take advantage of the diversity in channel gains and transmit powers, SI reduction decoding is performed at the receiver. Before decoding its own signal, a SIC receiver decodes other users' information one after another, according to a decoding order. The order of received signal powers determines the decoding order in SIC decoding, with the user with the highest received power being decoded first. The receiver subtracts one user's signal from the signal it has received once it has been decoded. With this process, the interference is removed one by one, resulting in a higher data rate. Inter-user interference (IUI) can be removed with the use of SIC. Compared with OMA techniques, superposition coding (SC) paired with SIC attains similar performance as the strong user attaining a weak user bound equal to that of single-user bound. Because of its high SNR, the strong user achieves the performance bond, but superposition coding allows it to receive a tiny amount of transmitted power, which causes minimal disturbance to weak users. To acquire single-user bound performance, an orthogonal technique should allot a similar DoF to the weak user, resulting in a more significant drop in strong user performance. As a result, NOMA achieves the capacity of general degraded broadcast channels compared to SC and SIC.

15.2 PERFORMANCE ANALYSIS OF NOMA WITH MULTIPLE ANTENNAS

Future wireless network standards are highly rigorous and demanding, as stated in the introduction. New technologies to be incorporated, in particular, to obtain a more significant improvement in the SE compared to the LTE baseline. The gain from NOMA with SIC had improved to 30%–40% approximately. Although they share some resources like frequency and time, the performance of SE is the same. Incorporating NOMA with multiple antennas solves this issue. In this session, possible extension of NOMA by applying multiple antennas through multiple degrees of freedom for further improvement of performance in 5G and beyond wireless communications are discussed. Despite the benefits of using numerous antennas in NOMA, several open-research concerns must be addressed. MIMO NOMA is more complicated to construct than single input single output (SISO) NOMA.

The critical issue of SISO-NOMA is identifying the best power allocation across users, whereas in MIMO-NOMA, the spatial DoF allows for beam formation to eliminate user interference in power and spatial domains. As a result, MIMO-NOMA's becomes difficult in designing. In general, multi-antenna NOMA introduces two primary problems. First, while MIMO-NOMA achieves comparable performance with OMA produces the best multi-user capacity region. MIMO-NOMA provides the best performance when the users' channels are quasi-degraded. On the other hand, the move from quasi-degraded to general channels is still a work in progress. Second, in MIMO-NOMA settings, user ordering is an NP-hard problem since elements of matrices/vectors define it. Since the users' channels are scalars in SISO, the SIC order is usually determined by channel gains. The receiver decodes the information of the weak user with less gain. On the other hand, this characteristic is not guaranteed with multiple antenna systems with distinct beamformers for each user.

The practical channel gains in MIMO-NOMA are linked to specific beam-forming designs, making beam-forming design and SIC ordering intertwined. In MIMO

NOMA [20,21] settings, different ordering algorithms are offered. The BS considers random beam formation, in which the users are ordered according to the quality of feedback channels. NOMA-based SC SIC technology maintains every cluster's consumers, which superposition codes several user signals within each beam. The fundamental distinction between these approaches is whether a single beamformer serves a single user or several users. It is worth noting that the decoding order utilized in this method is optimal, and different decoding orders result in higher rates.

15.3 COMPARISON OF OMA AND NOMA

The orthogonal multiple access (OMA) technique provides more straightforward receiver detection but fewer users' noise with reduced spectral efficiency. The NOMA method provides greater spectral efficiency with sophisticated connection density. It supports a moderate number of users with lower latency by supporting diverse QoS. As the number of users increases, the complexity of receivers also increases, thereby providing higher sensitivity to channel uncertainty.

15.3.1 NOMA and OMA Downlink System Model for N Users (Power Domain)

Consider the information related to the channel from the base station to the ith user. If N such channels, h1, h2,... hN are considered. Here assume that user 1, with channel $h_{1,}$ is farthermost from BS. Hence, the weakest user is user 2 is the following and so on. User N is the nearby/most robust user. Then the users' channel condition is organized: $|h1|^2<|h2|^2<...|hN|^2$.

Let x1, x2, ..., xN represent the messages that will be sent to the users. With these messages, the BS conducts superposition coding and sends the following NOMA signal into the channel:

$$x_{\text{NOMA superposition}} = \sqrt{P}\left(\sqrt{\propto_1}x_1 + \sqrt{\propto_2}x_2 + \ldots \sqrt{\propto_n}x_n\right. \tag{15.2}$$

Where P, the total transmitted power with α_1, α_1 .... α_N represent power allocation coefficients. Since the channels are ordered as $|h_1|^2 < |h_2|^2 < \ldots. |h_N|^2$., the power allocation factors are allocated as $\alpha_1 > \alpha_2 > \ldots. >\alpha_N$.

General formula for x after super position is

$$x_{NOMA\ super\ position} = \sqrt{P}\sum_{i=1}^{N}\sqrt{\propto_i}x_i \tag{15.3}$$

The received strength of the signal at user i is characterized as

$$y_i = x_{NOMA\ super\ position}h_i + w_i \tag{15.4}$$

W_i is the AWGN with mean zero and variance σ^2.

$$y_i = (\sqrt{P} \sum_{i=1}^{N} \sqrt{\propto_i x_i}) h_i + w_i \tag{15.5}$$

$$y_1 = \left(\sqrt{P} \left(\sqrt{\propto_1} x_1 + \sqrt{\propto_2} x_2 + \ldots \sqrt{\propto_n} x_n\right) h_1 + w_1 \tag{15.6}$$

From output signal y_1 user 1's signal is required, and then the first term that is $\sqrt{\propto_1} x_1$ is desired value and remaining is interference. So at user 1, the decoding rule should be followed to achieve the desired output. User 1 is weakest, so it was allocated with more power, the signal intended for use one will dominate the received signal. So user 1 can go for direct decoding, treating the other signals from remaining users as interference. Thus, user 1 can indirectly decode the signal by considering the SINR.

$$SINR\ at\ user1 = \frac{\propto_1 P|h_1|^2}{\propto_2 P|h_2|^2 + \propto_3 P|h_3|^2 + \ldots + \propto_n P|h_n|^2 + (\sigma)^2} \tag{15.7}$$

Date rate at user 1

$$R_1 = log_2(1 + SINR\ at\ user1) \tag{15.8}$$

Next at user 2

$$y_2 = \left(\sqrt{P} \left(\sqrt{\propto_1} x_1 + \sqrt{\propto_2} x_2 + \ldots \sqrt{\propto_n} x_n\right) h_2 + w_i \tag{15.9}$$

From output signal y_2 user 2's signal is required, and then the 2 × 2 term $\sqrt{\propto_2} x_2$ is desired value and remaining is interference. The following is the decoding rule for user 2.

Because user 1 is given the most significant amount of power, its message dominates the received signal y of all the other users. As a result, user 2 should decode user 1's message immediately before performing SIC to remove it from y2.

$$SINR\ at\ user2 = \frac{\propto_2 P|h_2|^2}{\propto_3 P|h_3|^2 + \ldots + \propto_n P|h_n|^2 + (\sigma)^2} \tag{15.10}$$

Repeat same thing for user 3,4 and so on up to N users.

For any user, the data rate or achievable rate is written as

$$R_i = log_2(1 + SINR\ at\ user) \tag{15.11}$$

Here, OMA is compared with NOMA.

In OMA transmission of N users, considering TDMA requires N time slots. The respective user's signal is transmitted at every time slot, assuming the same duration for every slot. The transmitted signal of a ith user in the ith time slot is given by,

$$x_{OMA} = \sqrt{P}x_i \tag{15.12}$$

From this the output signal $y_{i,\, OMA} = \sqrt{P}x_i hi + wi$ (15.13)

$$SINR\ at\ useri = \frac{P|h_i|^2}{(\sigma)^2} \tag{15.14}$$

Data rate or achievable rate for any user is given as

$$R_i = \frac{1}{N}\left(log_2(1 + SINR\ at\ useri)\right) \tag{15.15}$$

The complexity of the system increases with an increase in the number of users. The (BER) error rate in NOMA is also varied according to the number of users considered. Following are the results of the simulation of NOMA with a different number of users.

15.3.2 NOMA and OMA Uplink Signal Model for N Users (Power Domain)

Each user broadcasts their own signal x_i with a transmit power P_i in the uplink, resulting in a received signal at the base station depicted as

$$y_i = x_{NOMA\ super\ position}h_i + w_i$$

Where $x_{NOMA\ super\ position} = \sqrt{P}\left(\sqrt{\propto_1}x_1 + \sqrt{\propto_2}x_2 + \ldots \sqrt{\propto_n}x_n\right.$

General formula for x $x_{NOMA\ super\ position} = \sqrt{P}\sum_{i=1}^{N}\sqrt{\propto_i}x_i$

W_i is the AWGN with mean zero and variance σ^2.

$$y_i = (\sqrt{P}\sum_{i=1}^{N}\sqrt{\propto_i}x_i)h_i + w_i \tag{15.16}$$

$$y_1 = (\sqrt{P}(\sqrt{\propto_1}x_1 + \sqrt{\propto_2}x_2 + \ldots \sqrt{\propto_n}x_n)h_1 + w_1 \tag{15.17}$$

Here at the base station, SIC will be implemented. The first signal it decodes is a signal from the strongest, i.e., from the near user.

The signal-to-noise ratio for the near user, considering all remaining signals as interference, is

$$SINR\ at\ user1 = \frac{P|h_1|^2}{P|h_2|^2 + P|h_3|^2 + \ldots + P|h_n|^2 + (\sigma)^2} \tag{15.18}$$

Where P is the power transmitted by the consumer equipment or user 1 equipment

Generally, for the ith equipment, the SNR is

$$SINR\ at\ user\ i = \frac{P|h_i|^2}{\sum_{i=i+1}^{n} P|h_i|^2 + (\sigma)^2} \tag{15.19}$$

For any user equipment, the throughput is written as

$$R_i = W\ log_2(1 + SINR\ at\ useri) \tag{15.20}$$

Whereas in OMA, the SNR and throughput of users can be written as follows

$$SINR\ at\ user\ i_{oma} = \frac{P|h_i|^2}{(\sigma_i)^2} \tag{15.21}$$

$$R_{ioma} = W_i\ log_2(1 + SINR\ at\ useri_{oma}) \tag{15.22}$$

Where

$W_i = \frac{W}{N}$ n is the total number of users.

Uplink NOMA's optimal successive-interference cancellation decoding order, with the same transmit power, is the ascending order of channel gains, which is the inverse of downlink NOMA's. The transmission-power constraint for all uplink NOMA users can be eased when using a closed-loop uplink power control. Here, SIC algorithm is used at the receiver for perfect reception. The SIC flow will be.

SIC algorithm flow: It is an iterative method that decodes data in decreasing order of power levels. The data corresponding to the user with the highest power is decoded first, and then the data relating to the user with the second-highest power, and so on until all users' data has been deciphered. The steps involved in SIC decoding for the simple case considered is described below:

Step 1. Directly decode x to obtain the signal that is allocated the highest power.

For example, here, x_1 is allocated more power (i.e., $\alpha_1 > \alpha_2$).

Step 2. Multiply the decoded signal in step 1 by its parallel allocation factor and subtract it from x.

For example, if x_1 is decoded in the previous step, then subtract $\sqrt{a_1}x_1$ from x. This leads to$x - \sqrt{\alpha_1}x_1$

Step 3. Decode the signal obtained in step 2 to get the other signal allocated low power.

For example, decoding of $- \sqrt{\alpha_1}x_1$ obtained from the previous step would yield x_2.

15.3.3 Implementation of Downlink NOMA for Two Users

Figure 15.2 shows a general model for downlink NOMA with one base station and two users. The information of both users is transmitted from the base station, i.e., x1 and x2, in the same frequency band, with dissimilar transmit powers p1 and p2, respectively. The associated transmitted signal is denoted by

$$x = \sqrt{p_1}x_1 + \sqrt{p_2}x_2 \tag{15.23}$$

where p1 + p2 = 1 inhibits transmitted power. The user k received signal strength is given by

$$y_k = h_k x + n_k \quad where\ k \in [1,\ 2]$$

where h_k indicates the channel coefficient between transmitter and kth user, n_k denotes additive white Gaussian noise (AWGN)

Successive-interference cancellation (SIC) decoding is used at the user side in downlink NOMA systems. The optimal SIC decoding order is based on noise-normalized channel gains in the order of priority.

As a result of SIC, the following rates are easily achievable:

$$R_{1,2} = log_2\left(1 + \frac{p_2|h_1|^2}{p_1|h_2|^2 + (\sigma_1)^2}\right) \tag{15.24}$$

$$R_1 = log_2\left(1 + \frac{p_1|h_1|^2}{(\sigma_1)^2}\right) \tag{15.25}$$

$$R_2 = log_2\left(1 + \frac{p_2|h_2|^2}{p_1|h_2|^2 + (\sigma_2)^2}\right) \tag{15.26}$$

Where $R_{1,2}$ indicates the capacity at which user one still decodes the information of user 2, R_1 refers to the capacity at user 1 to decode the message of user 2, and R_2 refers to the rate for which user 2 to decode its own message. It is worth noting that obtaining the R1 data rate necessitates the following condition.

$$R_{1,2} \geq R_2$$

It indicates the likelihood of effective interference cancelation at the first user (near the user).

15.3.4 Implementation of Uplink NOMA for Two Users

Figure 15.3 shows a general system concept for uplink NOMA with one BS and two users. Both users are using the same frequency band and transmit power p to send their messages. At the BS, the received signal is provided by

$$y = \sqrt{p}\,h_1x_1 + \sqrt{p}\,h_2x_2 + n, \tag{15.27}$$

The base station transmits a downlink reference signal, and every user estimates the channel based on that signal. As a result, depending on the channel gains, transmitted power can be set to P_1 or P_2. User 1 is again considered the most robust consumer, with increasing channel gain than user 2, the weakest consumer. The base station executes SIC on each user signal after receiving the overlaid signal to decode the signals.

The user one signal is decoded first, with the user two signal being treated as noise by the base station. The signal that has been decoded is deducted from the received information signal. The excess signal is used to decrypt the user's two information. User 1 gets interference from user 2 in uplink NOMA, as seen from the above process. However, because user 1's signal is eliminated before decoding, there is no interference for user 2.

However, in downlink NOMA, user 2 encounters interference from user 1, and there will be no interference in user 1's signal because it is eliminated before decoding user 1.

The base station performs SIC decoding to obtain the messages s1 and s2 from the overlaid signal y in Uplink NOMA. The BS, for instance, decodes user 1's message before subtracting s1 from the overlaid signal y. The BS can then decipher user 2's message without the need for IUI. As a result, we may obtain the following individual data rates for users 1 and 2:

$$R_1 = log_2\left(1 + \frac{p|h_1|^2}{p|h_2|^2 + (\sigma)^2}\right) \tag{15.28}$$

$$R_2 = log_2\left(1 + \frac{p|h_2|^2}{(\sigma)^2}\right) \tag{15.29}$$

From Figures 15.4 and 15.5, user one is considered the far consumer, and user 2 is the nearest consumer to the base station. The powers are also allocated based on their distances. User 1 is allocated 75% of the total power, which is the highest and User 2 is allocated 25%. User 2 is allocated with the lowest power because it is nearer to the base station. So, the transmission of the signal to and from the BS is easier. The power allocation coefficients are shown below

$$a_1 = 0.75;\ a_2 = 0.25$$

From Figure 15.5, it is noted that the BER for NOMA systems is slightly greater than OMA systems. This is because, in NOMA systems, the data of both of the users is transmitted at a time.

In Figure 15.6 and Figure 15.7, user 1 is considered the far consumer, user 2 is the intermediate consumer and user 3 is the nearest consumer to the base station. The powers are also allocated based on their distances. User 1 is allocated 75% of the total power, which is the highest of all. User 2 is allocated with 20% and user 3 is allocated with 5% of the power. User 3 is allocated with the lowest power because it is nearer

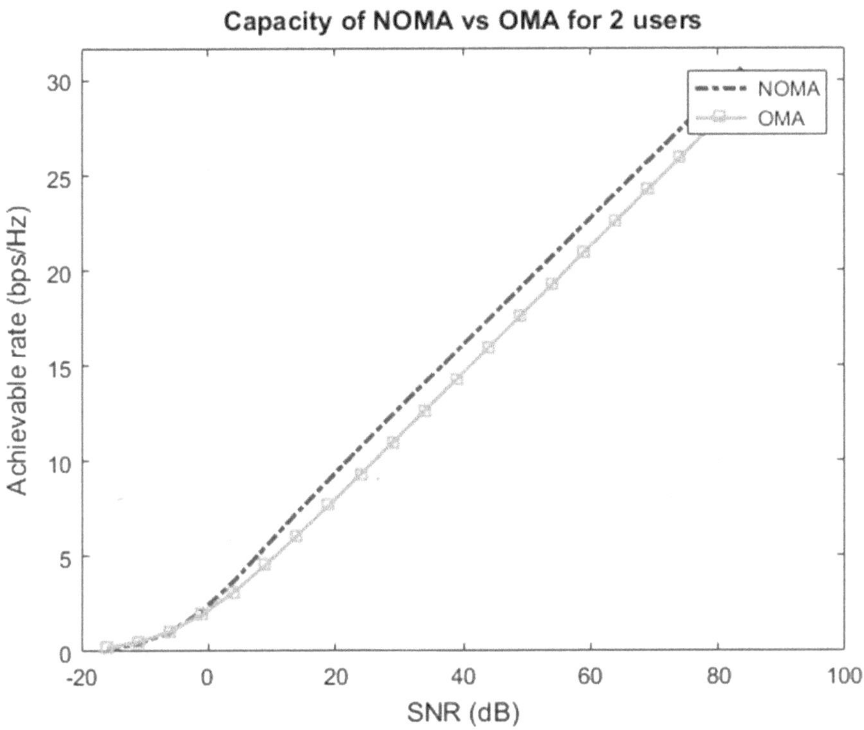

FIGURE 15.4 Capacity of NOMA vs OMA for two users.

to the base station. So, the transmission of the signal to and from the BS is easier. The power allocation coefficients are shown below

$$a_1 = 0.75;\ a_2 = 0.20;\ a_3 = 0.05$$

From Figure 15.7, it can be noted that user 3, irrespective of the lowest power given to it, transmits the signal with greater accuracy than the other two signals. The BER curve of user 3 proves this. This remains the same in the case of the NOMA downlink, which is shown in Figure 15.9.

Here is also shown the case where four users are transmitted at the same time. Following are the simulation results.

From Figure 15.8, it is noted that at a low signal-to-noise ratio, OMA is slightly better than NOMA in the rate of transmission. However, as the SNR increases, NOMA encompasses OMA and has a better rate of transmission.

In the above two results, user 1 is considered the far user, user 2 and user 3 are the middle users, and user 4 is the nearest user to the base station. The powers are also allocated based on their distances. User 1 is allocated 80% of the total power, which is the highest of all. User 2 is allocated with 16%, and user three is allocated with 3.2% of the power and user four is allocated with 0.8% of the total power. User 4 is allocated with the lowest power because it is nearer the base station. So, the

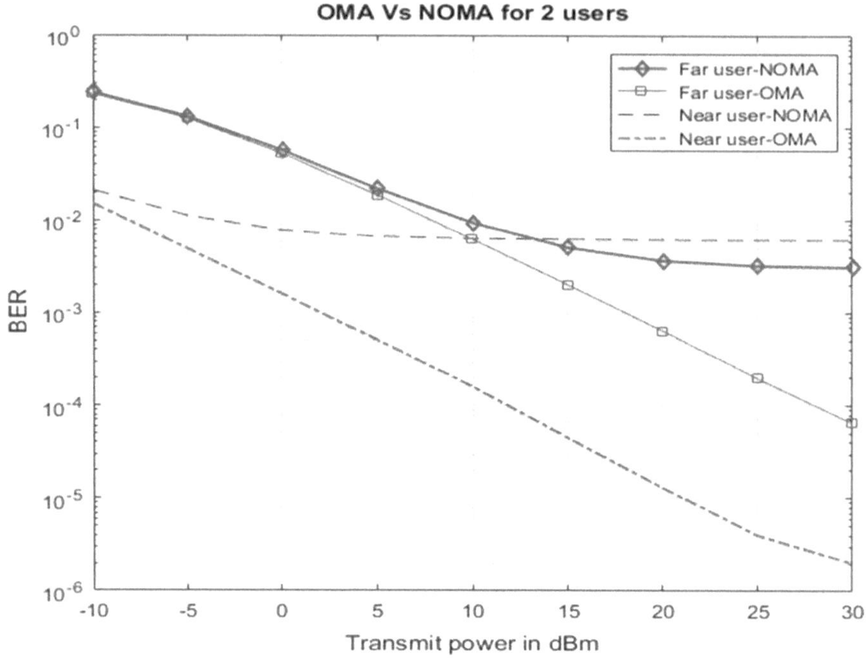

FIGURE 15.5 BER of NOMA vs OMA for two users.

transmission of the signal to and from the BS is easier. The power allocation coefficients are shown below

$$a_1 = 0.8;\ a_2 = 0.16;\ a_3 = 0.032;\ a_4 = 0.008$$

From Figure 15.9, it can be noted that user 4, irrespective of the lowest power given to it, transmits the signal with greater accuracy than the other two signals. The BER curve of user 4 proves this. So, compared to the OMA system, the bit-error rate in NOMA is slightly higher. This is because of the simultaneous transmission of the data in NOMA. This can be controlled by hiring sophisticated equipment at the receiver.

So, from all the results, it is concluded that the available rate of NOMA is greater than OMA, which leads to a higher rate of transmission. This results in a more incredible speed of data transmission. For all these simulations, the same power allocation factors and distances of the users from the base station are considered in the case of bit-error rate.

15.3.5 Spectral and Energy Efficiency of NOMA and OMA

Spectral efficiency can also be treated as bandwidth efficiency. It is defined as the amount of information transmitted over the communication channel for a given bandwidth.

Energy efficiency is referred to as the number of bits transferred for unit power consumption. It can be measured as bits per joule. The power required to transmit

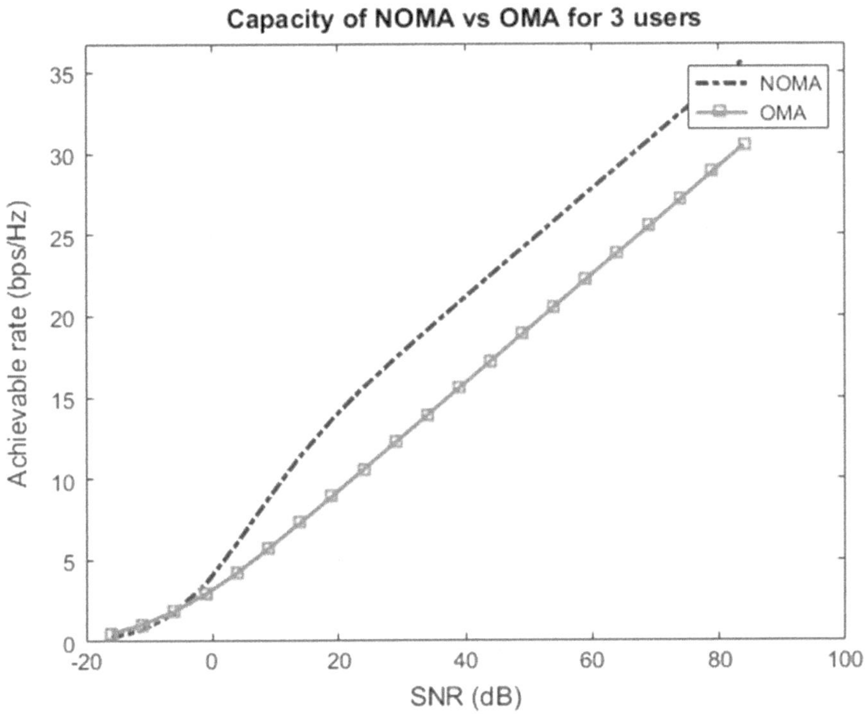

FIGURE 15.6 Capacity of OMA and NOMA uplink for three users.

data is the determining factor of energy efficiency for mobile devices. Spectral and energy efficiency are the vital factors that decide how efficiently the given resources are used. So, it is essential to assess these parameters for a multiple access technique to confirm its efficiency. The network's throughput performance has been a focus of most previous analyses. The energy efficiency (EE) and the spectral efficiency (SE) of NOMA systems are considered here to analyze the performance of the multiple access scheme. Taking the downlink into account, the total power consumed by the base station may be stated as

$$P_{total} = P_T + P_{static} \tag{15.30}$$

Where P_T is the total signal power, which is the power disbursed in the complete circuit.

Energy efficiency (EE) is referred to as the total rate concerning the total power of the BS [7].

$$EE = \frac{R_T}{P_{total}} = SE\frac{W}{P_{total}}(\text{bits/joule}) \tag{15.31}$$

Wherethespectralefficiency (SE) with bps/Hz.

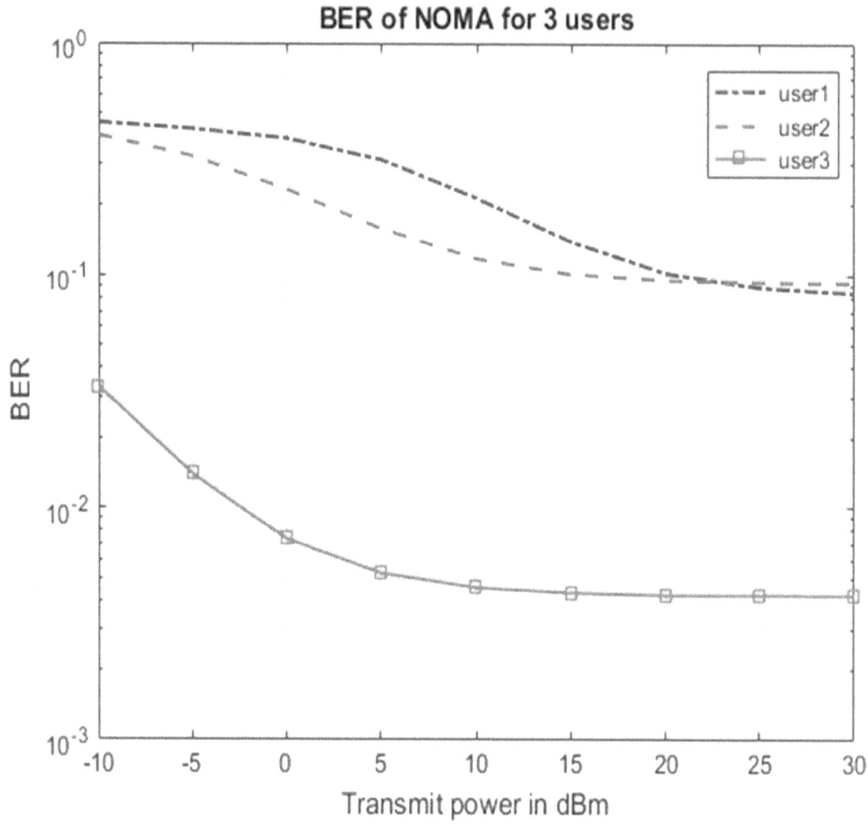

FIGURE 15.7 BER of NOMA downlink for three users.

Shannon's energy efficiency and spectral efficiency correlation (EE-SE) ignores the circuit's power consumption and is thus monotonic, with a more excellent SE continuously resulting in a lesser EE. WHEN CONSIDERING CIRCUIT POWER, the EE rises in the smaller SE region and falls in the greater SE region. The system's highest energy efficiency is at the peak of the curve. This point is referred to as the "green point." The EE-SE correlation is linear with a positive slope of for a fixed, with an improvement in SE resulting in EE.

RT is the sum capacity.

$$R_T = \sum_{i=1}^{i} R_i \tag{15.32}$$

Where 'i' denotes the number of users

A comparison of the spectral and energy efficiency for NOMA and OMA is presented here. Following are the simulation results of their comparison.

Figures 15.10, 15.11 and 15.12 represents the energy & spectral efficiency curves for 2, 3 and 4 users, respectively. The red curve represents OMA and the blue curve represents NOMA. It is noted that the spectral and energy efficiency for

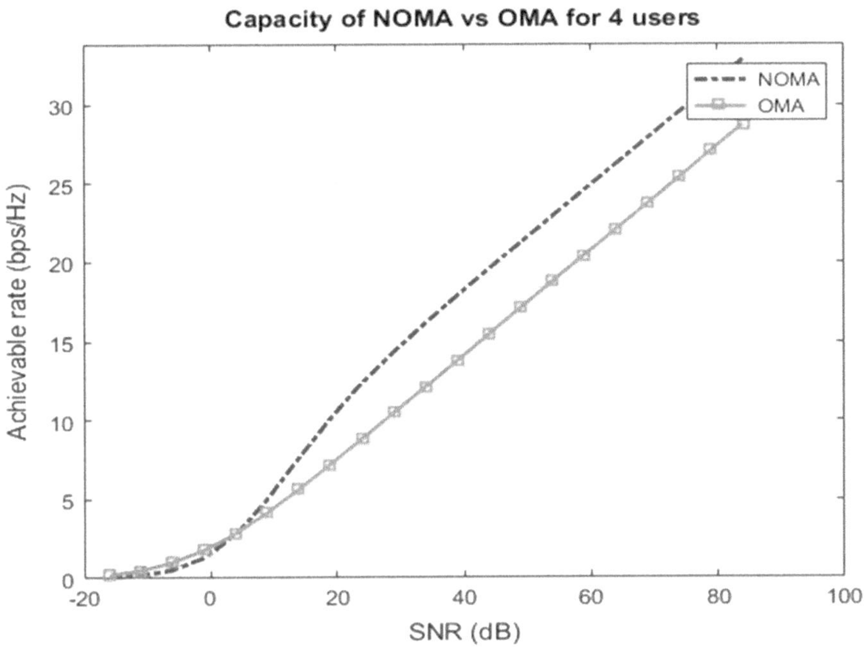

FIGURE 15.8 BER of NOMA uplink for four users.

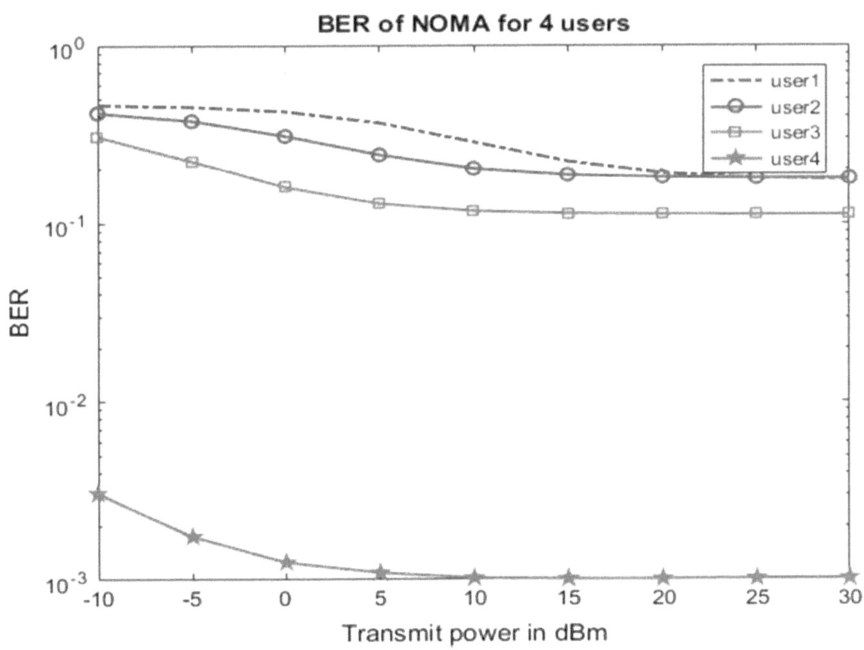

FIGURE 15.9 BER of NOMA downlink for four users.

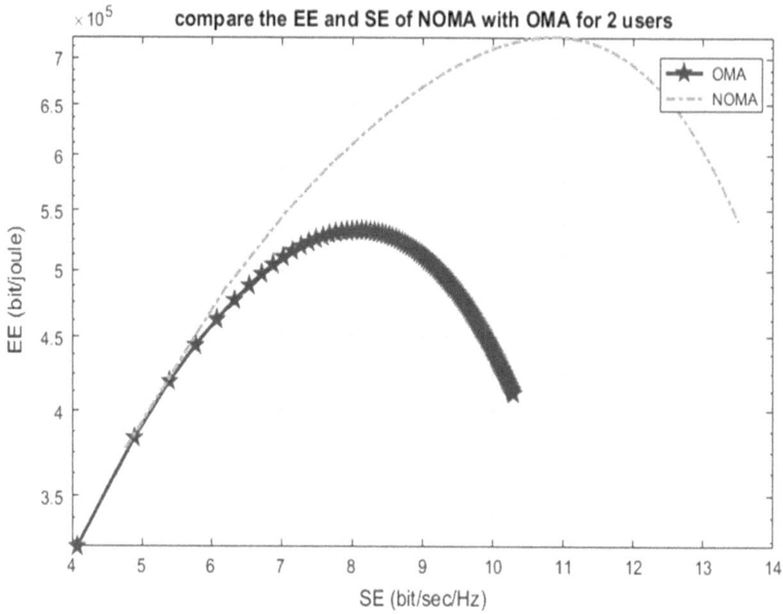

FIGURE 15.10 Spectral and energy efficiency comparison for two users.

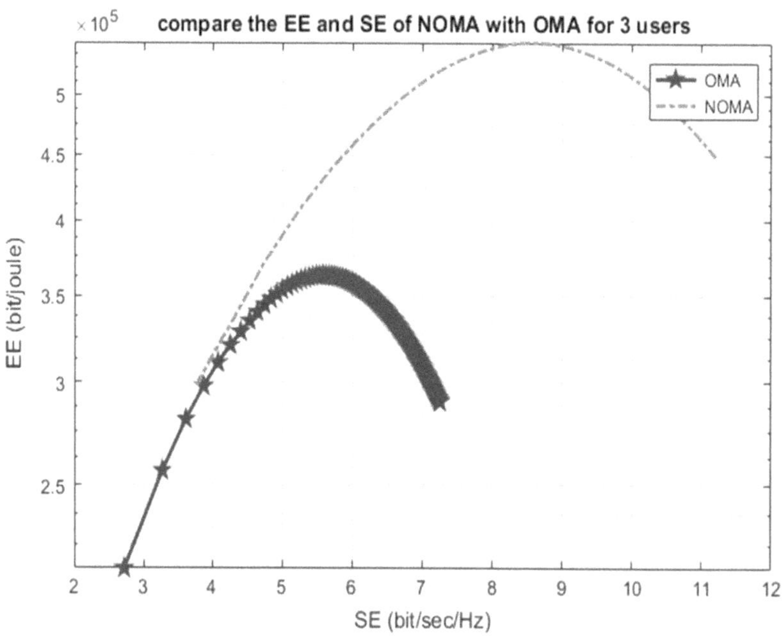

FIGURE 15.11 Spectral and energy efficiency comparison for three users.

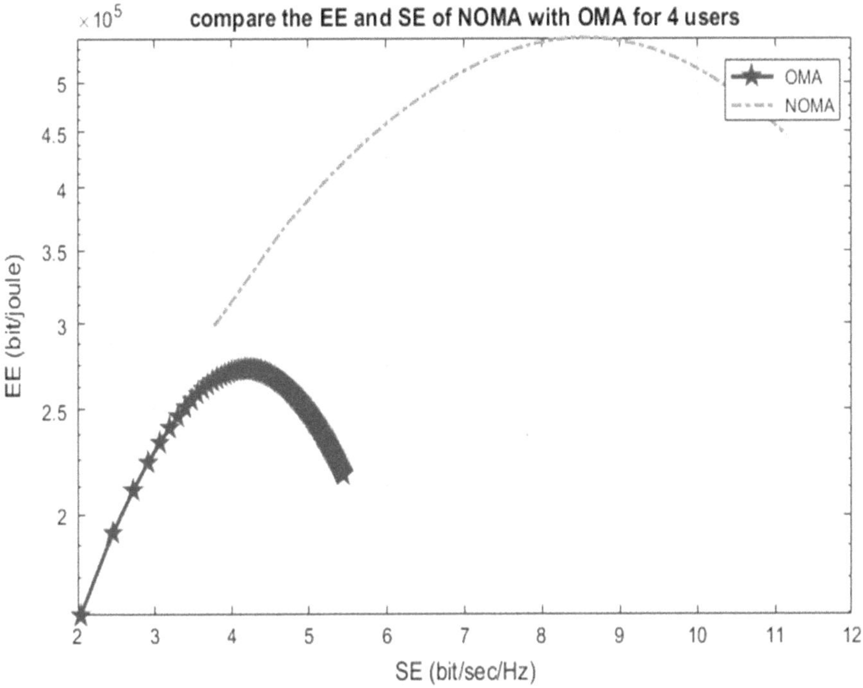

FIGURE 15.12 Spectral and energy efficiency comparison for four users.

NOMA, and all the three graphs is comparatively greater than OMA. Also, as the number of users increases, the spectral and energy efficiency of OMA keeps on decreasing. However, in the case of NOMA, it is nearly constant. This shows that NOMA has greater capability to use the given spectrum or the given power efficiently.

15.4 CONCLUSION

As time progressed, the number of users increased, and as the technology advanced, the need to increase the data rate emerged. The development of 5G provided solutions for all these requirements with actual data rates and massive connectivity. On the other hand, orthogonal multiple access techniques would raise the number of assets and expenditures in 5G. So, there is a need for that technology that provides all the requirements with existing resources. Power domain NOMA is a promising research topic that can meet most 5G communications' performance criteria. It can increase the system's spectrum efficiency, energy efficiency, and overall capacity. Also, NOMA methods outperform OMA schemes in terms of complexity and the cost of greater receiver complexity. From The simulation results, it has been observed that NOMA shows a noticeable part in shaping the future 5G wireless networks.

REFERENCES

1. L. Dai, B. Wang, Z. Wang, and S. Chen, "A Survey of Non-Orthogonal Multiple Access for 5G." *IEEE Communications Surveys & Tutorials*, vol. 20, no. 3, Thirdquarter 2018.
2. R. Chowdary Erpina and V. Reddy Gopireddy, "Performance of a Non-Orthogonal Multiple Access System with Full-duplex relaying over Nakagami-m Fading." *Master of Science thesis is submitted to the Faculty of Computing at Blekinge Institute of Technology*, January 2021.
3. A. Dhariwal, "Performance Evaluation of Energy Efficiency and Spectral Efficiency: NOMA vs OFDMA." *Telecommunications System & Management*, vol. 9, no. 3, 2020.
4. S. Ho, "NOMA for 5G Wireless Communication Systems." Thesis for Bachelor of Engineering (Electrical Engineering), School of Electrical Engineering and Telecommunications, UNSW Australia, 2017.
5. S. Srivastava, P. P. Dash, and S. Kumar, "Capacity Analysis of Non-orthogonal Multiple Access for Uplink and Downlink," *JMM*, February 2021. 10.13052/jmm1550-4646.171320
6. Z. Ding and H. Vincent Poor, "Design of MIMO-NOMA Downlink and Uplink Transmission." *IEEE Access*, vol. 4, 2016.
7. M. M. Barbé, "Development of a MATLAB application to generate NOMA-CAP 5G signals." Bachelor thesis, Universitat Politècnica de Catalunya, 2020.
8. N. Ye, H. Han, L. Zhao, and A.-H. Wang, "Uplink Nonorthogonal Multiple Access Technologies toward 5G: A Survey." *Wireless Communications and Mobile Computing*, vol. 1, no. 1, 2018.
9. R. C. Kizilirmak and H. K. Bizaki, "Non-orthogonal Multiple Access (NOMA) for 5G Networks, towards 5G Wireless Networks – A Physical Layer Perspective." vol. 1, no. 83, pp. 83–98, 2016.
10. H. Wymeersch and A. Eryilmaz, "Academic Press Library in Mobile and Wireless Communications.", vol. 2, no. 3, pp. 16–22, 2016.
11. S. Srinivasa, S. Vanka, Z. Gong, and P. Vizi, "Superposition Coding Strategies: Design and Experimental Evaluation." *IEEE Transactions on Wireless Communications*, vol. 11, no. 7, July 2012.
12. X. Su, H. F. Yu, W. Kim, C. Choi, and D. Choi, "Interference Cancellation for Non-orthogonal Multiple Access Used in Future Wireless Mobile Networks." *EURASIP Journal on Wireless Communications and Networking*, vol. 1, no. 1, pp. 1–12, 2016.
13. J. N. Patel and U. D. Dalal, "A Comparative Performance Analysis of OFDM Using MATLAB Simulation with M-PSK and M-QAM Mapping." In *International Conference on Computational Intelligence and Multimedia Applications (ICCIMA 2007)*, vol. 4, pp. 406–410, December 2007. 10.1109/ICCIMA.2007.142
14. R. Chataut and R. Akl, "Massive MIMO Systems for 5G and beyond Networks—Overview, Recent Trends, Challenges, and Future Research Direction." *Sensors*, vol. 20, no. 10, p. 2753, 2020. 10.3390/s20102753
15. J. Singh, "Generations of Wireless Technology." *International Journal of Engineering Research & Technology*, vol. 3, no. 10, April 2018.
16. K. Phlavan and M. Chase, "Spread-Spectrum Multiple-Access Performance of Orthogonal Codes for Indoor Radio Communications." *Ieee Transactions on Communications*, vol. 38, no. 5, May 1990.
17. Z. Wu, K. Lu, C. Jiang, and X. Shao, "Comprehensive Study and Comparison on 5G NOMA Schemes." *IEEE Access*, vol. 6, 2018.
18. L. Dai, B. Wang, Y. Yuan, S. Han, C. L. I, and Z. Wang, "Non-orthogonal Multiple Access for 5G: Solutions, Challenges, Opportunities, and Future Research Trends." *IEEE Commun. Mag.*, vol. 53, no. 9, pp. 74–81, 2015.

19. R. Xie, H. Yin, X. Chen, and Z. Wang, "Many Access for Small Packets Based on Precoding and Sparsity-Aware Recovery." *IEEE Transactions on Communications*, vol. 64, no. 11, pp. 4680–4694, November 2016.
20. J. Cui, Z. Ding, and P. Fan, "Outage Probability Constrained MIMO-NOMA Designs Under Imperfect CSI." *IEEE Transactions on Wireless Communications*, vol. 17, no. 12, December 2018.
21. H. K. Bizaki, "'Towards 5G Wireless Networks – A Physical Layer Perspective." *Intech Open Limited*, vol. 1, no. 2, pp. 16–22, 2016.

16 Advanced Wireless cum Digital Health Care and Artificial Human

Aaryan Sharma and Arvind Rehalia
Department of Instrumentation and Control, Bharati Vidyapeeth College of Engineering, New Delhi, India

CONTENTS

DOI: 10.1201/9781003230526-16

16.1 INTRODUCTION

Digital health or digi-health collaborates advanced biomedical science with modern digital technology such as artificial intelligence (AI) and machine learning (NL). This is the solution to the increasing need for medical personal and medical centers like hospitals, clinics, nursing homes, etc. This concept of digital health aims to help and cure many people with limited resources and who are in remote locations.

Digital health is made possible with the help of artificial intelligence engines and machine-learning algorithms. This will help medical centers to understand the root cause of the problems precisely. It will significantly improve the diagnosis and measurement of human body parameters. This will include hardware and software approaches, such as advanced microsensors and sensing algorithms with better calculation and elimination techniques. This would provide optimum efficiency for the machine and get fast and accurate results. This will become the backbone of the 'Telemedicine' concept and it will give birth to micro and personalized medicine for each individual [1–10].

With digital healthcare, we will aim and target a larger group of people suffering from all sorts of diseases. The AI and ML engine and algorithms will provide 'Prevention' and 'Prediction' of disease even before it reaches a later stage. In its initial years of implementation, this system will learn from past trials and improve itself for a better future (Figure 16.1).

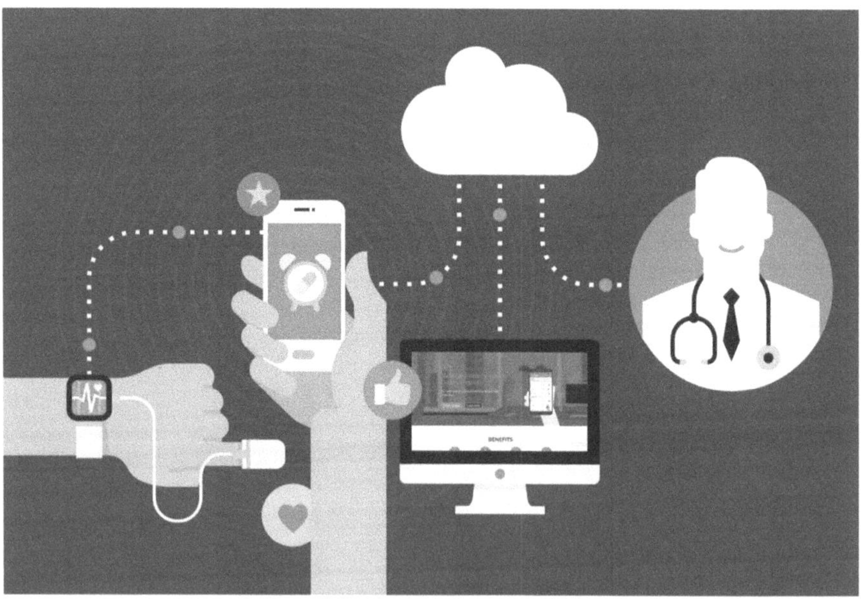

FIGURE 16.1 Digital healthcare [11].

16.2 NEED OF DIGITAL HEALTHCARE

In this modern and fast-growing world, everyone is on a pace to make progress. This hurry to make life better on this planet has badly hampered each individual's eating and bio-clock habits. Now, this much pressure and stress cause numerous types of diseases in the human body. This factor is making conditions even more complex day by day. With the increasing complexity of requirements, advanced science is needed, which cannot match the speeding complexity of diseases. On the other hand, due to the expanding population globally, we have a limited number of doctors and medical personal. This poor doctor-to-patient ratio makes things even more complex. With increasing deaths and worsening patients' conditions, something had to be done to keep everything in check.

Medical institutes, medical centers, doctors, entrepreneurs, engineering sectors came forward with a solution named 'digital health.' By this solution, we would reach new possibilities. This digital healthcare system would cover a vast population connected to a single grid named the 'internet.' This will significantly reduce the cost of healthcare and check-ups since we will have every important thing in our mobile phones (Figure 16.2).

This digital health system will take inputs from wearable technologies such as smartwatches, smart bands, fitness trackers, sleep trackers, etc. It will diagnose or predict any disease in the human body. With constantly learning and emerging algorithms, the system will continuously improve its diagnosis, accuracy, precision, error-correction, and disease searching capability in hospitals and remote care systems (Figure 16.3).

This local and remote patient care system will help in:

- Prediction
- Prevention
- Personalized medicine
- Human optimization

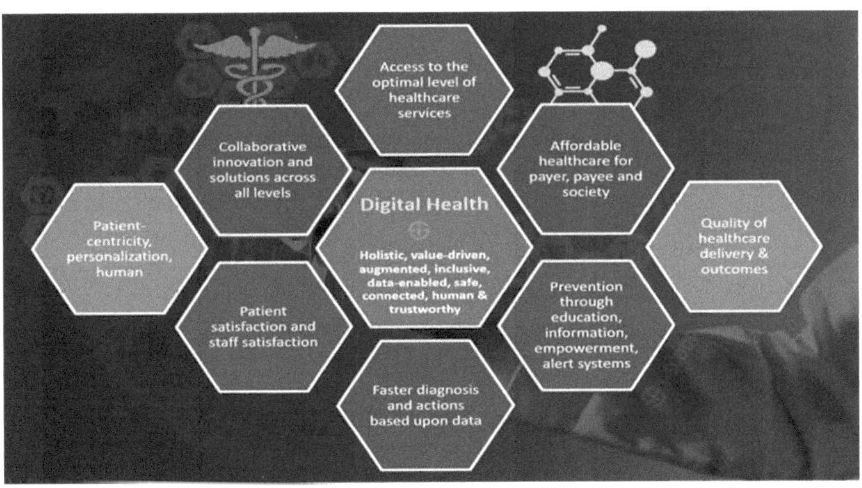

FIGURE 16.2 Digital transformation [12].

FIGURE 16.3 Personal-health-monitoring [13].

- Cure optimization
- Clinical assistance and decision making
- Engineering human factors
- Speech recognition

These favorable factors will provide a lot of assistance to doctors and medical personal in the fast and effective treatment of disease. This will reduce the immense amount of stress on doctors and medical staff as well.

In the future, when the situation is in control and everyone is exposed to this technology, predict upcoming disease just with the help of light wearable technology. Another possible factor is people suffering from chronic diseases like cancer, AIDS, etc. they need not visit hospitals regularly. They can be remotely monitored by doctors who provide essential instructions remotely. This will reduce rush at hospitals and medical centers and also, doctors in case of emergency can provide instructions from home. Rehabilitation can also be made easy with the help of digital healthcare. People who are given initial health treatment and are supported in their recovery need not regularly visit doctors or treatment centers. They would receive all day-to-day routines online, which they can do at home easily. As a feedback system, doctors will keep a check on patient's vitals and progress online.

This system will increase the overall life expectancy of human life across the globe, making the world a better place.

16.2.1 Digital Healthcare

16.2.1.1 Digital Revolution

In the early 2000s, when companies across the globe started making microsurgery instruments and machines, they were successful each time they had a goal. But the

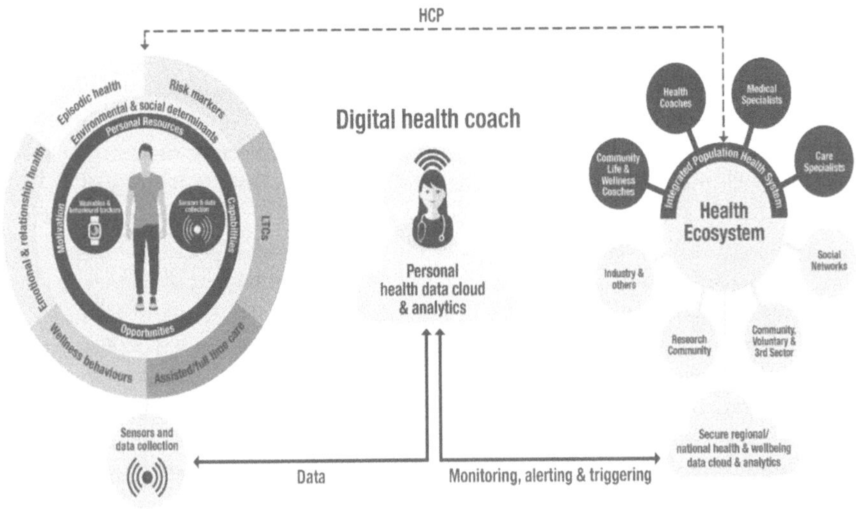

FIGURE 16.4 Digital health system [14].

problem was they did not have enough skilled people to know how to operate it. With collaboration with advanced technologies and engineers, they modified the systems controlled by AI &ML and used simple instructions. After this became fruitful, the medical and engineering sectors have been working together ever since. Everything can be made possible with the field of biomedical engineering. When we say digital revolution, it's a whole wide area to cover. But a few significant regions have proved themselves very useful, helping biomedical engineering to grow (Figure 16.4).

a. Image processing is one of them. It is an innovative program that can scan and detect all sorts of physical problems, such as a tumor, cancer, brain hemorrhage, fractures, nervous system damage, etc. Image processing on strong AI and ML algorithms can grow eventually. Over the years, both the operator and system will learn better with time. These systems will also significantly improve the accuracy and precision of the treatment and surgical procedures. A part of the already growing image-processing system is in the mobile application, which will tell the nutritional value of any food substance just by putting it under the mobile camera. This intelligent ML algorithm will make almost no need for food testing labs in the context of nutritional value checks. This would help remote care patients check what they should eat and what they should eat to maintain their health. With the help of each case the system solves, the learning algorithm becomes more and more reliable.
b. Micro medicine is another exemplary aspect of digital healthcare evolution. After a person is diagnosed with any disease, the system can be particular to what ratio and quantity of which salt or substance is needed. This will help in making personalized medicine for all individuals according to their needs. Also, the treatment will take micro medicine as it will be much smaller in size than usual.

c. Educating youth is another primary goal that mobile health apps and systems can fulfill. A significant chunk of the teenage population is addicted to drugs or alcohol in many world regions. This drug addiction is not only destroying the next generation but will also slow down the global progress of humanity. This leads to prevalent destructive causes like increased unemployment and crime rates. The 'eHealth's systems will keep them engaged in playful and peaceful activities, keeping them away from drugs, alcohol, and every possible thing which could ruin their lives. In countries like Africa, Brazil, some parts of America, etc., teenage girls indulge in sexual activities and get pregnant. This leads them to suffer, forces them to become teenage moms and even risk death. These factors alone are enough to ruin a girl's life for no reason. Thus, sex education is essential for both boys and girls. Significantly, one single mistake could ruin their life and destroy whatever they had dreamt of.
d. Safe barriers in technology comes along with all advancements in it. Safe barriers must be implemented in all AI and ML systems as there are possibilities of the system harming the patient instead of curing the patient. There are always chances of error in the program. There will surely be a time when system errors become nil, but that time is nowhere shortly (Figure 16.5).
e. Data privacy is a more profound concern when a person is connected to the grid, and there will be an electronic record of every individual. If that data gets hacked or leaked, it could be an entire fortune to that person.

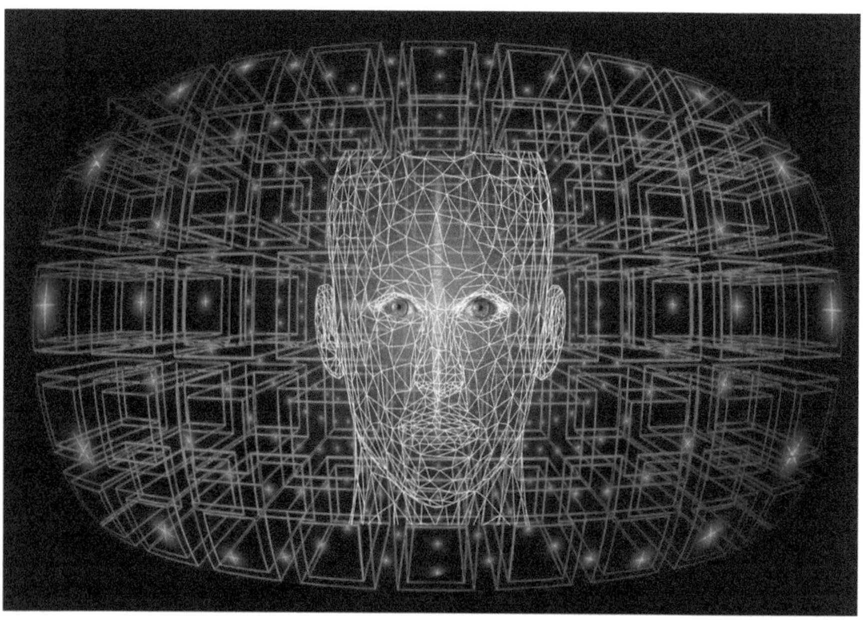

FIGURE 16.5 Image processing in healthcare [15].

FIGURE 16.6 Digital education [16].

Insurance companies would not insure a person if they come to know of their disease and that person would suffer to death simply because of poverty caused due to paying heavy bills. There can also be a group of people who could torture or bully or blackmail an individual if they come to know of one's disease or weakness. There can be all sorts of misuse of such confidential data (Figure 16.6).

16.2.1.2 Innovation Cycle

The innovation cycle is one of the most common ways to invent or discover something. It can be new technology, medicine, or treatment, depending upon the gravity of the situation. Its few crucial steps are discussed below:

a. Finding the problem is the beginning step, which is the most important step because any new virus or disease, just like COVID-19, can be very complicated to find. These new diseases can quickly go undetected if tested by the existing methods. Thus, every time a deadly virus is on the loose, brainstorming is essential.
b. Research on all different ways to tackle the root cause is fundamental. It enables scientists and doctors to invent new ways to stop this problem from spreading and keep people alive. They try all sorts of methods, old and new, to eliminate this disease. When the solution is ready, then comes the second approach of digital solution. If the problem is curable digitally, then with wearable technology, this is possible to treat a patient remotely.
c. Once the solution is ready and approved by the scientist and doctors, it goes further for testing and trial phases. In this phase, the answer to the problem goes on clinical trials. When it undergoes testing, many people are included in the test so that the solution is universal and effective on all human beings. This testing is evaluated by government bodies to give it a nod for public use and implementation.
d. Once the solution is ready, it is tested and approved. It now goes into implementation and mass production. This implementation can be digital or medical. In the case of digital performance, every program is checked.

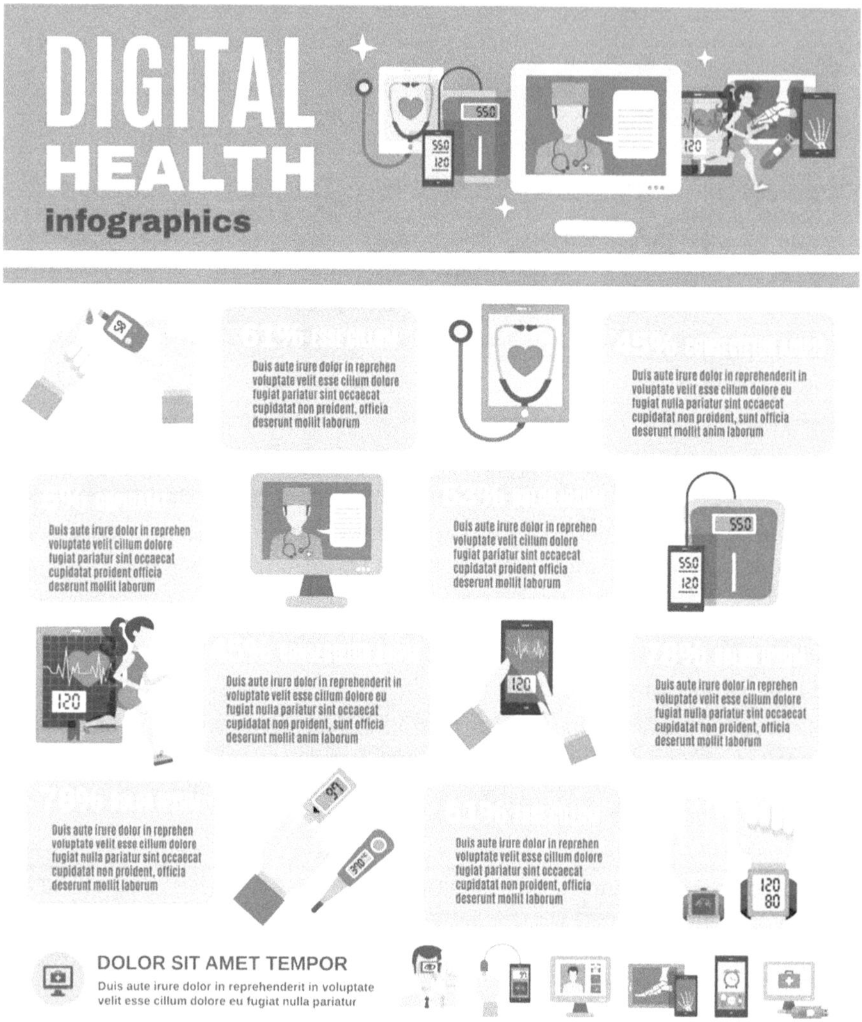

FIGURE 16.7 Digital health infographic [17].

Every bug is explored and all platforms are tested. Now, after all safety checks, the solution is digitally implemented. It may be using complex sensors like vibration or electrical therapy, or a BP controller, heartbeat tuning, or anything that is digitally possible (Figure 16.7).

16.2.1.3 Implementation of Digital Health

Implementation of a mission on such a large scale is complicated, especially in countries like Bharat, Pakistan, Bangladesh, Brazil, Philippines, etc., which are densely populated. Thus, the digital health mission is critical and complex to manage. It has started to spread its roots and will take some time, but, eventually, it will be a success. National healthcare programs are already making good progress in six countries around the world. It has recently been announced in 'Bharat' by Prime Minister Shri Narendra Modi.

Healthcare systems like in Canada are taking great shape and progress. In 2014, almost 75% of physicians in Canada used electronic records to examine patients, saving a tremendous amount of paper and other materials. They also have a reasonably strong e-prescribing medicine service and digitalized information system. This saves cost in many ways, making healthcare a bit cheaper. Canada's digital solution has committed $1.3 billion in the last six years. Implementation can be proven beneficial if focused on building core systems for diagnosis and clinical treatments. In some countries, like 'Uganda' and 'Mozambique,' mobile applications like 'mHealth' are constantly working with the local government to improve and almost instantly implement that system for better health and optimal solution. National health services (NHS) in the United Kingdom is making significant progress to make digital health solutions more effective and generate following generation medicines, next-gen devices and technologies. These next-generation technologies include gene therapy, genome sequencing, brain-computer interface, nanotechnology, mixed reality, etc. (Figure 16.8).

Huge organizations like United Nations, World Health Organization, and International Telecommunication Union constantly work together to create a global platform for digital health services and care. A few of them have already been implemented and are still in progress. One of them is the USA's 'Nightingale' program, which again develops core systems with the help of AI and ML technologies. These

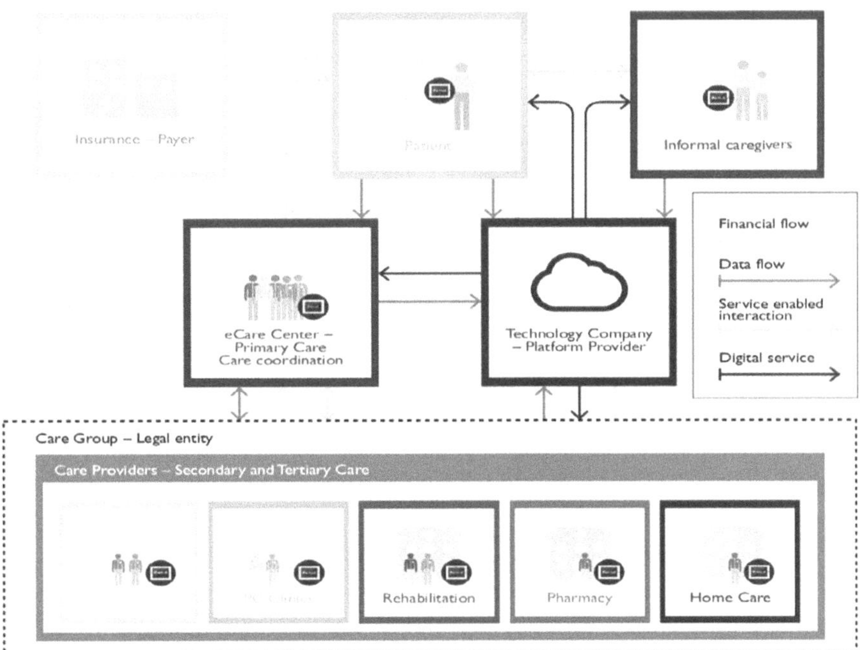

FIGURE 16.8 Digital health models [18].

systems in the future will be able to handle the countrywide population on digital servers and electronic records. Deep learning also plays a significant role in developing these systems as it allows the methods and algorithms to learn from past trials. Developing countries and third-world countries will be able to inspire their missions for digital healthcare from existing programs going on in the world.

16.2.1.4 Great Examples

With today's modern advancement, science is witnessing great inventions. Almost every day, there's a brilliant idea to change medical science and its capabilities like never before. One of the most incredible ideas of this era is 'NURALINK.' Its concept may be taken from some old sci-fi books, but today, it is the reality that is pretty successful so far. It aims to give an extra limb to significantly challenged brain conditions. The nuralink is nothing but a coin-sized brain implant that catches electrical signals running across the brain and transmits them to the external device for further use. This implant is fitted in brain cells with the help of a specially built neurosurgery robotic machine, which can put electrodes into the brain with the speed of 6 electrodes per second. These wires are 4–6 micrometer in width. With the help of this custom-made machine, they avoid any kind of blood loss or damage while operating. This type of interface is called the brain-machine interface. Nuralink implants are so micro-sized that it packs 1000+ wires in a coin-sized implant (Figure 16.9).

The signals from within this implant also help control many AI applications like computer control, Wi-Fi systems, etc. It picks up signals from a specific area of the brain that is responsible for body movement. This helps bionic limbs to work more efficiently as it stays in direct sync with the brain. People who have Parkinson's disease, where they cannot control the shaking and shivering of arms and legs, are

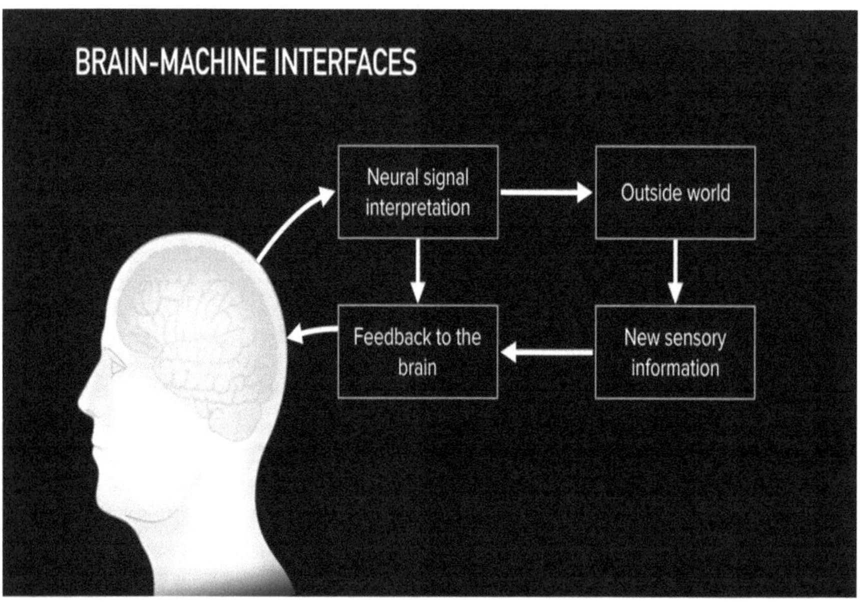

FIGURE 16.9 Brain microchip [19].

FIGURE 16.10 Health and cloud [20].

also helped by this technology. This acts as an external pathway for the signals and is directly received by the person's parts to function with. It aims to achieve in the long-term "symbiosis with artificial ntelligence." This is just done on a small scale, but this concept can help many in this world.

16.2.1.5 Future Scope

Digital healthcare has a massive future in the next 20–25 years and will always grow. The technology it has today is for sure going to be way more advance than what it is today. The medical industry can develop nanorobots that could enter the body and operate a surgery from within. AI cells that could help cancer patients, etc. It will surely aid millions and millions of unprivileged people both in and outside cities and remote areas (Figure 16.10).

Today, we can see a few nurse robots, and in the future, they will be everyday things found in typical homes. They, as the study says, will prove helpful for lonely senior citizens living by themselves. In countries like Korea and Japan, where loneliness is a national problem, robots have a future.

Bioprinting is another major invention. Using this technology, we will be able to 3D print organs and body parts. Printing small tissues are possible today. They can redefine the life of people missing legs or arms or another body organ. Even for people with organ failures, there's a chance of survival. On the other hand, the pharmaceutical field will improve significantly, especially after IT giants like Amazon and Google enter this field. E-pharmacy is not far from being ordinary. The United States spends $3 Trillion on just pharmacy. Biomedical engineering has a million possibilities, and it has a bright and wonderful future.

16.3 ARTIFICIAL HUMAN

Artificial human is the new human. There are currently hundreds of people around the world with bionic arms and legs. Some even with bionic eyes. Thus, the name artificial human. There are thousands of people who either have lost a limb or were born without one. Some are born blind or lose their sight for some tragic reason. Organ failure is another very serious happening. To help and aid them, healthcare systems are continuously evolving, especially digital healthcare systems. Digital health care systems also include bio-electric limbs and organs. These are mechanical arms and legs that are controlled by digital circuits and interfaces. On the other hand, organs are also electronically controlled but are not mechanically moving compared to arms and legs. These bio-electric limbs are called BIONIC limbs or body parts. In these recent years, there has been a lot of research and development in robotic and bionic limbs due to its growing demand, especially for the limbs controlled directly by the brain using an external pathway. This external communication pathway is made possible by the use of the Brain-Machine Interface (BMI).

16.3.1 Need of Artificial Human

This is a global need now. People from different parts of the world are demanding this facility. There are several reasons why this demand is increasing. People of all kinds and age groups are in need. Children need this resource because they are born deformed in some way or maybe because of genetic disorders. Many have met deadly accidents. Infections in some areas of the body can result in amputation. War-torn persons need donations of limbs. All of these people need the benefit of organs and stems, which can be a tremendous help, but the problem is that there are not enough donations to fulfill the current need.

Thus, bionic parts are so much needed. They complete a person to some extent and support them till a point which was never hoped for. These advanced machines can redefine prosthetic life. Today,

Artificial intelligent body parts are the most advance artificial bionics available today. They can help a person to stay alive and live one's almost entire natural life (Figures 16.11 and 16.12).

16.3.2 Bionic Limbs and Organs

16.3.2.1 Artificial Heart

People who suffer heart failure or gradually develop heart disease die. Sometimes, people who could have received a donated heart or could be revived in any manner die. To change this, biomedical engineering made an artificial heart. This artificial heart can stabilize a person for days until the person receives a donated heart. This artificial heart is a biomechanical copy of the human heart and is made from silicon. It produces the same blood pressure, flow, etc., which is done by a human heart. This artificial heart is electronically controlled (Figure 16.13).

16.3.2.2 Arms and Legs

This is the current mainstream focus. These bionic parts are very well developed today. These arms and legs can restore a person to their full working potential.

FIGURE 16.11 Artificial human [21].

FIGURE 16.12 Prosthetic limb [22].

Today, we have bionic arms with 8° of freedom and can send feedback to the brain about all that is happening. They can sense a touch of a thing or make the brain feel its movement. These systems are getting better day by day, evolving continuously.

Legs similarly are multitasking, and they too have degrees of freedom. They are strong and intelligent enough to carry a person's weight. Sportspersons participating in Paralympics or similar games use these technologies to assist them (Figure 16.14).

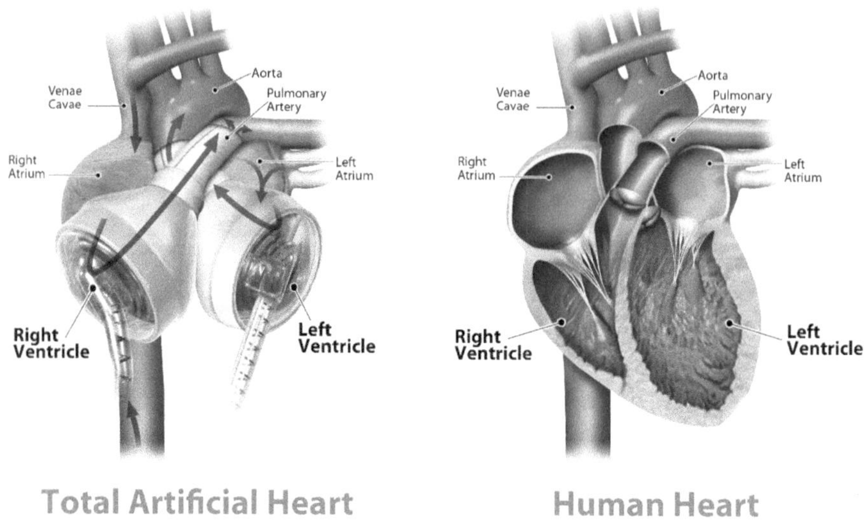

FIGURE 16.13 Artificial heart [23].

FIGURE 16.14 Bionic hands [24].

Two methods can control both arms and legs, either by myoelectric sensors that detect the muscle movement of the amputated part or by BMI. We prefer BMI as it is much more accurate and precise than other methods.

16.3.2.3 Bionic Eyes

Recently, the University of Melbourne in Australia has created the world's first fully functional bionic eye. This eye is a blessing to all those who are blind by birth or later lost their vision. Also, people who are color blind will be helped by this bionic eye. This eye mimics the human look and is very efficient in enabling humans to see. This has been possible by micro construction and the brain-machine interface (BMI). All the images captured by the eye are sent to the micro-implants in the brain, which relay these images to the brain, specifically to the section responsible for image recognition. This external pathway helps the person to see again. This interface is then further calibrated to one's body. This system is one-way communication.

16.3.2.4 Hearing Aids

In the United States itself, 300,000 people are assisted with hearing aids and systems that enable them to listen. Hearing problems or disabilities have been a long-time problem for people. These advanced hearing aids, with the help of technology, have evolved from large machines to small boxes and finally to microchips. These machines, too, use BMI as the bridge to communication with the brain. This system is also one-way communication (Figure 16.15).

16.3.2.5 Bionic Bodies

This is a future possibility wherein a person with multiple organ and limb failure can get various replacements. This will be a step closer to the future of bionic technology and how humanitarian assistance can improve.

16.3.2.6 Exoskeletal Systems

This exoskeletal system is a metallic and hydraulic frame adjusted to one's condition, which helps the skeletal system of the body to work correctly. The exoskeletal system is fixed with the body to keep it in alignment and support the weaker joints. People who have suffered genetic disorders or anything severe with the bones usually tend to help walk again. This machine works by detecting soft

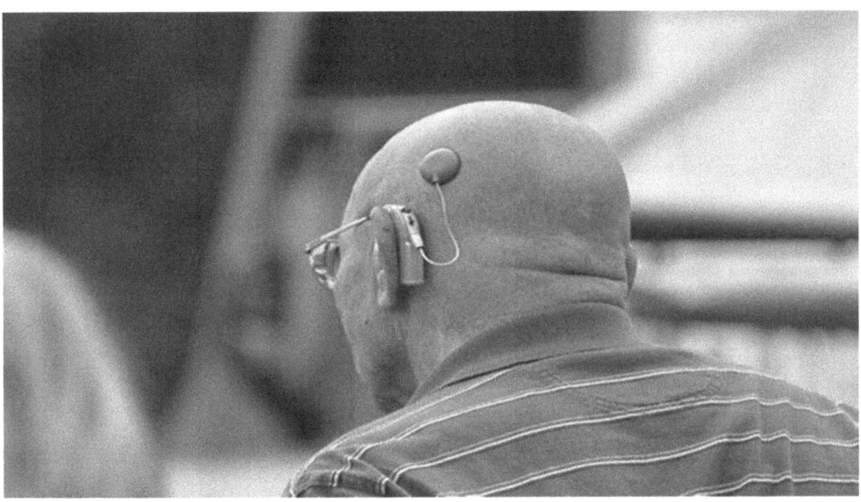

FIGURE 16.15 Hearing aids [25].

movements and then supporting that particular movement with increased strength. It can also be configured with BMI.

These systems are even developed by the militaries of powerful nations for warfare usage and creating super intelligent and strong armies.

16.3.3 Technology Used

16.3.3.1 Brain-Machine Interface (BMI)

This idea of linking the human brain with external devices dates back to 1970, when someone created an external pathway to control instruments by using the brain and not doing anything physically (Figure 16.16).

We have different parts of the brain, each responsible for other operations like thinking, sight, speech, image recognition, performance, etc. These parts are studied and then enabled with chipsets for various requirements. For example, the bionic eye helps people to see with its sophisticated design and circuits. But it is useless unless the brain can see through it. It is essential to connect with the right part of the brain to synchronize both the brain and the eye. Similarly, BMI used for bionic arms and legs also used different parts of the brain. Interfacing of feeders has more electrodes embedded in our brain than those used for legs. This is because arms

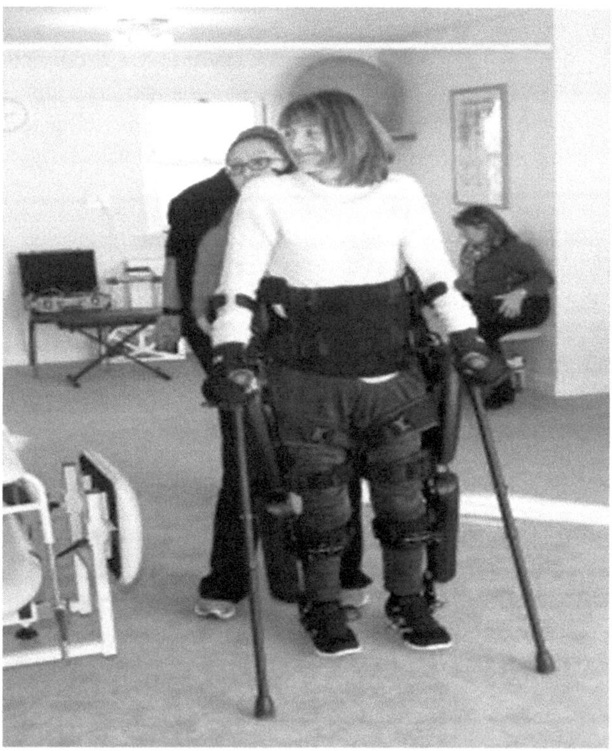

FIGURE 16.16 Exoskeleton [26].

have much more movement than legs. Modern bionic arms and legs even have feedback systems that also send sensation to the brain about the presence of the limbs when this gets calibrated and adjusted. After this is done, our brain can sense the completion of actions instructed by the brain, once the person gets used to it. These bionic arms or legs can also be used continuously, adjusting to mimic the natural human tendency of movement.

BMI can also be developed for any mechanical organ or body part, depending upon the requirement of one need. BMI can also be used to control external devices such as computers, home automation systems etc. (Figure 16.17).

16.3.3.2 Myoelectric Sensors

Myoelectric sensors are implantable sensors that are micro-sized and are implanted in muscles to detect movement. These sensors are placed at the position of which we want to know trends. Human body muscles generate electrical signals whenever there is movement in the cardiovascular system of the body. The implanted myoelectric sensors pick up these signals and identify the angle of movement made and the degrees of freedom the patient wants. These sensors use electrical coils, which pick up current to detect motion. These signals are further passed on to the bionic arm or leg, which executes the instruction. Sometimes this is less accurate than BMI, but it is cheaper since there is no surgery involved. In BMI, we have to operate the brain to fix implants like 'NURALINK.'

This system is also efficient in calibrating the system to one's needs. Every person has a different body, and every person needs a separate calibration of the muscles and body to get maximum precision and accuracy.

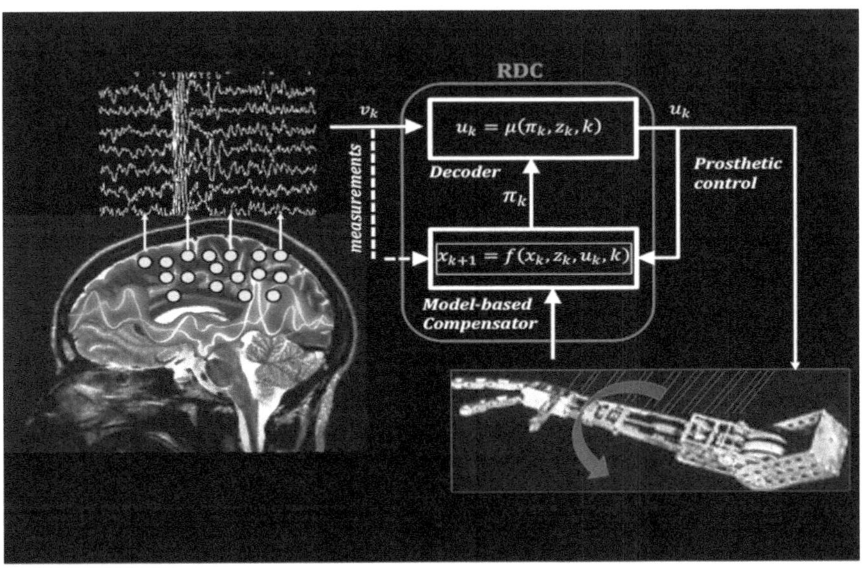

FIGURE 16.17 Brain-machine interfaces [27].

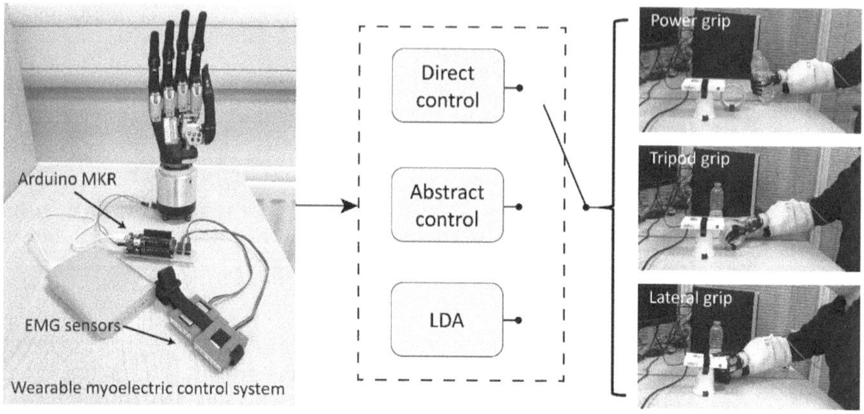

FIGURE 16.18 Schematic of the implantable myoelectric sensor system [28].

16.3.4 Advantages of AI-Enabled Bionic Systems

These systems will cure a person in many ways and, in some cases, give superhuman abilities. With their sophisticated design and capability, these systems provide a person abilities they never had before (Figure 16.18).

This upgraded the human body in multiple ways, enabling them to walk longer, work harder, think better and more. Systems like exoskeletal systems are a tremendous benefit. Retina implants improve one's sight significantly. People suffering from some types of eye problems find this as the ultimate solution. This retina implant is an electronic eye sensor that assists the retina in capturing images and sending them to our brains. With this assistance, now the person can see even better-than-average eyesight. This enhances the vision and eliminates most of the chances of future disease.

In the future, these systems in nano form will surely help human immunity and various systems of the human body. We surely can expect better disease-free lives shortly. This could also help us reconstruct faces or bones for those who have suffered war damages, especially grenade-hit soldiers. People who have suffered catastrophic accidents or have genetic deformation will live everyday life once again.

16.3.5 Osseointegration

It is the direct structural and functional connection between living bone and the surface of a load-bearing artificial implant.

It is a procedure and technique which helps amputees to walk again. It is used for bone and dental purposes. This implant is directly connected to the bone without the interface of any tissue or ligament. It is famous for knee replacements because, after a certain age, the bone starts to degenerate and the lubrications dry out, which causes immense pain upon moving (Figure 16.19).

When a person loses a leg to an accident or cancer or infection, the portion has to be amputated, which forces the person to walk on support. In osseointegration, the

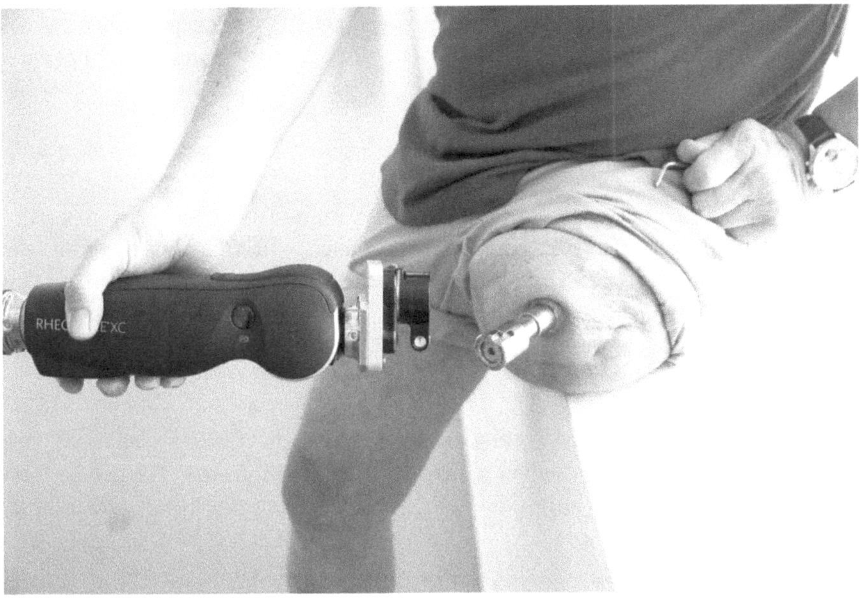

FIGURE 16.19 Osseointegration [29,30].

bone is hollowed from inside and is fitted with a high tensile titanium rod, which extends out of the body at the amputated cross-section. This extension is one part of a joint to connect to other external limbs or attachments. This enables a person to wear or remove the prostatic leg. Once a person is equipped with this, it takes only a few days to adapt and walk again. The titanium inside the body is entirely harmless. The procedure is done with much precision, such that the fixed rod does not move and is set by the body at a place, which means that the body fills up the micro gaps while healing the wound. If this fixed rod moves more than 150 micrometers, it could make a scar ligament or leave marks and scars. This procedure is quite successful nowadays, especially in Australia, with the most advanced and leading technology for such systems. Also, they have the most significant number of procedures done and the highest success rate in prostatic procedures.

With the help of this bionic tool, a person can walk again, but with the fusion of automated bionic legs and osseointegration, there is a lot of advancement. The person can run, walk or go to the gym, and the best part is that the person can feel their legs with the help of a sensory feedback system responding to the brain.

16.4 ONGOING PROJECTS AND RESEARCHES

16.4.1 Open Bionics (Bristol, UK)

A company situated in the UK is developing and manufacturing 3D-printed arms. They offer all the primary technology in the arm and are low in cost because they are 3D printed, a cheaper alternative to metallic-made components. This will allow a large number of people to get help. They can manufacture cheap bionic arms at a much faster rate. They are also developing 3D plastic bones to help people differently.

16.4.2 Touch Bionics (Livingston, Scotland)

They are more technology-based than cost. Their product 'i-limb' is a generation advanced bionic hand. They have developed products with full 8° of freedom (axis on which arm moves). Their products are faster, stronger, and more accurate than their competitors.

16.4.3 SynCardia (Arizona, US)

They are situated in the United States. They focus on bionic organs and their technologies. Their current product is the battery-powered hearts. These hearts can keep a person alive for a long time. They are usually made up of silicon-based materials.

16.4.4 OSSUR (California, US)

They are the first company to allow thought-controlled bionic legs. These legs are highly advance and use BMI for their functioning. They were the first to tap brain signals and utilize them to understand the mechanics of our body movement system. This enables the patient to move legs and feel them with BMI. It is two-way communication between the legs and the brain. It needs a brain implant just like 'neuralink' to work.

16.4.5 ReWalk Robotics (Maryland, US)

This company focuses on exoskeletal systems. They aim to help people suffering from conditions like paralysis and muscular degeneration. The exoskeletal is a metallic and hydraulic frame that supports the skeletal system's functioning. This design also helps people to recover from spinal injuries and enable them to walk again freely.

These were some significant companies working continuously for human betterment and advancements. Shortly we will see better versions of what we have today. Better designs and sophisticated technologies. The cost will also reduce with improvement, enabling more and more people for treatment and afford facilities.

16.4.6 Future Possibilities

Our future has a million possibilities, below listed are some possibilities upon which companies and medical institutes are working:

1. Human brain replica
2. 3D printing of arteries and veins (Figure 16.20)

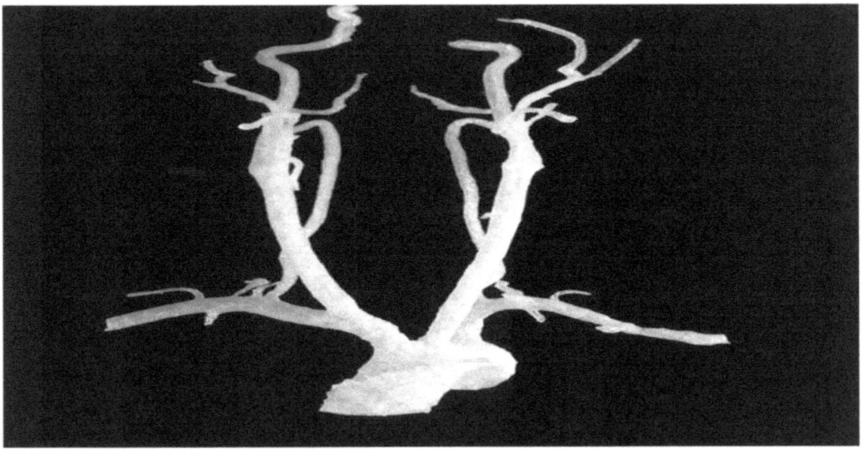

FIGURE 16.20 3D printed carotid and vertebral arteries.

3. Artificial body senses
4. 3D printing of complex organs
5. Improving human senses
6. Reconstruction of the human body
7. Better personalized limbs
8. Better appearance, etc.

Most humans want to evolve spiritually, mentally, and in many more ways. Most want to become the species of 'homo sapiens' to another level, which is an extremely tough job. Thus, they are developing assistance for the human body in any way possible. Most importantly, if we can recreate the arteries and veins, it will be a milestone for medical and engineering sciences.

16.5 CONCLUSION

Through all of these facts and researches, we can conclude that life will improve dramatically with the help of digital health care. Great inventions are either present or they are yet to come. Anyhow, life will be disease-free and healthy. Food fuel for this body will go back to its natural and non-adulterated state, making disease less in the human body. It will increase average expectancy and loneliness will not be a problem anymore. Brain-machine interfaces will become routine for challenged people. The world will be a better place to live.

REFERENCES

1. https://en.wikipedia.org/wiki/Digital_health
2. https://www.fda.gov/medical-devices/digital-health
3. https://www.who.int/health-topics/digital-health#tab=tab_1
4. https://www.digitalhealth.gov.au/get-started-with-digital-health/what-is-digital-health

5. https://www.news-medical.net/health/What-is-Telemedicine.aspx
6. https://www.livescience.com/48001-biomedical-engineering.html
7. https://neuralink.com/
8. https://www.wired.com/story/neuralink-is-impressive-tech-wrapped-in-musk-hype/
9. https://www.youtube.com/watch?v=q8RVVCQUe0c
10. https://www.youtube.com/watch?v=hzaH2tG19_g
11. https://www.galendata.com/digital-healthcare-future-heathcare/
12. https://www.i-scoop.eu/digital-transformation/healthcare-digital-health/
13. https://www.orbit-rri.org/blog/2018/03/30/ethical-analysis-personal-health-monitoring-uk/
14. https://medium.com/healthbeyondthefog/the-story-in-graphics-e504a73dd1f2
15. https://www.skyfilabs.com/blog/free-image-processing-project-ideas-for-beginners
16. https://www.hometownsource.com/sun_current/news/local/somali-group-combats-terrorism-by-educating-youth/article_258a66b5-3949-52e9-ad07-fcbe4c5d65c2.html
17. https://upload.wikimedia.org/wikipedia/commons/thumb/4/49/Digital_HealthInfographic.jpg/330px-Digital_Health_Infographic.jpg
18. https://www.mobihealthnews.com/content/boehringer-and-lilly-tap-inovalon-analytics-and-more-digital-health-deals
19. https://www.businessinsider.in/science/biology/elon-musk-finally-took-the-wraps-off-his-new-brain-microchip-company-that-plans-to-connect-peoples-brains-to-the-internet-by-next-year/articleshow/70268767.cms
20. https://medium.com/healthbeyondthefog/the-story-in-graphics-e504a73dd1f2
21. https://www.forbes.com/sites/bernardmarr/2020/02/17/artificial-human-beings-the-amazing-examples-of-robotic-humanoids-and-digital-humans/#33a2f9fa5165
22. https://www.theguardian.com/technology/2018/nov/15/being-bionic-how-technology-transformed-my-life-prosthetic-limbs
23. https://healthmanagement.org/c/cardio/news/successful-heart-transplant-with-experimental-artificial-heart
24. https://openbionicslabs.com/blog/open-bionics-and-the-nhs-launch-worlds-first-trial-of-3d-printed-bionic-hands-for-children
25. https://www.ibtimes.co.in/obesity-linked-to-hearing-loss-physical-activity-wards-off-defect-525437
26. https://www.hobbsrehabilitation.co.uk/rewalk-exoskeleton.htm
27. https://sarmalab.icm.jhu.edu/research/brain-machine-interfaces/
28. https://www.researchgate.net/figure/Schematic-of-the-implantable-myoelectric-sensor-system8_fig14_236981286
29. https://www.iconinmotion.com/osseointegration/
30. https://commons.wikimedia.org/wiki/File:3D_Printed_Carotid_and_Vertebral_arteries_20151204.jpg

17 Data Security on Internet of Things Devices Using the Public-Key Cryptography Method

Rutvik Patel
Department of Electronics and Communication Engineering, Institute of Computer Science and Technology (Ganpat University) Ahmedabad, Gujarat, India

CONTENTS

17.1 INTRODUCTION

The Internet of Things (IoT) is a future communication technology in which everything in our environment is implanted with a microcontroller, allowing them to connect with other objects and people [1]. IoT, also known as machine-to-machine communication, uses the internet as its connectivity to give real-time information without distance limits or human involvement. Smart systems, such as smart homes, smart energy, smart governance, and smart infrastructure are examples of IoT development.

The expanding use of IoT brings with it new issues, one of which is security. Good security is one of the biggest challenges in building IoT. The sensor, network, and application layers are the three levels that make up the IoT framework [2]. To guarantee

DOI: 10.1201/9781003230526-17

data integrity, availability, and confidentiality, each layer must have its own security protocol, especially when the IoT system is attacked. The botnet, Man-in-the-Middle, Data & Identity Theft, Social Engineering, and Denial of Service are all common forms of assaults [3,4].

If you look at how the IoT works, the microcontroller will send and receive data from the server via a public network or the internet. The data sent is still in plaintext, which means that anyone may read the data format. If a data & identity theft attack occurs, the attacker can read the device's data by sniffing the network connected by the microcontroller. Confidentiality of data is no longer assured.

Data-encryption technologies are one way for securing data on the Internet of Things. Cryptology, the science of hiding and keeping messages secret, includes data encryption. This field of study is concerned with cryptography and cryptanalysis [5,6]. The initial message, known as plaintext, is encrypted using a secret key and a specific algorithm in cryptography. Messages that have been encrypted will be called ciphertext. This message cannot be read by someone who does not know the algorithm and key to open it. The process to change a message from ciphertext to plaintext so that it can be read back is called the decryption process. Strong encryption can ensure data confidentiality.

This study will develop and analyze public-key cryptography methods in IoT devices for data security, focusing on encrypting data passed from the microcontroller to the server. The RSA technique was employed to encrypt the data. Following the implementation of this strategy, an evaluation of the effectiveness of processing time and memory consumption of the devices will be conducted.

17.2 INTERNET OF THINGS

IoT is a revolutionary technology in wireless communication technology that is still evolving. The primary premise behind the Internet of Things is that any object or device, such as RFID tags, sensors, actuators, smartphones, microcontrollers, and so on, can communicate with one another to achieve a specific objective. Each item will have a different marker address to identify it from the others [1]. (Figure 17.1).

The sensor, network, and application layers are the three layers that make up the Internet of Things. The sensor layer is the IoT's outermost layer, and it is physical and touchable. This is where RFID tags, sensors, actuators, and other electronic devices are kept. Wireless or wired networks, network media, and the primary network are all part of the network layer. The network layer is the data transmission path from the sensor layer to the application layer. The application layer is a system with an interface that allows the user to see all of the data transformed into information [2].

IoT is a complicated system that is prone to cyber threats. Attacks are possible at all layers of the IoT, especially at the sensor layer. Attackers can easily target these devices connected to a public network or the internet. One issue is data and identity theft, which occurs when the attacker takes data that is temporarily transmitted from

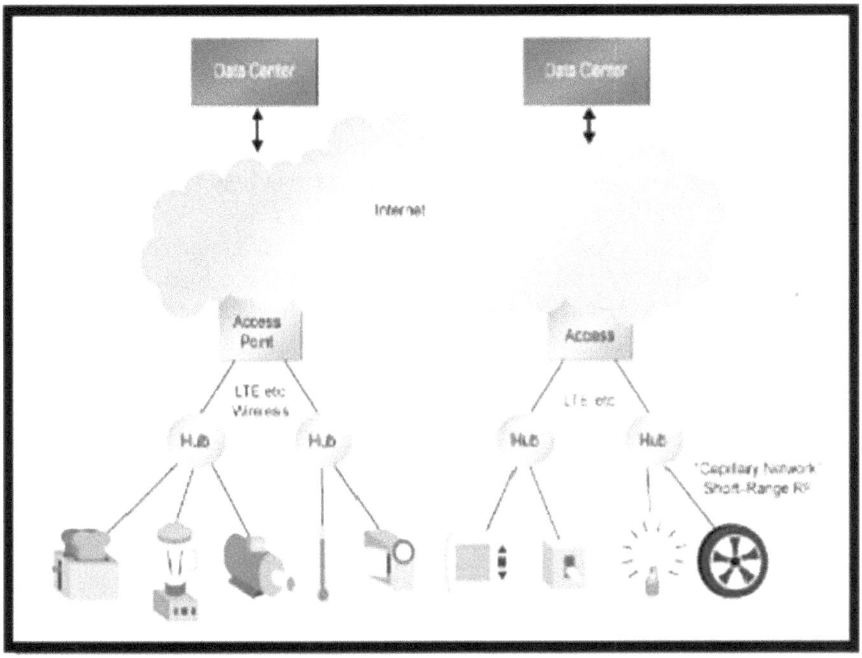

FIGURE 17.1 Layer Internet of Things.

the device to the data server and vice versa. Because the data was not encrypted before being delivered, the attacker can read the stolen data.

17.3 PUBLIC-KEY ENCRYPTION

A public-key infrastructure (PKI) is a set of functions, protocols, and techniques for generating, distributing, utilizing, saving, and revoking digital certificates and managing public-key encryption. The general public-key issue in the IoT ecosystem is the requirement for a verified exchange of public keys. The public key infrastructure (PKI) is a set of elements for safely distributing public keys, widely utilized in today's networks.

Interconnected devices must offer trustworthy information to consumers and services in an IoT environment, but sustaining confidence over an extensive network is challenging. The IoT devices are prone to assault, and communications between nodes are often insecure. The PKI system has created a trustworthy environment in current financial systems, cellular stations, and mobile networks. As a result, the PKI is a potential Internet of Things (IoT) option [2].

a. When it comes to assurance and validation, PKI is the way to go.
b. Scale: While some PKI deployments can manage millions of certificates, the vast majority do it at a far lower level.

c. Technological issue: The Internet of Things would be populated by computers that are extraordinarily low-power and low-cost. Traditional encryption, which is computationally costly and mathematically intensive, is not designed for these conditions.

Another issue is the development of credentials. It is not easy to make good keys, and doing so in large quantities can quickly become a bottleneck.

Low-power crypto algorithms and quick essential generation techniques already exist and have been thoroughly tested. We will go through the fundamentals of PKI before diving into the specifics of how it works. The certification authority (CA) is a reputable third party that issues a certificate limiting an entity's public key and identity. The following are the essential elements:

a. Signature: the algorithm for generating a signature.
b. Subject PKI: the algorithm that generated the public key, as well as the algorithm's identity.
c. Validity: the period for which the certificate is valid.
d. Issuer: CA's identifier.
e. Signature value: The value of the issuer's signature on the hash of the primary elements.

To check the certificate, you will need access to the CA's public key. This takes us back to the original problem. Root CAs have self-signed certificates and are at the top of the trusted hierarchy. Furthermore, root CAs are pre-installed in applications, such as by browser vendors.

In the world of cryptography, two encryption methods can be used to maintain the confidentiality of data, namely symmetric and asymmetric encryption methods. Symmetric encryption methods are often called single-key cryptography and asymmetric encryption methods are often called public-key cryptography.

Since symmetric encryption employs the same key for encryption and decryption, it is referred regarded as single-key cryptography. When using this method to send data, the recipient must be notified with the message's key to decrypt it. The key determines the encryption security of this approach. If someone else has the key, they can encrypt and decrypt the message. There are many different forms of symmetric encryption methods, with the DES algorithm being the most prevalent. The critical distribution process and sender authentication are the two fundamental flaws in the symmetric encryption approach. Therefore, to overcome this problem, an asymmetric encryption method was created.

The asymmetric encryption approach is commonly referred to as public-key cryptography since the keys used for encryption and decoding are distinct. The key is split into two parts in asymmetric encryption: the public and private keys. A public key is a key that is created for public distribution and is used for encryption. Since this private key will be needed to decrypt the data, it must be kept hidden. The RSA algorithm is the most often utilized in this manner. The fact that this method exists solves two significant issues with the symmetric method. However, this strategy has a flaw: it is inefficient in terms of time. Data encryption and decryption are often slower than the symmetric technique because the process utilizes huge numbers and involves extensive power calculations.

17.4 RSA ALGORITHM

The RSA algorithm belongs to the asymmetric encryption methods category. The acronym RSA is an abbreviation of the algorithm's three creators' names, Ron Rivest, Adi Shamir, and Len Adleman, who discovered the algorithm in 1977. The RSA algorithm is the world's most popular asymmetric encryption algorithm. Digital signatures can also be created using the RSA algorithm. The difficulty of factoring massive numbers into prime factors is what makes the RSA algorithm safe. The private key for decrypting the message is obtained by factoring. The key generation, message encryption, and decryption processes are part of the RSA algorithm's operation.

17.4.1 Key Generation

The essential generation procedure involves the creation of public and private key pairs. This process is carried out on side A.

a. Choose p and q as two huge prime numbers.
b. Find $N = p * q$ and phi () = (p-1) for $N = p * q$. (q-1).
c. Determine the coprime of phi and an integer e between one and phi (1 e).
d. Count from d to $d * e$ 1 (mod).

(n, e) is the public key, and its value will be sent to B, while the private key is (d, p, q). The values of d, p, q, and phi should be kept hidden at all times. The length of an RSA key that may be considered safe has been 1024 bits or more significant.

Example of a 1024 bit n key in hexadecimal format

```
0A 66 79 1D C6 98 81 68 DE 7A B7 74 19 BB 7F B0
C0 01 C6 27 10 27 00 75 14 29 42 E1 9A 8D 8C 51
D0 53 B3 E3 78 2A 1D E5 DC 5A F4 EB E9 94 68 17
01 14 A1 DF E6 7C DC 9A 9A F5 5D 65 56 20 BB AB
```

17.4.2 Message Encryption

On side B, the message encryption process is carried out with the public key of A.

a. B receives the public key from A. (n, e).
b. A hands over the public key to B.
c. A positive number m (1 m n) is generated from the plaintext message.
d. The ciphertext is determined using the formula c = me mod n.
e. B transmits A the ciphertext c.

17.4.3 Message Decryption

The decoding of the communication is done on side A with A's private key.

a. From B, A receives ciphertext.
b. Calculate m = c d mod n using the private key (n, d).
c. Use m to reshape plaintext.

Even utilizing the brute-force technique, which will take decades to uncover the appropriate key combination, the ciphertext communication cannot be decrypted by the other party, including the B side, as long as the private key stays safe and secret. However, as compared to other encryption methods, RSA's drawback is its prolonged computing process. As a result, RSA is more suited to encrypting short communications.

17.5 PUBLIC-KEY CRYPTOGRAPHY ON IoT DEVICES

The IoT devices discussed in this study focus more on microcontroller devices. In most cases, the data sent from the microcontroller to the server is still in plaintext, which means anyone may read it. To protect the data, encryption methods should be utilized. The symmetric encryption method is the most suited since it is faster and more memory efficient than asymmetric encryption. However, the symmetric encryption key saved on the device is still subject to third-party discovery. In this situation, that person accesses the device memory's source code.

As a result, the asymmetric encryption method is more appropriate for the application. Even if the other person has access to the device's public key, they will not decipher the stolen ciphertext message. The RSA algorithm is used as an asymmetric encryption technique.

17.5.1 Prototype IoT Device

The Arduino UNO R3 board was used as the basis for the microcontroller in this study's prototype. The ESP8266 wireless module is used to connect to the internet network in this prototype.

The static random-access memory (SRAM) of this prototype is only 2 KB. This includes RAM allocation, which the ESP8266 wireless module will use. This prototype only has a clock speed of 16 MHz when it comes to computation processing speed. (Figure 17.2) (Table 17.1).

17.5.2 Implementation of the RSA Algorithm

The efficiency of memory consumption and the length of the encryption procedure are the challenges in implementing RSA encryption in this prototype. To meet the situation of the device's resources, the data-encryption operation will be brief. Although the encryption is made simple, the encryption's strength is preserved. The

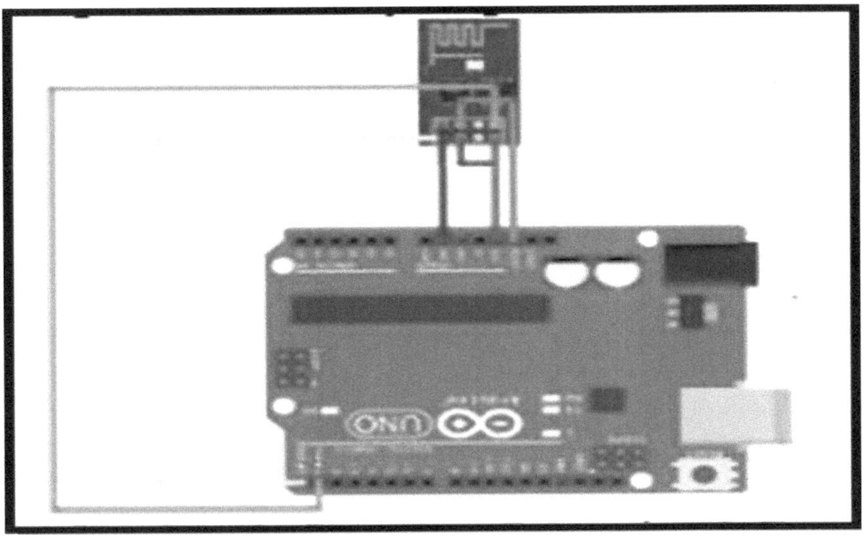

FIGURE 17.2 Prototype IoT device.

TABLE 17.1
Hardware Specification

Hardware	Type/Large
Microcontroller	ATmega328P
Flash Memory	32 KB
SRAM	2 KB
EEPROM	1 KB
Clock speed	16 MHz

IoT device will be referred to as the client side in the following system architecture, while the data server will be referred to as the server side. Clients and servers use the TCP protocol to interact over the internet network.

The phases of the RSA encryption procedure on the system architecture are as follows:

a. The client connects with the server to obtain the public key (n, e).
b. The client sends the server's public key.
c. The client is provided with a public key.
d. The client uses the public key to encrypt plaintext.

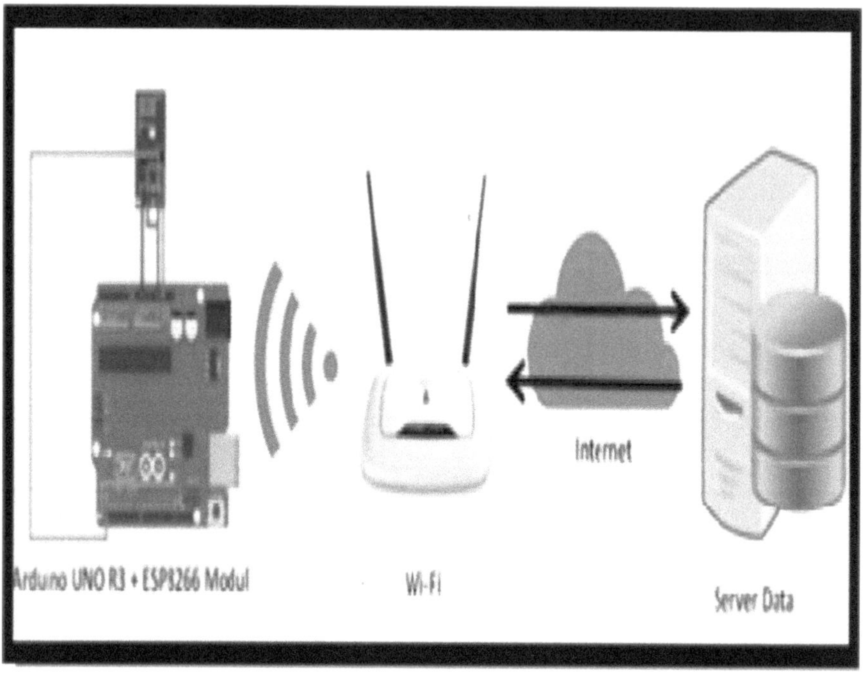

FIGURE 17.3 System architecture.

e. The client gives the server the encrypted ciphertext.
f. The server receives the ciphertext and uses the private key to decrypt it (n, d). (Figure 17.3).

Distributing public keys from the server to the client is divided from Stage 1 to 3. This stage is only performed once after the device has been turned on. The process of transmitting data from the client to the server takes place in stages 4 to 6. This step will recur as long as the device's time interval is configured.

17.5.3 Encryption Process Analysis

Only the client-side or IoT device is subjected to analysis. The efficiency of the encryption process will be measured by the quantity of memory used, while the efficacy will be determined by the time it takes to conduct encryption. The RSA public key that was used during the testing process is just 16 bits long. The encrypted data is 18 bytes in length.

The given data takes 1.97 seconds to decrypt a plaintext message using the RSA technique. As much as 1018 bytes of memory are consumed. Memory allocation for supporting variables and encryption process computation is included. The encryption results will be placed into the HTTP header format and transmitted to the server after encryption. (Table 17.2)

TABLE 17.2
Memory Allocation Specification

Key n	B3 46 (16 bit)
Key d	6A (8 bit)
Plaintext message (m)	marcel until a slang (18 bytes)
Hypertext Message (c)	C9 7A BA 06 0A DC D5 0B 7A 29 56 65 BF 7A 31 7A 01 F5 (18 bytes)
Encryption time	1.97 seconds
Memory Usage	1018 bytes

HTTP header format for the request to save data to a server

```
POST /save.php HTTP/1.0
Host: iot.marselsampeasang.web.id
Content-type: application/x-www-form-urlencoded
Content-Length: 28

C9 7A BA 06 0A DC D5 0B 7A 29 56 65 BF 7A 31 7A 01 F5
```

The analysis results of the time and memory usage above are pure for the RSA encryption-calculation process. This does not include the ESP8266 wireless module, the server connection, or any other variables that are not needed in the encryption process. The amount of time spent on request and response data from or to the server, particularly for a connection to the server, cannot be monitored. The amount of time it takes to complete the operation is determined by the internet speed used.

It can be observed that utilizing a key length of 16 bits and a data length of 18 bytes, RSA encryption can be calculated relatively quickly. However, keep in mind that a secure RSA key has a length of at least 1024 bits. It can be concluded that the memory in the device used in this investigation is insufficient to conduct calculations of this magnitude.

Program Implementation

The public key will be available to everyone for communication, while the private key will be unique to each user. The goal is to figure out a single person's public and private keys. Also, describe how this public key and private key will be used to encrypt data. (Figure 17.4).

17.6 CONCLUSION

This study used the RSA technique to implement public-key cryptography on IoT devices such as a microcontroller. The test results suggest that the RSA algorithm's encryption procedure may work successfully despite requiring a lot of time and memory. The encrypted data is 18 bytes long, while the RSA encryption key is 16 bits long.

```
#include<iostream>
#include<math.h>
using namespace std;
// find gcd
int gcd(int a, int b) {
    int t;
    while(1) {
        t= a%b;
        if(t==0)
        return b;
        a = b;
        b= t;
    }
}
int main() {
    //2 random prime numbers
    double p = 3;
    double q = 11;
    double n=p*q;//calculate n
    double track;
    double phi= (p-1)*(q-1);//calculate phi
    //public key
    //e stands for encrypt
    double e=7;
    //for checking that 1 < e < phi(n) and gcd(e, phi(n)) = 1; i.e., e and phi(n) are coprime.
    while(e<phi) {
        track = gcd(e,phi);
        if(track==1)
            break;
        else
            e++;
    }
    //private key
    //d stands for decrypt
    //choosing d such that it satisfies d*e = 1 mod phi
    double d1=1/e;
    double d=fmod(d1,phi);
    double message = 9;
    double c = pow(message,e); //encrypt the message
    double m = pow(c,d);
    c=fmod(c,n);
    m=fmod(m,n);
    cout<<"Original Message = "<<message;
    cout<<"\n"<<"Encrypted message = "<<c;
    cout<<"\n"<<"Decrypted message = "<<m;
    return 0;
}
```

```
C:\Users\rutvik\Desktop\prac 8.exe
Original Message = 9
p = 13
q = 11
n = pq = 143
phi = 120
e = 7
d = 0.142857
Encrypted message = 48
Decrypted message = 9
--------------------------------
Process exited after 0.08862 seconds with return value 0
Press any key to continue . . .
```

FIGURE 17.4 Program implementation for encrypt data.

This key is still a long way from the safe length of an RSA key, which is roughly 1024 bits and higher. Even if the microcontroller's performance is pushed to be very slow, the device employed is unlikely to complete computations for a key of such size. On the other hand, because the device must consistently deliver real-time information, the microcontroller's memory and implementation must be simplified. In the future, it is suggested that a hybrid encryption approach be used for the scenarios mentioned above. The hybrid encryption approach combines symmetric and asymmetric encryption techniques. On IoT devices, encryption can assure data security.

REFERENCES

1. Y. Wang, Public-Key Cryptography Standards: PKCS. 2012, https://arxiv.org/abs/1207.5446
2. S. Li, IoT Node Authentication, Springer, pp. 69–95, 2017.
3. L. Atzori, A. Lera, and G. Morabito, The Internet of Things: A Survey, *Comput. Netw*, vol. 54, no. 15, pp. 2787–2805, 2010.
4. M. O'Neill, Insecurity by Design: Today's IoT Device Security Problem, *Elsevier Journal of Engineering*, vol. 2016, no. 2, pp. 48–49, 2016.
5. R. Wilson, A Clearing Picture of the Internet of Things, *Altera Corporation*, 2013. https://www.embedded.com/a-clearing-picture-of-the-internet-of-things/
6. D. Ireland, RSA Algorithm, *DI Management*, 2010 , http://di-mgt.com/rsa_alg.html. Retrieved 04-06-2014

18 PAPR Reduction in MIMO-OFDM System Using Artificial Neural Network

Khushboo Pachori
Department of Electronics and Communication Engineering, Oriental Institute of Science and Technology, Bhopal, India

Harpreet Kaur
Department of Electronics and Communication Engineering, GNI Technical Campus, Hyderabad, India

Sarabpreet Kaur
Department of Electronics and Communication Engineering, CEC Jhanjeri, Mohali, India

Manpreet Kaur
Department of Electronics and Communication Engineering, GNI Technical Campus, Hyderabad, India

CONTENTS

DOI: 10.1201/9781003230526-18

18.1 INTRODUCTION

The need for multimedia wireless communication is increasing at a breakneck speed today, and this trend is projected to continue. Wireless communication devices such as mobile phones are being used primarily for voice communications, but the main attention is now diverting to devices capable of many more capabilities like data communication (such as pictures, steaming videos etc.). The several existing wireless standards for orthogonal frequency division multiplexing include a multicarrier interface based on OFDM [1]. The main difficulty for communication engineers is to obtain high data rates in a noisy and crowded environment without expanding bandwidth or transmiting power. The recent successful and viable solution to achieve high data rates is the use of multiple antennas, i.e., antenna arrays at the receiver or transmitter end, which are generally termed as MIMO systems. The combination of OFDM and MIMO are used to obtain diversity and/or capacity gains of a wireless system under fading channels.

18.1.1 Specific Applications

OFDM was introduced in 1958, but was first implemented in 1960 [2]. Multicarrier modulation methods were often used in utilised in high-frequency military radios at the time viz. KINEPLEX, ANDEFT, KATHRYN [3], etc.,. A patient for OFDM was granted in 1970s [4]. Parallel data-communication systems were the first to use the discrete fourier transform (DFT) [5] in 1971 as a result of modulation and demodulation protocol. In 1980, Hirosaki *et al.* [6,7] proposed an equalisation method for reducing intersymbol, as well as intercarrier interference caused by timing and frequency errors in the channel impulse response. In the 1990s, OFDM was exploited for wireless communication systems [8,9], international wireless protocols [10] as an example. Digital-video-broadcasting (DVB), HIPERLAN-2, Digital-audio-broadcasting (DAB), IEEE 802.15.3a (UWB), IEEE 802.11a (WLAN), IEEE 802.20 (MBWA), and IEEE 802.16e (WiMAX) have all adopted OFDM.

18.1.2 Development in MIMO-OFDM

MIMO techniques can improve capacity by a factor of the minimum broadcast and receive antennas for narrow-band channels or flat-fading. It is natural to combine OFDM with spatial-temporal processing or STC (space-time coding) for wideband transmission to cope with frequency selectivity of wireless channels and gain diversity and/or capacity improvements [11–15]. As a result, MIMO-OFDM has turned into a significant better solution for future high-data-rate transmission across broadband wireless channels. MIMO-OFDM was initially suggested to employ OFDM for determining a MIMO systems with minimal ISI. From the past few decades, MIMO-OFDM is popular in various wireless communication applications viz., third generation (3G) systems, 3GPP long-term evolution (LTE), fourth-generation (4 G) systems, etc. [16,17].

18.1.3 Challenges and Solutions in MIMO-OFDM

The research focuses on issues or obstacles in MIMO-OFDM systems, also including synchronisation, high PAPR (peak-to-average power ratio), equalization, and channel estimation. These issues introduce ICI (inter-carrie-interference), and ISI due to which performance of a system degrades.

Another difficult challenge in wireless systems is channel estimation. The broadcast signal is subjected to a number of harmful effects on its way to the receiver, which distort the signal and often limit the system's performance. The signal is spread in frequency, duration, and angle due to multipath propagation, mobility, and local scattering. Channel performance are directly related to these statistics. Various pilot-assisted schemes, blind and non-blind channel-estimation schemes are available in literature to estimate MIMO-OFDM channels [18–20].

Most wireless systems in communication employ a HPA (high-power amplifier) at the transmitter's output to achieve adequate transmit power for wide-area coverage. To ensure maximum power efficiency, the HPA is generally run at or close to saturation. When high peak power signals travel through an HPA, they cause peak clipping and non-linearity. The BER increases as a result of the additional interference. In the linear region of amplifier operation, the distortion can be reduced. However, due to its inefficiency and high cost, this linear amplifier is not recommended. As a result, the PAPR must be lowered for an OFDM system to achieve increased power efficiency, broad area coverage, and low BER. [21].

In the literature, various PAPR reduction strategies for MIMO-OFDM systems have been discussed [22–26]. PAPR reduction procedures are divided into four categories: pre-distortion methods, signal distortion, probabilistic (scrambling) approaches, and coding methods. By non-linearly changing the OFDM signal, signal-distortion methods reduce the PAPR. Filtering and clipping, peak widowing, and non-linear companding are examples of approaches. For the PAPR reduction, the coding methods used some error-correcting codes. Scrambling each OFDM signal with multiple scrambling sequences and selecting the sequence that gives the lowest PAPR is the basis of probabilistic methods. Before performing IFFT, the pre-distortion approaches rely on re-orienting or distributing the energy of the data symbol. Aside from these many PAPR reduction techniques, the literature also has certain hybrid methods [27–29] that reduce PAPR while increasing complexity.

18.1.4 Paper Organization

The following is how the rest of the survey is organised: Section II reviews the fundamental ideas of a conventional OFDM-MIMO system, as well as the PAPR measure. Section III discusses the criteria that go into evaluating the effectiveness of PAPR-reduction approaches. Section IV discusses the numerous PAPR-reduction approaches that have been published in the literature, as well as the most recent relevant bibliography. Finally, Section V concludes and summarises the survey.

18.2 MIMO-OFDM SYSTEM MODEL

A (1) MIMO-OFDM system is presented using two antennas denoted as *Nt* and *Nr* used for transmitting and receiving, respectively. The figure's space-time processing can employ any space-time techniques developed for flat fading channels. Despite the fact that it is still referred to as space-time processing, MIMO-OFDM processes signals in both the space and frequency domains.

Space-time processing, as illustrated in the diagram, multiplexes the transmitted symbol or data stream $\{x_n\}$ into N_t substreams $\{x_k^{\langle m \rangle}\}$, along with transmission through multiple antennas and OFDM modulation. The demodulated signal at each receive antenna is a superposition of signals from many transmit antennas if only a multipath of wireless channels is considered and the cyclic prefix is long enough and then it can be represented as: [30]

$$y_k^{\langle i \rangle} = \sum_{m=1}^{N_t} H_k^{\langle i,m \rangle} x_k^{\langle m \rangle} + z_k^{\langle i \rangle} \tag{18.1}$$

for $i = 1, \ldots, N_r$, where $H_k^{\langle i,m \rangle}$ represents the kth subchannel frequency response and $z_k^{\langle i \rangle}$ is channel noise having an influence at the kth sub-channel of ith antenna used for receiving, it is independent of different k or i, with mean equal to zero and Gaussian. The Equation (18.1) can alternatively be expressed as a matrix form

$$y_k = H_k x_k + z_k \tag{18.2}$$

where y_k, H_k, x_k, and z_k are representing receive signal vector, kth subchannel of channel matrix, transmitted signal vector, and noise vector, respectively, and which is referred as

$$y_k = \left(y_k^{(1)} \ldots\ y_k^{(N_r)}\right)^T$$

$$x_k = \left(x_k^{(1)} \ldots\ x_k^{(N_t)}\right)^T$$

$$z_k = \left(z_k^{(1)} \ldots z_k^{(N_r)}\right)^T$$

$$H_k = \begin{pmatrix} H_k^{(1,1)} & \ldots & H_k^{(1,N_t)} \\ \vdots & \ddots & \vdots \\ H_k^{(N_r,1)} & \ldots & H_k^{(N_r,N_t)} \end{pmatrix}$$

The PAPR of $x(t)$, which is a continuous time signal can be determined by the ratio of the maximum instantaneous power to the average power. The PAPR is calculated from L – times oversampled time domain signal samples and is equal to

$$PAPR = \frac{1}{E\left[|x|^2\right]} \max_{0 \le k \le LN-1} |x_k|^2 \quad (18.3)$$

where $E[\cdot]$ represent expectation.

18.3 PERFORMANCE CRITERIA OF PAPR REDUCTION TECHNIQUE

To illustrate the efficacy of a PAPR reduction programme, the several key variables are typically employed.

18.3.1 Complementary Cumulative Distributive Function

The most useful statistic for evaluating the PAPR is the empirical CCDF. The amount of CCDF reduction obtained is used to determine PAPR-reduction capabilities. The CCDF determines the likelihood of the OFDM signal's envelope exceeding a specific PAPR threshold during the OFDM symbol and it is given as

$$CCDF\left[PAPR\left(y^n(t)\right)\right] = prob\left[PAPR\left(y^n(t) > \lambda\right)\right] \quad (18.4)$$

where $PAPR(y^n(t))$ is the PAPR of the nth MIMO-OFDM symbol and λ is the required threshold.

18.3.2 Bit-Error Rate

The necessary signal-to-noise ratio (SNR) to achieve a particular bit-error rate is how a modulation technique's performance is measured (BER). The main objective of PAPR-reduction strategies is to reduce the CCDF, which is a positive attitude, by raising the BER.

18.3.3 Power Spectral Density

The highest peak is limited by a power amplifier (PA), resulting in an increase in both distortions at in-band and out-of-band. The second causes an unfavorable rise in power of the PSD side lobes of the MIMO-OFDM signal. Spectral spreading generates increased contact between the subbands of the MIMO-OFDM signal as

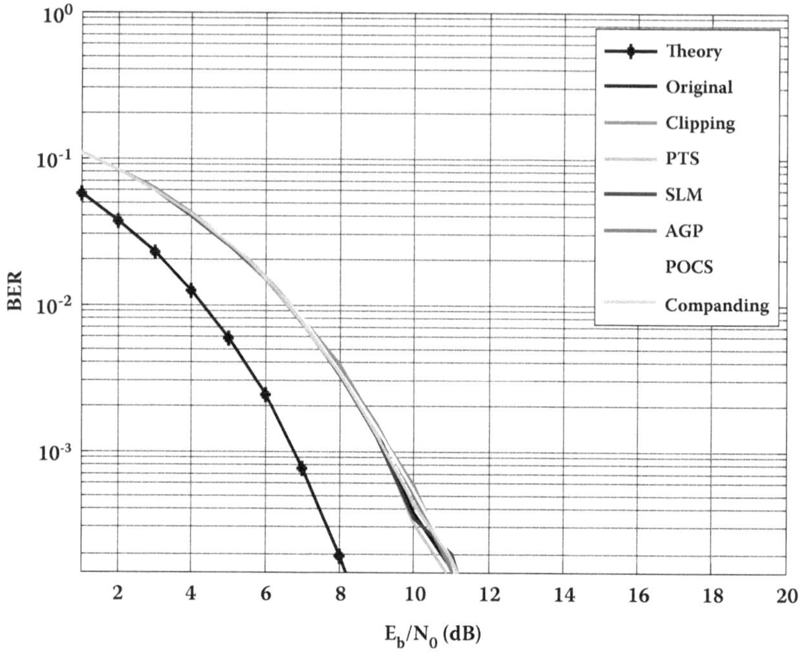

FIGURE 18.1 Comparison of BER with subcarriers for MIMO-OFDM system.

shown in Figure 18.1, unless the frequency difference between neighboring subcarriers is simultaneously enhanced to ensure orthogonality.

18.3.4 Loss of Data Rate

Several techniques result in some data-rate loss due to the added bandwidth required to transmit side information. Other strategies may necessitate the use of non-information symbols to manage PAPR as shown in Figure 18.2. If the information data rate must be the same as it was before the approach was used, a bandwidth expansion will be the consequence.

18.3.5 Computational Complexity

Another significant factor to consider when choosing a PAPR reduction technique is computational complexity. Complex methods, on the other hand, demand more processing power, time, and hardware. In reality, both hardware and processor complexity should be maintained to a minimum to allow real-time system operation and save money.

18.3.6 Other Factors

The effect of non-linear equipment in the transmitter, such as the DAC, mixer, transmit filter, and PA, should be considered when reducing PAPR. The cost of

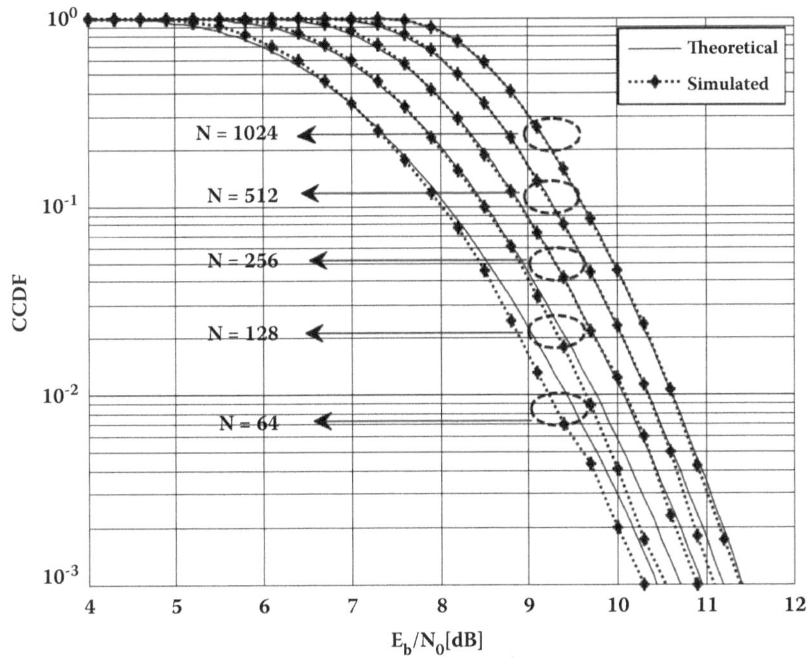

FIGURE 18.2 Comparison of CCDF with various values of N.

these devices, as well as the nonlinearity they introduce, are two crucial elements in the system design process.

18.4 PAPR-MINIMIZING METHODS

Many approaches for reducing PAPR have been proposed in the literature. Signal-distortion techniques, pre-distortion methods, probabilistic approaches, and coding techniques are the primary types of these techniques. In this section, we will concentrate on various techniques used to minimize PAPR for multi-carrier transmission.

18.4.1 Signal Distortion

Signal distortion methods reduce the PAPR through distorting the broadcasted MIMO-OFDM signal prior passing it through the PA. Peak windowing [31,32], companding [33–35], clipping, and filtering [36,37], and peak cancellation are the most often used signal-distortion methods. These methods reduce PAPR significantly, and they often create in-band and out-of-band distortion, increasing BER.

18.4.1.1 Clipping and Filtering

Amplitude clipping [23] may be the easiest way for PAPR reduction. A clipper is used in this approach to limit the signal envelopes to a predetermined clipping

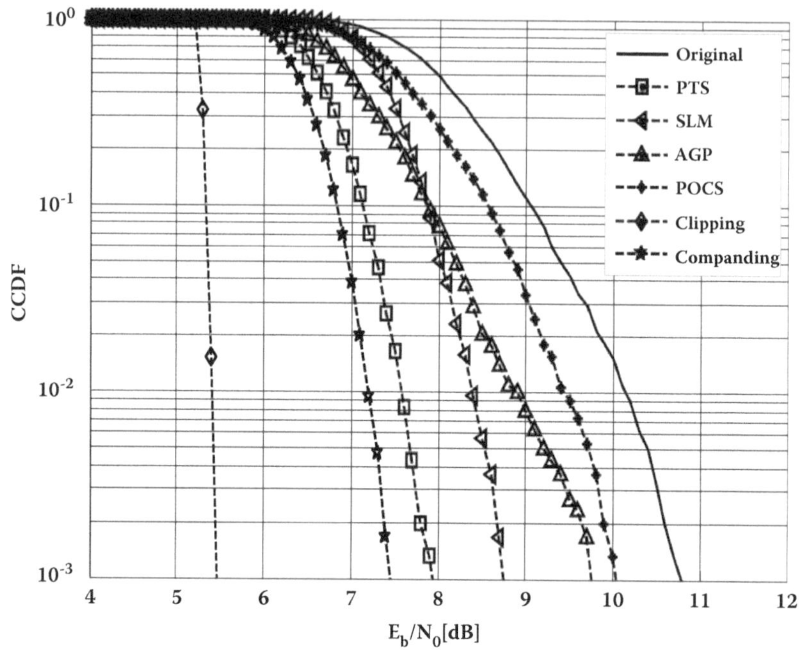

FIGURE 18.3 Comparision of CCDF with subcarriers for MIMO-OFDM (2 × 1) system.

threshold (A) if the signal exceeds that level; otherwise, the signal is unaffected. [38], given as

$$B[y(n)] = \begin{cases} y[n] & if|y[n]| \leq A \\ Ae^{j\phi y[n]} & if|y[n]| > A \end{cases} \tag{18.5}$$

where $y[n]$ is the OFDM signal, A is the clipping level and $\phi y[n]$ is the angle of $y[n]$. According to [37], clipping is a non-linear phenomenon that creates distortions both in and out of the band distortions. Filtering a clipped MIMO-OFDM signal improves BER by retaining spectrum efficiency and minimising distortion in out-of-band signal, and it can also cause regrowth of peak power as shown in Figure 18.3. Authors of [39] devised an improved technique for repetitive clipping and filtering that employs convex optimization to provide a frequency response filter that is acceptable. The filter's goal is to minimise signal distortion to a point where the PAPR is below a particular level. Also, the method achieves desired PAPR at the expense of increasing computing complexity.

18.4.1.2 Peak Windowing

The primary objective of peak windowing is to decrease radiation in out-of-band signal by attenuating peak signals with narrow band windows such as the Gaussian window. In fact, any time-domain window is narrow with significant spectral qualities can be implemented [40]. In 2008, [32] published an enhanced peak-windowing approach that solves the disadvantages of the traditional peak-windowing method. It successfully suppresses the peak signals to the proper critical threshold when the successive peaks happen within 50% of the window length.

18.4.1.3 Non-linear Companding

Non-linear companding is a one-of-a-kind clipping method that decreases PAPR while increasing performance of BER with no effort and no bandwidth expansion. Because the companding transform uses a rigorous monotone growing function to compand the original signals, the companded signals can be successfully retrieved at the receiver by inverting the companding transform. In the literature, there are numerous companding approaches [41]. Most companding techniques work on the principle of transforming a Rayleigh-scattered OFDM signal into a uniformly distributed signal.

18.4.1.4 Peak Cancellation

Using this approach, a peak-elimination waveform was generated, resized, shifted, and deducted from the OFDM signal during segments having high peaks. The waveform created is restricted to a few peak cancellation tones that aren't utilised in transmitting data.

18.4.2 Pre-distortion Methods

Before performing IFFT, the pre-distortion approaches rely on re-orienting or distributing data symbol's energy. The pre-distortion system includes constellation shaping, DFT spreading, and pre-coding or pulse shaping. Techniques like TI (tone injection), ACE (active constellation extension), and TR (tone reservation) are used to shape constellations.

18.4.2.1 ACE (Active-Constellation-Extension)

Active constellation extension (ACE) is indeed an approach in which the modulation constellation over active subcarriers in an OFDM data block is modified to reduce the data block's PAPR without significantly lowering BER performance [42]. During the modulation process, a few of the outer constellation points are dynamically extended toward the outside of the original constellation.

The ACE technique concurrently reduces PAPR while marginally lowering BER. Furthermore, there is no need for side information, resulting in no loss in data rate. Also the method, on the other hand, increases the average transmitted signal strength and is only suitable for modulation schemes with large constellation sizes [43].

The amount of algorithms that are discussed in literature to handle this problem efficiently, including first projection onto convex sets and approximation gradient project. A new approach is provided in a literature to handle this problem efficiently using first POCS (projection-onto-convex-sets) and second AGP (approximation gradient project). Both algorithms are similar in some ways; however, POCS has a sluggish convergence difficulty, and it's tough to identify the right clipping level number. The steepest gradient path for solving to minimise the peak value after ACE constraints is the approximate gradient. An AGP is created by considering the clipped signal *Cclip*. defined in Equations in 18.6 and 18.7

$$\bar{y}[n] = y[n] + c_c lip[n] \tag{18.6}$$

where

$$c_{clip}[n] = \begin{cases} 0 & if\ |y[n]| \leq A \\ (A - |y[n]|)e^{j\phi y[n]} & if\ |y[n]| > A \end{cases} \tag{18.7}$$

To enforce all constellation-extension constraints, rebuild all interior spots to their original values by projecting exterior points into the region of increased margin. Figure 18.4 depicts the QPSK extension, in which all points move in the same direction. The difference between AGP and POCS stems from how each subchannel's

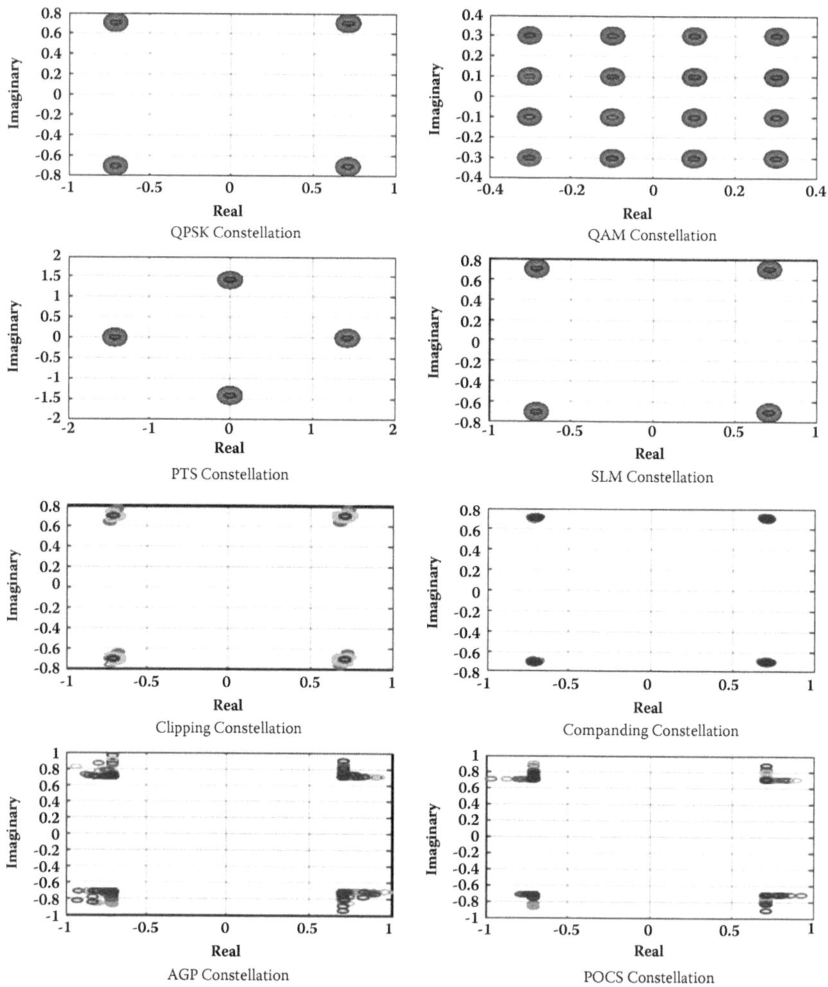

FIGURE 18.4 Constellations of various CNN.

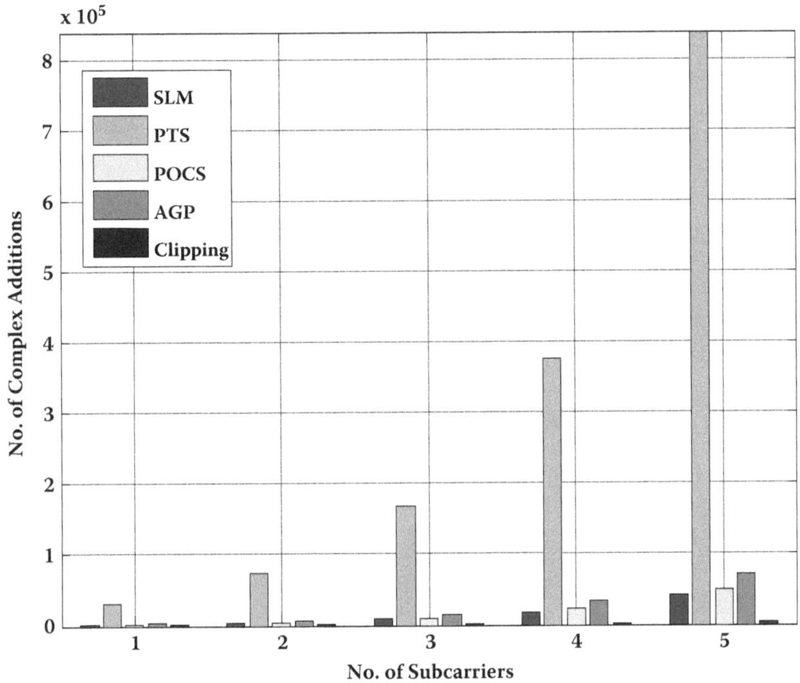

FIGURE 18.5 Number of complex additions.

allowed extension directions are determined and stored. Apply ACE limitations after switching from time to frequency domain. In addition, only maintaining components of $C_c lip$ are the extension directions that are acceptable for the supplied subchannel constellations as shown in Figure 18.5, also the rest is set to zero. Use an IDFT to get c so that step size can be calculated by equations

$$x^{(i+1)} = x^i + \mu c \tag{18.8}$$

where μ is gradient step size, that creates the difference between the approaches and the POCS. This process end with acceptable PAR or a maximum iteration count has reached. This method ends when the PAR is acceptable or when the maximum number of iterations has been achieved as shown in Figure 18.6.

18.4.2.2 Tone Reservation

Subset tone is a crucial notion in reservation. For the linear addition of reserved tone, frequency-domain processing is performed. For PAPR optimization, the statistical vector is being added to the OFDM symbol and given as follows

$$\hat{y} = y + c = IDFT\,(Y + A) \tag{18.9}$$

where Y represent OFDM symbol and A is reserved tone.

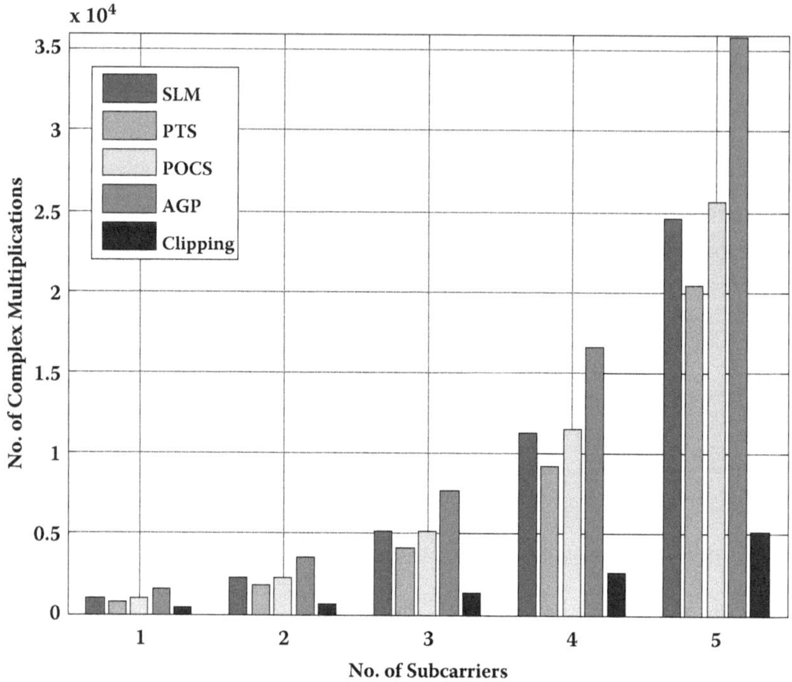

FIGURE 18.6 Number of complex multiplications.

PAPR is evaluating a variable

$$\min_{c} \|y + c\| = \min_{c} \|y + IDFT(A)\|_{\infty} \tag{18.10}$$

18.4.2.3 Tone Injection

Tone injection philosophy is the mapping of constellations into expansion. The extended constellation's new points are equivalent to the old constellation's points, resulting in more permutations that can be utilised to reduce PAPR.

18.4.3 Probabilistic Methods

The probabilistic approaches scramble each OFDM symbol with several scrambling orders and select the one with the lowest PAPR. While out-of-band power has little effect, as the number of subcarriers grows, so does spectral efficiency and complexity. Because side information is required at the receiver, it contributes to the overhead as well.

18.4.3.1 Selective Mapping

In SLM, the input data sequences are multiplied by each of the phase sequences to generate alternative input symbol sequences. The supplied data is divided into

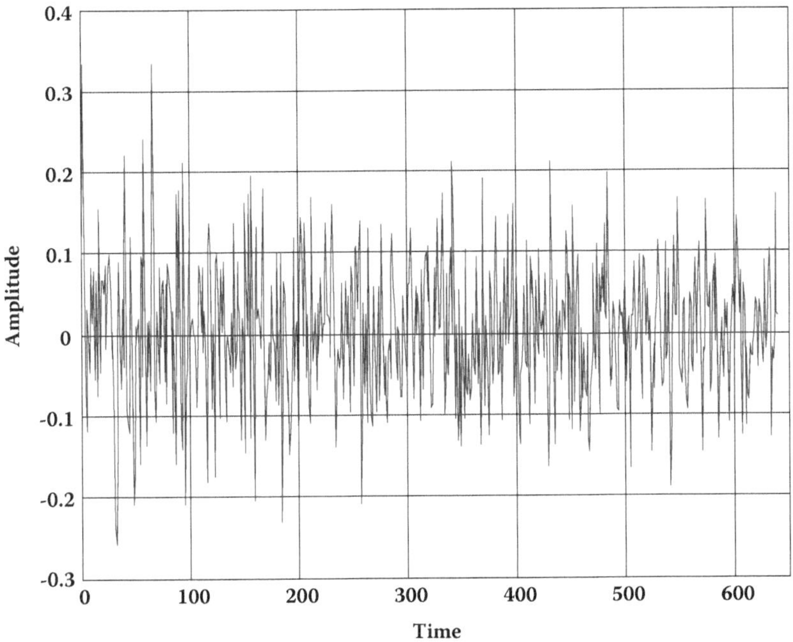

FIGURE 18.7 Channel signal.

Ydata blocks of length N. Then, using phase sequences W as shown in Figures 18.7 & 18.8, these data blocks are multiplied element by element

$$W^{(v)} = [w_{v,0},\ w_{v,1}, \cdots, w_{v,N-1}]^T,\ for\,(v = 1, 2, \cdots, V), \tag{18.11}$$

resulting into V modified data blocks

$$Y^{(v)} = [Y_{v,0},\ Y_{v,1}, \cdots, Y_{v,N-1}]^T, \tag{18.12}$$

where $Y_{v,k} = Y_k.\ W_k^v$ for $k = 0,\ 1,\ \cdots,\ N - 1$. After being operated by the Alamouti SFBC, the alternative signal $\mathbf{Y}^{(v)}$ is encoded into two vectors $\mathbf{Y}_1^{(v)}$ and $\mathbf{Y}_2^{(v)}$ as [44]

$$\begin{aligned} Y_1^{(v)} = [&W^v(0)Y(0),\ -W^{v*}(1)Y(1)^*, \cdots, W^v(N-2)Y(N-2),\ -W^{v*}(N \\ &- 1)Y(N-1)^*], \\ Y_2^{(v)} = [&W^v(1)Y(1),\ W^{v*}(0)Y(0)^*, \cdots, W^v(N-1)Y(N-1),\ -W^{v*}(N \\ &- 2)Y(N-2)^*]. \end{aligned} \tag{18.13}$$

The optimal set with the minimum PAPR of the two signals is chosen as

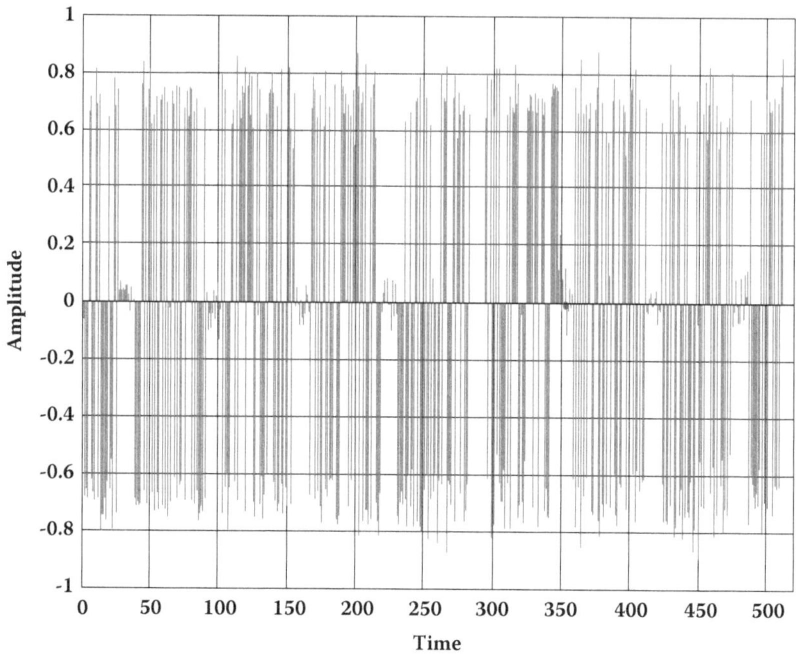

FIGURE 18.8 Received signal.

$$\hat{v} = arg \min_{0 \le v \le V-1} \left(\max_{i=1,2} \max_{0 \le n \le N-1} |y_i^v(n)| \right). \tag{18.14}$$

Generally, the V phase rotation sequences $\mathbf{W}^v$ should be transmitted to the receiver as the SI with $log_2 V$ bits.

18.4.3.2 Partial Transmit Sequence

The original frequency-domain data sequence is separated into numerous discontinuous sub-blocks, which are then weighted by a collection of phase sequences to produce a set of candidates using the PTS approach as shown in Figure 18.9. The input data is divided into sub-blocks that do not overlap. The weighting of the sub-carriers in each sub-block is determined by phase rotations that are chosen to minimise the PAPR [45]. The final signal is formed by combining the sub-carriers with the lowest PAPR. The original sub-carriers are retrieved during reception using inverse phase rotations. The PTS algorithm is described as follows:

- Define data symbol with N sub-carriers

$$x = (x_0, x_1, \cdots, x_{N-1}). \tag{18.15}$$

- The vector X is partitioned into V disjoint sub-blocks $\{x_v, v = 1, 2, \cdots, V\}$ with $x = \sum_{v=1}^{V} X_v$.

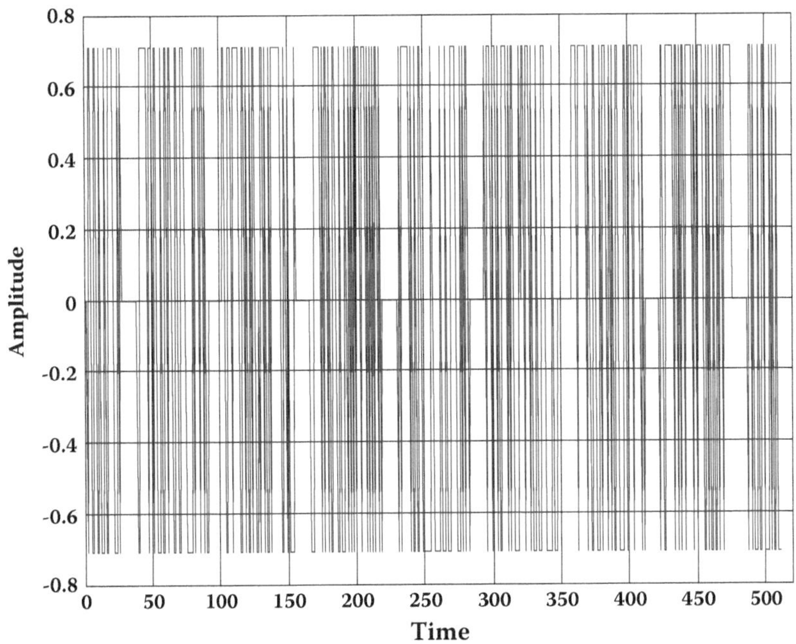

FIGURE 18.9 Transmitted signal.

- The combined V sub-blocks are as follows:

$$x' = \sum_{v=1}^{V} b_v x_v \tag{18.16}$$

where $b_v = exp(j\varphi_v)$ is the phase rotation factor and $\varphi_v = \epsilon\,[0,\ 2\pi]$, which are called side information.

18.4.4 Coding Methods

The purpose of coding is to find such a codeword to reduce PAPR. This procedure is straightforward because all that is required is the discovery of acceptable code. Some of the codes are mentioned below

- Block coding
- Golay complementary sequence
- Reed Muller code
- Hadamard code, etc.

Block coding can be used to convert 3-bit data to 4-bit data. However, selecting appropriate code necessitates a thorough search. Encoding and decoding still necessitate massive search tables. If the number of subcarriers is big, this might

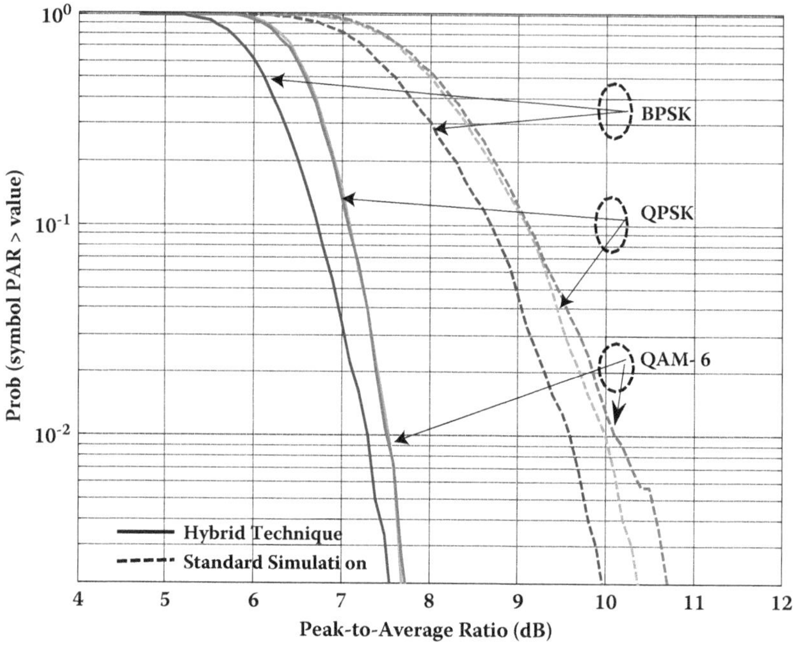

FIGURE 18.10 PAPR comparsion.

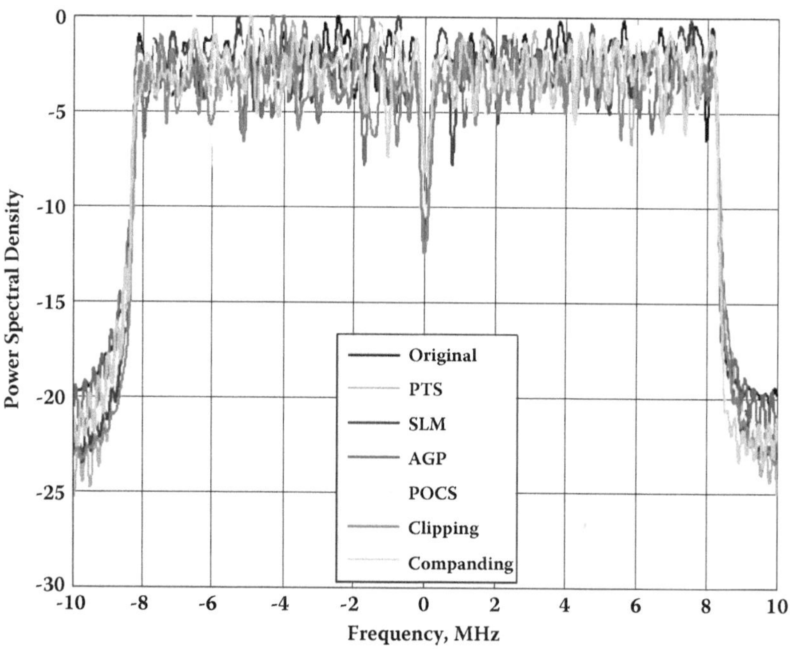

FIGURE 18.11 PSD comparison.

become difficult. The Golay complementary sequence is another well-known example. When phase-shift keying (PSK) modulation is utilised, PAPR is about $3dB$.

18.5 CONCLUSION

In this paper, a hybrid technique is used to reduce the PAPR of a MIMO-OFDM system utilising an artificial neural network-based approach. The AGP-NN module, as well as the PTS scheme, are combined in this hybrid approach. In terms of PAPR reduction, BER performance, and out-of-band radiations, the suggested solution outperforms conventional methods. The proposed method also has the benefit of being less computationally complex. When applied to a large number of subcarriers, the suggested approach delivers greater PAPR reduction in MIMO OFDM systems as shown in Figures 18.10 & 18.11.

REFERENCES

1. L. J. Cimini, Jr., "Analysis and simulation of a digital mobile channel using orthogonal frequency division multiplexing." *IEEE Trans. Commun.*, vol. COM-33, no. 7, pp. 665–675, July. 1985.
2. R. W. Chang, "Synthesis of band-limited orthogonal signals for multichannel data transmission." *Bell Syst. Tech. J.*, vol. 45, pp. 1775–1797, December 1966.
3. M. S. Zimmerman and A. L. Kirsch, "The AN/GSC-10 (KATHRYN) variable rate data modem for HF radio." *IEEE Trans. Commun. Technol.*, vol. COM-15, no. 2, pp. 197–205, April. 1967.
4. R. W. Chang, "Orthogonal frequency division multiplexing." U. S. Patent 3 488 445, 6 January 1970.
5. S. Weinstein and P. Ebert, "Data transmission by frequency-division multiplexing using the discrete Fourier transform." *IEEE Trans. Commun. Technol.*, vol. COM-19, no. 5, pp. 628–634, October 1971.
6. B. Hirosaki, "An analysis of automatic equalizer for orthogonally multiplexed QAM systems." *IEEE Trans. Commun.*, vol. COM-28, no. 1, pp. 73–83, January 1980.
7. B. Hirosaki, S. Hasegawa, and A. Sabato, "Advanced groupband data modem using orthogonally multiplexed QAM technique." *IEEE Trans. Commun.*, vol. COM-34, no. 6, pp. 587–592, June 1986.
8. Y. G. Li and N. R. Sollenberger, "Clustered OFDM with channel estimation for high rate wireless data." *IEEE Trans. Commun.*, vol. 49, no. 12, pp. 2071–2076, December 2001.
9. A. R. S. Bahai and B. R. Saltzberg, *Multi-carrier Digital Communications: Theory and Applications of OFDM*, New York: Plenum, 1999.
10. I. Koffman and V. Roman, "Broadband wireless acess solutions based on OFDM access in IEEE 802.16." *IEEE Commun. Mag.*, vol. 40, no. 4, pp. 96–103, April 2002.
11. G. J. Foschini, "Layered space-time architecture for wireless communication in a fading environment when using multi-element antennas." *Bell Labs Tech. J.*, vol. 1, no. 2, pp. 41–59, April 1996.
12. V. Tarokh, N. Seshadri, and A. R. Calderbank, "Space-time codes for high data rate wireless communication: Performance criteria and code construction." *IEEE Trans. Inf. Theory*, vol. 44, no. 2, pp. 744–764, March 1998.
13. A. Wittneben, "A new bandwidth efficient transmit antenna modulation diversity scheme for linear digital modulation." In *Proc. IEEE Conf. Commun.*, pp. 1630–1634, 1993.
14. S. M. Alamouti, "A simple transmit diversity technique for wireless communications." *IEEE J. Sel. Areas Commun.*, vol. 16, no. 8, pp. 1438–1451, October 1998.

15. J.-C. Guey, M. P. Fitz, M. R. Bell, and W.-Y. Kuo, "Signal design for transmitter diversity wireless communication system over Rayleigh fading channels." *IEEE Trans. Commun.*, vol. 47, no. 4, pp. 527–537, April 1999.
16. H. Yang, "A road to future broadband wireless access: MIMO-OFDM based air interface." *IEEE Commun. Mag.*, vol. 43, no. 1, pp. 53–60, January 2005.
17. H. Sampath, S. Talwar, J. Tellado, V. Erceg, and A. Paulraj, "A fourth generation MIMO-OFDM broadband wireless system: Design, performance, and field trial results." *IEEE Commun. Mag.*, vol. 40, no. 9, pp. 143–149, September 2002.
18. S. Shahbazpanahi, A. B. Greshman, and J. H. Manton, "Closed-form blind MIMO channel estimation for orthogonal space-time block codes." *IEEE Trans. on Signal Processing*, vol. 53, no. 12, pp. 4506–4517, 2005.
19. J. Via, I. Santamaria, and J. Perez, "Code combination for blind channel estimation in general MIMO-STBC systems," *EURASIP Journal on Advances in Signal Processing*, vol. 2009, no. 3, 1–12, 2009.
20. R. Kumar and R. Saxena, "Performance Comparison of MIMO-STBC Systems with Adaptive Semiblind Channel Estimation Scheme." *Wireless Pers. Commun.*, vol. 2013, no. 72, pp. 2361–2387, 2013.10.1007/s11277-013-1154-4
21. T. Jiang and Y. Wu, "An overview: peak-to-average power ratio reduction techniques for OFDM signals." *IEEE Trans. on Broadcasting*, vol. 54, no. 2, pp. 257–268, June 2008.
22. J. Armstrong, "Peak-to-average power reduction for OFDM by repeated clipping and frequency domain Itering." *IEEE Electronics Letters*, vol. 38, no. 5, 246–247, 2002.
23. R. O'Neill and L. B. Lopes, "Envelope variations and spectral splatter in clipped multicarrier signals." In *Proceeding of IEEE PIMRC*, Toronto, Canada, pp. 71–75, 1995.
24. B. S. Krongold and D. L. Jones, "PAR reduction in OFDM via active constellation extension." *IEEE Trans. on Broadcasting*, vol. 49, no. 3, pp. 258–268, 2003.
25. X. Zhu, "A low-BER clipping scheme for PAPR reduction in STBC MIMO-OFDM systems." *Wireless Pers. Commun.*, vol. 13, no. 65, pp. 335–346, 2012. 10.1007/s11277-011-0259-x
26. X. Jr. and L. J. Cimini, "Effect of clipping and filtering on the performance of OFDM." *IEEE Communication Letters*, vol. 2, no. 25, pp. 131–133, 1998.
27. C. Wang, S. Ku, and C. Yang, "A low-complexity PAPR estimation scheme for OFDM signals and its application to SLM-based PAPR reduction." *IEEE Journal of Selected Topics in Signal Processing*, vol. 4, no. 3, pp. 637–645, 2010.
28. A. Ghassemi and T. A. Gulliver, "PAPR reduction of OFDM using PTS and error correcting code sub-blocking." *IEEE Transctions on Wireless Communication*, vol. 9, no. 3, pp. 980–989, March 2010.
29. B. M. Lee, Y. Kim, and R. J. P. de Figueiredo, "Performance analysis of the clipping scheme with SLM technique for PAPR reduction of OFDM signals in fading channels." *Wireless Pers. Commun.*, vol. 63, no. 2, pp. 331–344, March 2012.
30. Y. G. Li and G. Stuber, *Orthogonal Frequency Division Multiplexing for Wireless Communications*, Boston, MA: Springer-Verlag, January 2006.
31. D. Kim, D. Shi, Y. Park, and B. Song, "New peak - windowing for PAPR reduction of OFDM systems." In *Proc. Asia-Pacific Conference on Wearable Computing System (APWCS)*, pp. 169–173, August 2005.
32. S. Cha, M. Park, S. Lee, K. J. Bang, and D. Hong, "A new PAPR reduction technique for OFDM systems using advanced peak windowing method." *IEEE Transactions on Consumer Electronics*, vol. 54, no. 2, pp. 405–410, May 2008.
33. H. Xiao, L. Jianhua, C. Justin, and Z. Junli, "Companding transform for the reduction of peak-to-average power ratio of OFDM signals." In *Proc. IEEE Vehicular Technology Conference (VTC)*, pp. 835–839, 2001.

34. X. Huang, J. Lu, J. Zheng, K. B. Letaif, and J. Gu, "Companding transform for reduction in peak-to-average power ratio of OFDM signals." *IEEE Trans. Wireless Commun.*, vol. 3, no. 6, pp. 2030–2039, November. 2004.
35. S. Cha, M. Park, S. Lee, K. J. Bang, and D. Hong, "A new PAPR reduction technique for OFDM systems using advanced peak windowing method." *IEEE Transaction on Consumer Electronics*, vol. 54, no. 2, pp. 405–410, May 2008.
36. K. R. Panta and J. Armstrong, "Effect of clipping on the error performance of OFDM in frequency selective fading channels." *IEEE Trans. Wireless Commun.*, vol. 3, no. 2, pp. 668–671, March 2004.
37. H. Ochiai and H. Imai, "On the clipping for peak power reduction of OFDM signals." In *Proc. IEEE Global Communication Conference (GLOBECOM)*, San Francisco, USA, pp. 731–735, 2000.
38. J. Heiskala and J. Terry, *OFDM Wireless LANs: A Theoretical and Practical Guide*. Indianapolis Indiana: Sams Publishing, 2001.
39. Y. C. Wang and Z. Q. Luo, "Optimized iterative clipping and filtering for PAPR reduction of OFDM signals." *IEEE Trans. Commun.*, vol. 59, no. 1, pp. 33–37, January 2011.
40. R. Van Nee and R. Prasad, *OFDM for Wireless Multimedia Communication*. London: Artech house Publisher, 2000.
41. M. J. Omidi, A. Minasian, H. Saeedi-Sourck, K. Kasiri, and I. Hosseini, "PAPR reduction in OFDM systems: Polynomial-based compressing and iterative expanding." *Wireless Personal Communications*, vol. 2014, no. 75, pp. 103–118, 2014, 10.1007/s11277-0.13-1350-2
42. B. S. Krongold and D. L. Jones, "PAR reduction in OFDM via active constellation extension." *IEEE Trans. Broadcast.*, vol. 49, no. 3, pp. 258–268, September 2003.
43. S. H. Han and J. H. Lee, "An overview of peak-to-average power ratio reduction techniques for multicarrier transmission." *IEEE Wireless Commun.*, vol. 12, no. 2, pp. 56–65, April 2005.
44. T. Jiang, C. Ni, and L. Guan, "A novel phase offset SLM scheme for PAPR reduction in alamouti MIMO-OFDM systems without side information." *IEEE Transaction of Electronics*, vol. 20, no. 4, pp. 383–386, 2013.
45. S. H. Muller and J. B. Huber, "OFDM with reduced peak-to-average power ratio by optimum combination of partial transmit sequences." *IEEE Electronics Letters*, vol. 33, no. 5, pp. 368–369, 1996.

Index

FSC
www.fsc.org
MIX
Papier | Fördert gute Waldnutzung
FSC® C083411

Kolibri 360 GmbH
Ferdinand-Jühlke-Straße 7
99095 Erfurt, Deutschland
produktsicherheit@kolibri360.de